Dictionary of Scientific Literacy

Dictionary of Scientific Literacy

Richard P. Brennan

Wiley Science Editions

John Wiley & Sons, Inc.

New York ■ Chichester ■ Brisbane ■ Toronto ■ Singapore

In recognition of the importance of preserving what has been written, it is a policy of John Wiley & Sons, Inc., to have books of enduring value published in the United States printed on acid-free paper, and we exert our best efforts to that end.

Library of Congress Cataloging-in-Publication Data

Brennan, Richard P.
Dictionary of scientific literacy / by Richard P. Brennan.
p. cm. -- (Wiley science editions)

ISBN 0-471-53214-2
1. Science--Dictionaries. 2. Technology--Dictionaries.
I. Title. II. Series.
Q123.B68 1991
503--dc20 91-4307

Printed in the United States of America

For Carolyn F. Brennan
With Love

Contents

Foreword

Our home library is full of reference books on just about every subject imaginable. Next to our old Merriam-Webster's, you can find dictionaries of slang, aphorisms, names and quotations, and various encyclopedias and atlases. Then there are the flower, restaurant, country inn, CD, and how-to-save-on-your-tax guides, and all the handbooks on fence-building, do-it-yourself-upholstery, and choosing a college. Shelf space is at a premium, and yet something very critical is missing from this collection.

Life is complicated, and getting more-so, it seems, by the hour. One aspect of this—perhaps the central one of our times—is that science and technology have left the lab and the factory and are no longer the sole concern of scientists and engineers and their clients. Those fields of thought and action are now part of every one of our lives, no matter what our occupation, no matter where we live. Science, both basic and applied, has much to do with the nature of the work we do, how we travel and communicate, diseases we are prone to and those that no longer threaten us, what we eat, how we wage war, what we do to our surroundings and to other living creatures, and much else. Scientific literacy is the ticket for admission to such a world.

Scientific literacy enables us to share in the joy of discovery and to make informed decisions about public and personal matters involving science and technology. It makes life more interesting and more responsible.

Most of us can achieve the necessary understanding by reading widely. Even in the unlikely event that we were blessed with an outstanding education in science in high school or college, the fact is that in order to read contemporary articles and books on science we need a good dictionary at hand—I would say, right next to Roget, Partridge, and Bartlett, not to mention our *Oxford Companion to English Literature* and Bierce's *Devil's Dictionary*.

I suggest that the most useful science dictionary for most of us will have to meet three very tough criteria. One, there should not be too many terms. The focus should be on the most common, current vocabulary. Two, the terms selected should cover basic science, the applications of science, and science-related technology. This makes the selection of terms all the more crucial if the dictionary is not to grow unwieldy. And three, the clarity of the definitions must be superb. It also helps if the definitions are to the point and intriguing enough to cause us to sample new terms at random after locating the answer to our initial query.

This *Dictionary of Scientific Literacy* meets those standards, and then some. In every respect—coverage, level, felicity—I have found this book to be in harmony with *Science for All Americans,* the landmark standard for literacy in science, mathematics, and technology, developed under the guidance of the American Association for the Advancement of Science and published by Oxford University Press.

For me, the only problem now is figuring out how to fit this book next to all the other valuable reference works on our crammed and sagging shelves. One thing is clear, though: Whether I have to sacrifice my *Dictionary of Films and Filmmakers,* or my wife has to give up her *Encyclopedia of Flowering House Plants,* we will surely find a place for this excellent and important book.

James Rutherford, Chief Education Officer
American Association for the Advancement of Science

Preface

What does it matter if a large proportion of our population thinks that Chernobyl is a ski resort, DNA a food additive, a megabyte an orthodontal problem, and protons something you put on salads?

This book is for those who know that it does matter.

It has been estimated that there are more new words in a science course today than there are in an introductory foreign-language class. Because of the burgeoning role of science and technology in our lives, all of us are in effect taking a crash course in technical terminology. The problem is learning how to "hang in there" in this fast-moving class.

Purpose of This Book

The aim of this book is to aid you in your quest for scientific literacy by providing a handy, quick-reference guide to the vocabulary. And vocabulary is the key to the game. Jon Miller, director of Northern Illinois University's Public Opinion Laboratory, has conducted extensive surveys of American adults at all educational levels by asking a set of questions about elementary scientific concepts. He has estimated the national level of scientific literacy at only 6 percent and has concluded that only 17 percent of American college graduates have even a rudimentary knowledge of general science. He defines a scientifically literate person as one who understands the vocabulary well enough to follow public debates about issues involving science and technology. What are the ways in which the scientifically literate achieve and maintain their literacy? The answer, according to Miller's polls, is that these people take in a lot of information—they watch science programs on television, they go to museums, they read magazines and newspapers. Reading especially distinguishes them from the rest of the population because, unfortunately, a lot of people have lost the ability to read and understand moderately complex material.

In my book *Levitating Trains and Kamikaze Genes* I offered what one reviewer called "a layman's survival guide to our high-technology world." Written for nonspecialists, *Levitating Trains and Kamikaze Genes* is intended to help the intelligent layperson *catch up* with the technologies of the 1990s. This *Dictionary of Scientific Literacy* is intended to help the reader *keep up* with the dynamic, rapidly changing world of science and technology.

As public policy issues become increasingly technical today, good citizenship requires a working knowledge of science and technology. The alternative, the more pessimistic Cassandras tell us, is an apathetic, scientifically illiterate constituency that is much more likely to breed, bomb, gas, or pollute themselves to extinction. Are they overstating the case? Possibly, but numerous recent studies have made it abundantly clear that most Americans are not scientifically literate and that as a nation we are suffering from an epidemic of scientific illiteracy.

Achieving scientific literacy is more than just social duty. Dennis Flanagan, the longtime editor of *Scientific American,* has said that for him, following the work of modern science as a spectator can be one of life's deepest pleasures, fully comparable with the pleasures of literature, painting, and music. As another spectator in this grand mind-stretching adventure, I enthusiastically concur.

Who Should Use This Book?

The target user for this dictionary is the intelligent nonscientist who recognizes a possible technological deficiency, realizes that a fair amount of reading will be required to play keep-up, and who needs an easy-to-understand reference book to better comprehend newspaper and magazine articles on scientific and technological subjects—articles that are often written on the assumption that everybody knows what a *superconductor* is or the difference between *neutrinos* and *neurons,* or *quarks* and *quasars.*

How to Use This Book

Imagine that as you are reading your daily newspaper you come across a story on the dangers of *radon* gas. Now what the heck is *radon* and is it something you have to worry about? Or, to take another example, you see the term *algorithm* in a news item on the latest in computers and you are not completely sure you understand the terminology. Still another example might be Z *particle* in an article on the new developments in physics. All of these terms are examples of the rapidly expanding vocabulary of science and technology with which we are confronted every day. All of the terms printed in italics above are included in this dictionary.

This book is intended as a reference source for busy readers. The entries are arranged alphabetically without regard to subject classification (i.e., astronomy, biology, chemistry), because it is easier to use that way. Each entry consists of a concise definition followed by a sentence or two relating the technical term to something in our daily lives or expanding on the concise definition in some way to make it more easily understandable or memorable. I call this approach *Definition +*.

To further expand the information given in each entry and to indicate relationships between various entries, cross-references are used. Set in all CAPITAL letters, cross-references appear as part of the Definition + or are listed at the end of the entry.

Some of the entries, acronyms for example, have unusual pronunciations and therefore require a pronunciation guide. This guide appears in parenthesis after the entry word.

The following sample entry illustrates the approach taken throughout this dictionary:

Radon A colorless, radioactive gaseous ELEMENT produced by the natural decay of radium and present in soil everywhere. Emissions of sufficient strength are considered a hazard to health. According to the U.S. Environmental Protection Agency (EPA), radon-induced lung cancer is one of today's most serious health issues. The EPA's recommended maximum allowable concentration of radon is 4 picoCuries per liter of air. A Pico is one-trillionth of something (see NUMBERS: BIG AND SMALL) and a CURIE (named after Pierre Curie, the co-discoverer of radium) is a unit of radiation approximately equal to the amount of radiation produced by one gram of radium per second. The estimated average U.S. indoor-radon concentration is 1 picoCurie per liter of air.

Emitted by rocks in the soil, radon enters buildings

largely through cracks in the foundation. Two 1990 EPA surveys showed that about one out of five homes in the United States exceed a reading of 4 picoCuries. This concentration of radon poses about the same lung-cancer threat to householders as smoking half a pack of cigarettes daily or receiving 200 to 300 chest X rays annually. By further polluting indoor air, smokers elevate radon risks both to themselves as well as to nonsmokers. See **RADIATION, NATURAL BACKGROUND.**

This volume is not a comprehensive encyclopedia of technical terms—such reference documents exist usually in multivolume form. The goal here is a selective (as opposed to exhaustive) collection of those scientific and technical terms thought to be most useful in understanding today's high-tech world presented in an easy-to-use, quick-reference format. This approach is subjective and there will be those who quarrel about which terms should be included or excluded. My criteria in choosing terms for inclusion has been to select those technical terms that are necessary for a broad understanding of a major science or those terms that nonspecialist readers will most likely want to know a little more about.

I am grateful to Carolyn F. Brennan for the illustrations that enhance the text as well as for her critical reviews of the manuscript and her invaluable assistance and support throughout this challenging project.

Richard P. Brennan

Dictionary of Scientific Literacy

Absolute Zero Theoretically, the lowest possible temperature. At close to this temperature, substances possess minimal energy. Often stated as "zero Kelvin," absolute zero is equal to about -273°C or -460°F. All substances, whether in gas, liquid, or solid form, are collections of molecules and temperature affects the velocity of these molecules. The higher the temperature, the faster they move and the more volume they require (i.e., they expand). The lower the temperature, the more slowly they move and the less volume is required (i.e., they contract). As temperature decreases, the molecules eventually possess too little energy to move at all. In other words, the substance is frozen solid. Although physicists have come within a millionth of a degree, absolute zero itself is unattainable. The quest for knowledge about the odd behavior of material or substances in the neighborhood of absolute zero is called CRYOGENICS. See also **TEMPERATURE.**

Acceleration The rate of increase in velocity. Acceleration can be expressed as:

$$\frac{\text{increase in velocity}}{\text{time}}$$

Acceleration is often measured in *g* forces. The normal force of gravity at the Earth's surface is called 1 *g*. An acceleration force that in effect doubles the body's weight is 2 *g*, a force that triples the body weight is called 3 *g*.

Accelerator Short for particle accelerator. Sometimes called an *atom smasher*. A device, such as an electrostatic generator, cyclotron, or linear accelerator, that speeds charged particles or nuclei to high velocities and high energies useful for subatomic particle research. See **COLLIDER.**

Acid Rain The term encompasses all types of precipitation (rain, snow, or sleet) whose acidity exceeds the

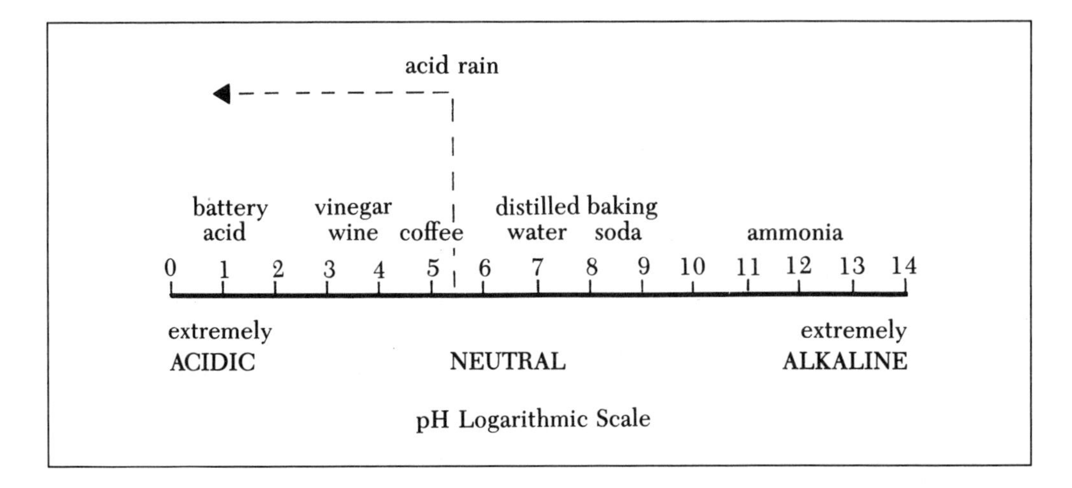

ACID RAIN and the pH Scale. The lower the pH number, the more acidic the substance. The range is between highly acidic battery acid at pH 1.0, to highly alkaline ammonia at pH 12. Distilled water, at pH 7.0, is neutral. Acid rain is pH 5.6 or lower. The pH scale is logarithmic, so water with a pH of 5 is ten times more acidic than water with a pH of 6.

normal (pH 5.6). On the pH scale, the lower the number, the more acidic the substance. Rain is made unnaturally acidic by emissions of sulfur dioxide and nitrogen oxides from the burning of fossil fuels—primarily coal.

There is no evidence that acidic rain is harmful to us directly, but when it falls into lakes and streams it acidifies them enough to kill fish and other water life. Acid rain also damages trees. Tests have also shown that highly acidic rain—pH 3.0 or lower—does damage some crops. Because of prevailing westerly winds, the northeastern part of the United States and the southeastern part of Canada suffer from acidic precipitation originating in the American Midwest. The amount of pollution emitted by coal-burning power plants can be minimized by technological means—either by cleaning the coal before it is burned or adding scrubbers to the smoke stacks—but at an increase in the cost of generating electricity.

The problem is that acidity levels are worsening each year and some sort of corrective action, however costly, will eventually have to be undertaken. The acid rain situation is a clear case of paying for the corrective action now or paying for the consequences later.

Aerodynamics The dynamics of the interaction of moving objects within the atmosphere, specifically in the design of vehicles to reduce the resistance to motion. Improved aerodynamic design has contributed a major part in reducing the resistance to motion, and consequently improving the fuel efficiency, in current passenger automobiles. See **DRAG COEFFICIENT**.

Age and Aging Aging is a normal process in all animals, including humans. Its effects vary greatly among individuals. In general, muscles and joints tend to become less flexible, bones and muscles lose some mass, energy levels diminish, and the senses become less acute. The aging process in humans is also associated with external influences such as disease and injury, diet, mutations arising and accumulating in the cells, wear and tear on tissues such as weight-bearing joints, and exposure to harmful substances. The accumulation of injurious agents such as fatty deposits in arteries, damage to lungs from smoking, and radiation damage will adversely affect the normal life span.

In July of 1990, a research team at the University of Virginia announced the preliminary results of a study showing that a recombinant form of human growth hormone may boost muscle mass and skin thickness in some people over age 60 and perhaps improve their strength and endurance. Scientists caution that this experimental treatment must undergo much more testing before it can be declared effective and safe. The preliminary studies are exciting but they are not a cure for old age. See LIFE EXPECTANCY.

AIDS (Acquired Immune Deficiency Syndrome) A deadly disease in which some of the body's white blood cells (lymphocytes) are destroyed and, consequently, the body's natural defense system is impaired. For people with AIDS, survival means fighting one illness after another, as normally meek microbes slip past crippled immune defenses. After a decade of this almost invariably fatal disease, the total U.S. death count from AIDS by March 1991 was 102,803. The Pan American Health Organization estimates that more than 3 million people in the Western Hemisphere will be infected with the AIDS virus by the mid-1990s. In other parts of the world, most specifically Africa, the disease is taking an enormous toll and the full impact of the AIDS epidemic has yet to be felt. AIDS in the United States remains concentrated among male homosexuals and needle-sharing drug users. There are, however, small signs of a "breakout" of the disease into groups that were previously little affected: teenagers, notoriously careless in their sexual practices; and women apparently infected by bisexual partners.

Scientists have made some advancements in the decade since the epidemic was first identified. The viral origin and means of transmission of AIDS have been identified, and a rejected cancer drug, AZT (Zidovudine), has been adapted to AIDS treatment. It is no cure, but AZT has lengthened the lives of patients who can tolerate its strong side effects. See AZT, and HIV.

Air Bags One type of automatic crash protection equipment now available on many new cars. Air bags have proven to be extraordinarily effective safety devices. Built into the steering wheel or dashboard, air bags are activated when a serious frontal crash occurs—equivalent to hitting a stationary object at a speed greater than 12 miles per hour. Within 1/25 second after impact, the bag is inflated to create a protective cushion between the driver/occupant and the dashboard or windshield. Immediately on inflation, the bag starts to deflate and is out of the way almost at once.

The solid chemical sodium azide initiates the inflation process, and nitrogen, which composes 78 percent of the

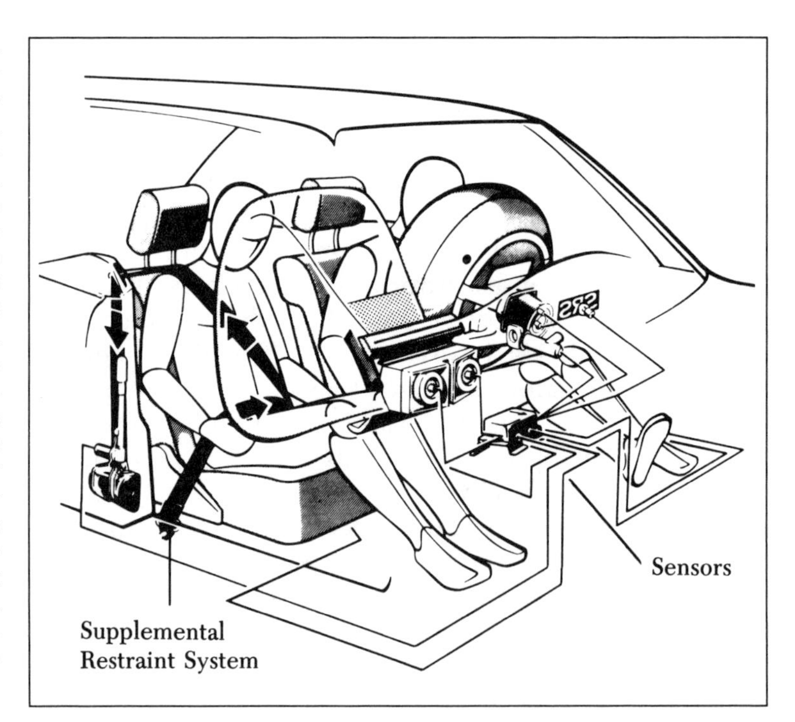

AIR BAGS cushion the driver/passenger during high-energy frontal crashes. The neoprene-coated nylon air bags inflate with harmless nitrogen gas. Deployment takes only 40 milliseconds (1/25 of a second) and, after absorbing the forces of a crash, deflation begins immediately. Some automobiles today include a supplemental restraint system that tightens the seat belt harness during an emergency stop.
Source: Mercedes-Benz.

air we breathe, inflates the bag. Air bags are not designed to inflate on a side impact crash or rollover, or if the vehicle is struck from behind.

The National Highway Traffic Safety Administration predicts that air bags will save 2,400 lives and prevent 29,000 serious injuries in the six years ending December 31, 1995. The federal safety agency also expects that 18 percent of passenger cars in use at the end of 1995—about 25 million vehicles—will be equipped with inflatable restraints.

Albedo, Earth's The ratio of light reflected by Earth to that received by it. The amount of light that a planet or satellite reflects is called its albedo. The maximum possible albedo, when all the incident light is reflected away again, is 1, and the minimum, when no light is reflected, is 0.

Albedo plays a major role in climate change. Ice covers about 10 percent of Earth's land area and reflects light more efficiently than bare rock or soil; ice reflects 90 percent of the light that falls on it, whereas bare soil reflects less than 10 percent. A slight increase in ice cover would mean that more sunlight would be reflected and less absorbed, so that the average temperature on Earth would drop. Similarly, if Earth's temperature rises even slightly, the ice fields would retreat—less sunlight would be reflected and more absorbed, accelerating the warming trend. This positive feedback effect makes the albedo an important factor in long-range climate changes. See FEEDBACK CONTROL.

Algorithm A set of instructions for accomplishing a particular task or solving a specific problem in a finite number of steps. Synonymous terms are *method,* or *procedure.*

An example is a step-by-step recipe in a cookbook. In mathematics the term refers to any recursive computational procedure. An algorithm intended for execution by a computer is also called a program.

In programming, an algorithm must be precise. Each step of the algorithm must specify exactly what action is to be carried out. There is no room for vagueness. Each step must be stated explicitly. None can be *understood* or *assumed.* Because it is difficult to achieve the necessary precision in English, a number of algorithmic languages, including programming languages, have been devised. These languages are analogous to the notations used to express technical ideas in mathematics, music, or chemistry more precisely than is possible in English. See **COMPUTER**.

Aliens See **EXTRATERRESTRIAL INTELLIGENCE**, **SETI** (Search for Extraterrestrial Intelligence), and **UFOs**.

Alpha Centauri A triple-star system that includes Proxima Centauri, the nearest star to Earth after our own Sun. Alpha Centauri is so distant from us that it takes light, traveling at 186,000 miles per second, 4.3 years to reach Earth (See **LIGHT-YEAR**).

Interstellar distances are so immense that they are difficult to imagine, but, if we had a spacecraft that could travel at a speed of a million miles per hour (40 times faster than any present-day spacecraft), it would still take us nearly 3,000 years to reach the vicinity of Alpha Centauri.

With the naked eye, an observer can see about 6,000 stars in the night sky. Alpha Centauri is located low in the southern skies and is not visible north of the latitude of Tampa, Florida. From Earth it appears to be the third brightest star in the heavens. See **ASTRONOMY**, and **COSMOLOGY**.

Alpha Particle A positively charged composite particle, indistinguishable from a helium atom nucleus and consisting of two **PROTONS** and two **NEUTRONS**. See **ATOMS**.

One of three types of radioactive emission (the other two being **BETA PARTICLES** and **GAMMA RAYS**), a stream of alpha particles is called alpha rays or alpha radiation.

Alzheimer's Disease The most common neurodegenerative (brain damaging) disease in the United States, afflicting two in ten people over the age of 70, and a total of four million people in the country annually. Usually beginning in late middle age, Alzheimer's disease is characterized by memory lapses, confusion, emotional instability, and progressive loss of mental ability. The cause of this disease is not yet known but scientists have identified a specific brain protein associated with the disease. Identification of this protein may, in the future, allow earlier diagnosis and earlier treatment. See also **PARKINSON'S DISEASE**.

American Standard Code for Information Interchange. ASCII (pronounced *aśkey*) Analogous to the Morse code wherein combinations of dots and dashes represent the letters of the alphabet as well as numerals and punctuation, ASCII is a character code in which two symbols, 0 and 1 (called **BITS**) are used in combinations to process information in character form. The ASCII code uses seven bits to represent each character and can represent 128 characters in all. These include the upper- and lowercase letters of the alphabet, the digits, the punctuation marks, a few mathematical symbols, and other signs. The seven-bit ASCII character will fit into an eight-bit unit called a **BYTE**. (The eighth bit is used as part of the internal machine code.)

As an example, the name Carolyn is shown here in ASCII code:

C	1000011
a	1100001
r	1110010
o	1101111
l	1101100
y	1111001
n	1101110

Amino Acids Organic acids that are the building blocks from which proteins are constructed. There are 20 common amino acid types and the properties of every protein depend on exactly how—in what order—all the amino acids are arranged in the molecular chain. **PROTEIN** molecules are long, usually folded chains made up of the 20 different kinds of amino acid molecules. The function of each protein depends both on its specific sequence of amino acids and the shape the chain takes. See **CELLS**, **DNA**, and **RNA**.

Amniocentesis Medical procedure involving the withdrawal and analysis of a small amount of watery fluid in which the human embryo is suspended in the womb. This procedure is usually performed during the 16th or

17th week of pregnancy and provides a good deal of information about the fetus. Fluid is withdrawn by needle usually after an **ULTRASOUND** examination has located the position of the fetus in the womb. The amniotic fluid is then analyzed for a number of proteins and enzymes. Chromosomal analysis can reveal any genetic abnormalities.

This procedure is now required in some countries (Canada, for example) for pregnant woman over 40, who are statistically more likely to give birth to genetically abnormal children. The procedure remains controversial in the United States. Amniocentesis and chromosomal analysis is 99 percent effective in predicting birth defects. For this reason, many couples opt to abort the pregnancy and avoid the emotional and financial hardships involved in raising handicapped children.

Critics of abortion point out that if a pregnancy can be aborted because of some devastating neurological defect or physical malformation, this technique may lead to abortion because of some relatively minor disorder, or even because of gender. Amniocentesis is an example of how modern technology can sometimes raise ethical and moral questions. See **SONARGRAPHY**.

Ampere The basic unit of electrical current. A flow of electricity can be likened to water flowing in a pipe. Voltage can be thought of as the pressure that pushes the water through the circuit. Resistance then is the size of the pipe restricting the flow of current through the circuit, and the amount of water that makes up the current is measured in amperes. See **OHM'S LAW**, and **VOLT**.

Analog Usually used in reference to a type of data or equipment, analog means analogous to or representative of some reality, as opposed to *digital,* which means having to do with numbers. Analog information is continuous, as in a standard radio or TV signal—a waveform, as contrasted with discrete groups of information broadcasts by satellites to earth stations in a digital communication system. See **DIGITAL**.

Andromeda Galaxy A large system of stars held together by gravitational forces. One of the billions of

GALAXIES in the universe, Andromeda is the nearest galaxy similar in size to our own MILKY WAY. It is 2.2 million light-years away. Like the Milky Way, Andromeda is a large spiral galaxy. The Andromeda galaxy, its several satellite galaxies, the Milky Way, and a galaxy called the Large Magellanic Cloud form what astronomers call the Local Group. Because it is gravitationally bound to the Milky Way, the Andromeda galaxy is currently approaching us, rather than receding as is the case for other galaxies. See ASTRONOMY, and COSMOLOGY.

ANDROMEDA GALAXY, a large system of stars held together by gravitational force, is the nearest galaxy to our own Milky Way.
Source: NASA photo.

Angioplasty Sometimes called balloon therapy—more technically known as *percutaneous angioplasty* (*percutaneous* meaning through the skin, and *angioplasty* meaning blood vessel repair)—this technique is an alternative to heart surgery. A catheter with a tiny balloon is threaded through an artery and guided through the body until it reaches the blocked area. There the balloon is inflated, compressing the fatty material (called *plaque*) against the artery wall, thus increasing the diameter of the blood vessel.

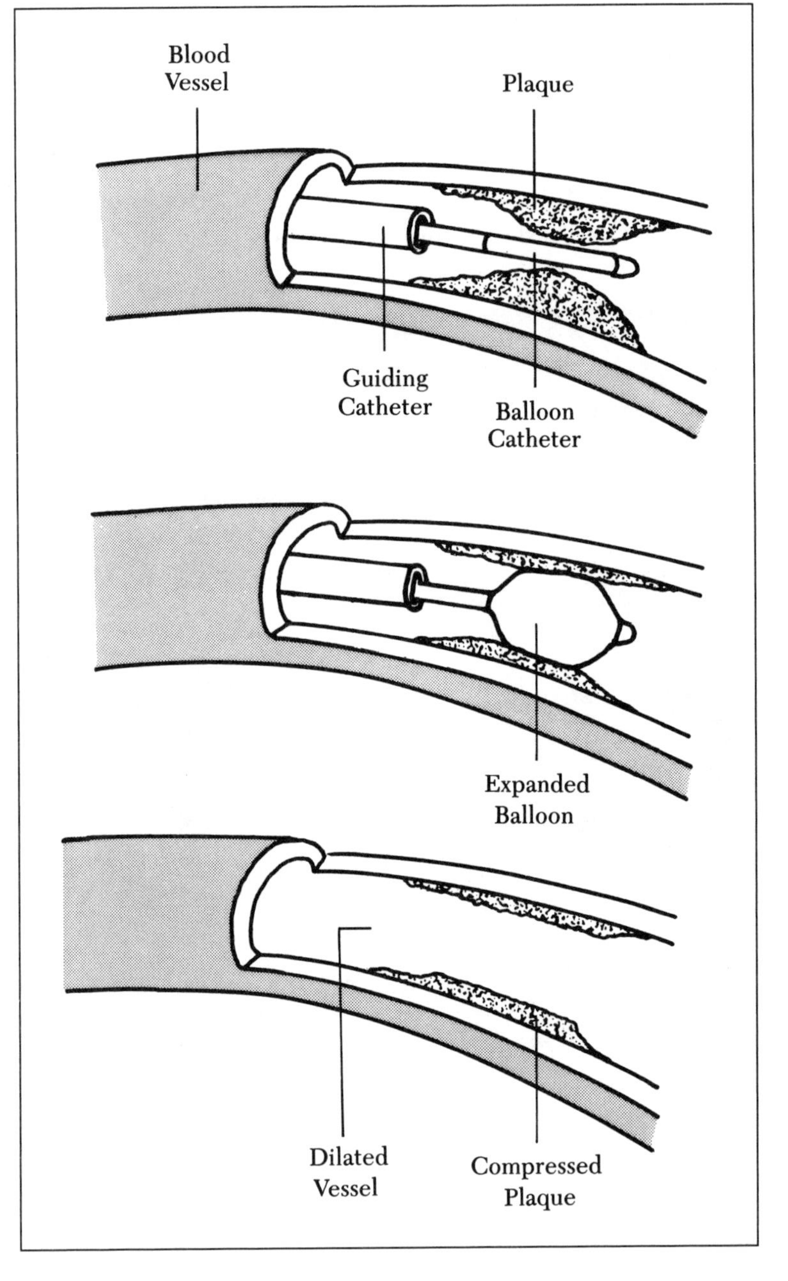

ANGIOPLASTY, or balloon therapy, involves threading a catheter with a tiny balloon through an artery to a blockage area. The balloon is inflated, compressing the plaque against the artery wall and thus increasing the diameter of the blood vessel.

Angstrom A unit of length used for measuring extremely small amounts. An angstrom is equal to one hundred-millionth (10^{-8}) of a centimeter (see **NUMBERS: BIG AND SMALL**). Named after Swedish physicist Anders

Angstrom, the unit is especially used to specify radiation wavelengths at the very high frequency end of the ELECTROMAGNETIC SPECTRUM.

The angstrom is the unit of distance commonly used by scientists because the diameter of most atoms is 1 to 4 angstroms. One angstrom is equal to 100 trillionths of a meter, 4 billonths of an inch, or about a millionth the diameter of a human hair.

Animalia One of the five taxonomic kingdoms of living organisms. Animalia comprise all multicelled animal life. The other four classifications are FUNGI, MONERA, PLANTAE, and PROTISTA. See TAXONOMY.

Animal Research About 20 million animals are used in medical research annually (down 40 percent since 1968). The American Association of Medical Colleges claims that many benefits to humankind result from this research, including: vaccines for polio, diphtheria, rabies, measles; drugs such as insulin, antibiotics, painkillers; treatments for cancer, diabetes, cystic fibrosis, high blood pressure, epilepsy, hemophilia; advances in viral research, angiograms, artificial limbs; and surgical advances such as blood transfusion and organ transplants.

Animal rights groups, however, claim that lab animals are mistreated and abused. They say most animal research is unnecessary and should be replaced by methods such as using tissue and cell cultures, computer models, and clinical studies.

Antibiotics Microorganisms that destroy BACTERIA. An example is penicillin, which has become one of the most important drugs in the practice of medicine. Penicillin is only one of the many antibiotics that have been developed in the past 50 years. Taken together, antibiotics have helped to conquer many infectious diseases. The antibiotics, however, have had little success against viruses that reproduce themselves by a different mechanism than bacteria. See VIRUS, BIOLOGICAL.

Antibody Any of various proteins in the blood that are generated in reaction to foreign proteins to neutralize them or produce immunity to them. As a protective protein molecule, an antibody can combine with a foreign virus protein and inactivate the virus. Cells in the immune system of the human body are capable of making more than a million different kinds of antibodies. See **IMMUNE SYSTEM**.

Antigen Any foreign substance, such as a virus, that when entering the body's chemistry stimulates the production of an **ANTIBODY**. Antigens commonly invade the human body on the surfaces of bacteria and pollens. The body protects itself by making antibodies.

Antimatter According to the **STANDARD MODEL**, the name given to the set of theories that attempts to describe the nature of matter and energy, each of the family of particles that make up all matter has a counterpart or mirror image called an antiparticle. Antimatter is composed of antiparticles. The existence of antimatter is the subject of theoretical conjecture. The only known examples of antimatter are the particles created when subatomic particles are slammed together in an accelerator/collider and then they don't last long. As soon as antimatter encounters its equivalent matter, the two mutually annihilate, converting both back to energy. Antiparticles produced experimentally in colliders have charges opposite to those of their corresponding particles. Antiprotons and **POSITRONS** are examples of antimatter. A **PROTON** is positively charged, but an antiproton is negatively charged. The counterpart of a negative electron is the positive positron. Why so little antimatter is found in nature is one of the mysteries that particle physicists hope to solve when the giant **SUPERCONDUCTOR SUPERCOLLIDER** (SSC) becomes operational sometime around the year 2000. Antimatter propulsion systems are a favorite of science fiction writers and the U.S. Air Force has funded some research on the subject. See **SUBATOMIC STRUCTURE**, **BIG BANG**, **COLLIDER**, and **MATTER**.

Aphelion The point on an elliptical orbit at which a planet or comet is farthest from the Sun. The Earth, for instance, was at aphelion on July 3, 1990, at which time it was 94,508,105 miles from the Sun. This is not to be confused with apogee, the point in the orbit of any celestial body or man-made satellite at which it is farthest from Earth. See also **PERIHELION**.

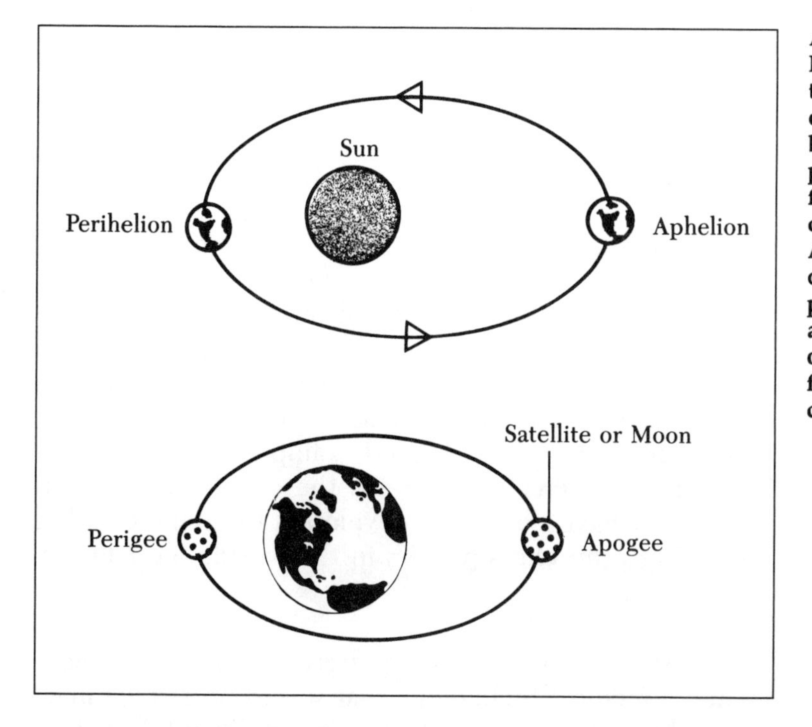

APHELION and PERIHELION are those two points on an elliptical orbit at which a planet or comet is farthest from or closest to the Sun. APOGEE and PERIGEE are those two points on an orbit at which a satellite or celestial body is farthest from or closest to the Earth.

Aquaculture The cultivation of fish and plants in natural or controlled underwater environments. Fish farming has soared in the United States in recent years. In 1980, fish farms provided only about 1 percent of all fish consumed. By 1990 that figure had reached 10 percent, and the U.S. Commerce Department has predicted that the figure will rise to 20 percent by the turn of the century. Aquaculture is an economically attractive alternative to commercial fishing for some species. Species being farmed today include striped bass, shrimp, salmon, sturgeon, red-fish, oysters, clams, crayfish, catfish, and trout.

Artificial Intelligence (often called AI) Advanced computer systems capable of complex problem solving, pattern recognition, and decision making that can now only be accomplished by human beings. The objectives of AI are to imitate, by means of electronic machines, as much human mental activity as possible. Eventually, it is the hope of AI engineers to improve upon human decision making.

AI covers a broad spectrum of technologies that includes such fields as speech recognition, robotics, problem solving, and pattern recognition. Much of the attention so far has been focused on the development of *expert systems*, in which the essential knowledge of a particular specialty—medical, legal, financial, and so on—is programmed into a computer system. Expert systems are intended to mimic the knowledge, procedures, and decision-making ability of real people—sort of an electronic clone of some human decision maker.

The core concept of artificial intelligence is the use of **HEURISTICS**, sometimes called the art of good guessing. Heuristics allows computers to deal with situations that cannot be reduced to mathematical formulas and may involve many exceptions. A good example of machines exhibiting intelligent behavior are the chess-playing computers that have now been developed to mimic the skill level of the top chess players in the world. See **NEURAL NETWORKS**.

Artificial Organs Predictions about advancements in medical technology that would permit artificial organs have proven unrealistic. Experts now tell us that no permanent artificial organs are likely to replace worn-out hearts, lungs, livers, or kidneys by the year 2000. Researchers underestimated the problems associated with blood clotting, infection, miniaturizing power sources, and reliability. See **BIONICS**, and **ORGAN TRANSPLANTS**.

Artificial Reality Sometimes called *virtual reality*, this computer-based concept involves combining three-dimensional simulations of reality with tools that can sense a user's eye, hand, and body movements. Changes in the visual model are then coordinated with the user's real movements by means of computer software. The overall effect is startling: The user sees three-dimensional moving pictures that are constantly adjusted as the user "moves" through the scene by turning his or her head or moving his or her hand or body. For example, a user can perform repair work remotely while a robot does the real repair in some dangerous or inaccessible location. Another application might involve learning surgery. A simulated scalpel can be moved through a patient's simulated body. Or a distant spacecraft or undersea robot can be controlled with precision while the operator is safely on the ground. Artificial reality has great potential for designers of new aircraft or automobiles, or any other complex mechanism, because the designer can manipulate the design as if it were real.

We currently perceive and experience reality by means of light, or soundwaves, or motion, which our brains interpret as real. Thus if a computer can send signals to mimic those sent by real objects, the result will be virtually indistinguishable from the real thing. This technology is still in its infancy, and the creation of mock-ups of reality that send the right signals back to the user is expensive.

Asteroids Relatively small (in comparison to the planets), solid objects that orbit the Sun and shine by reflective light. Most asteroids are located in what is called the *asteroid belt* between the orbits of Mars and Jupiter. Although asteroids number in the hundred thousands, their total mass is but a small fraction of Earth's. They range in diameter from less than a mile to 600 miles. Some asteroids are made largely of rock, others of pure nickel-iron alloy, and others are made of a carbon compound something like tar.

The U.S. GALILEO spacecraft will take a close look at an asteroid named Gaspra in 1991 and fly by a second asteroid, Ida, in 1993. It has been proposed that we land astronauts on an asteroid for firsthand scientific study. Because asteroids have little or no gravity, landing and takeoff would be technologically simpler than it was on the Moon. Asteroids also present a possible hazard to Earth. A small, approximately 30-foot-wide asteroid came within only 106,000 miles of Earth on January 17, 1991. The Moon is nearly 239,000 miles from Earth and therefore this asteroid came within about half that distance. Astronomers say that if this asteroid (called Asteroid 1991 BA) was composed of iron and fell to Earth, it would have been able to withstand the heat of plunging through the atmosphere and would have made a 300-foot crater perhaps 100 feet deep. It has been estimated that an asteroid crashes into Earth about once every 250,000 years, and that such a collision would have disastrous results. Scientists tell us that if an asteroid landed in the middle of the Atlantic, it would propel a wall of water hundreds of feet high onto the adjacent continents. One theory about the extinction of the dinosaurs and most other forms of life on Earth 65 million years ago involves the collision with a large asteroid and the resulting drastic change in Earth's climate. Today, technologies are available to prevent such a collision. Space scientists tell us that it should be possible to send a piloted mission to a threatening asteroid and plant a few large rockets on it that are powerful enough to nudge the massive object off course just enough to miss Earth entirely. See **ASTRONOMY**, **ASTROPHYSICS**, **COMETS**, and **METEOROIDS, METEORS, AND METEORITES**.

Astigmatism Distorted vision caused by an uneven curvature of the cornea, the outside front portion of the eye. Astigmatism sometimes occurs in conjunction with farsightedness or nearsightedness. It is usually present from birth and does not grow worse with age. Glasses, contact lenses, or surgical techniques are used to correct astigmatism. See **KERATOTOMY**, and **LASERS, MEDICAL USE OF**.

Astrology The belief, unsupported by any creditable scientific evidence, that human affairs and people's characters and personalities are influenced by the positions of the planets. Astrology is a popular pseudoscience—the astrologers of the world far outnumber the astronomers—but it should not be confused with the real science of **ASTRONOMY**. The revelation (in Donald Regan's book, *For The Record*) that the former First Lady, Mrs. Reagan, had an astrologer, by whom "virtually every major move and decision the Reagans made" had to be approved in advance came as a shock to many, but not to all. In a review of the Regan book in the *New Yorker*, Francis FitzGerald said, "if at least one of the Reagans did not believe in astrology the couple would be a statistical improbability in California." See also **ZODIAC**.

Astronomical Cycle Scientists believe astronomical cycles touch off changes in the ocean-atmosphere system that drives the world's climate. Glacial cycles (ice ages) are set in motion by (1) periodic wobbles in the tilt of Earth's rotation, (2) changes in the tilt of its axis, and (3) the shape of its orbit, occurring over tens of thousands of years. By altering the angles and the distances from which the Sun's energy reaches Earth, the three overlapping cycles control the timing of global warming and cooling, and the long-term advance and retreat of glaciers. See CLIMATE.

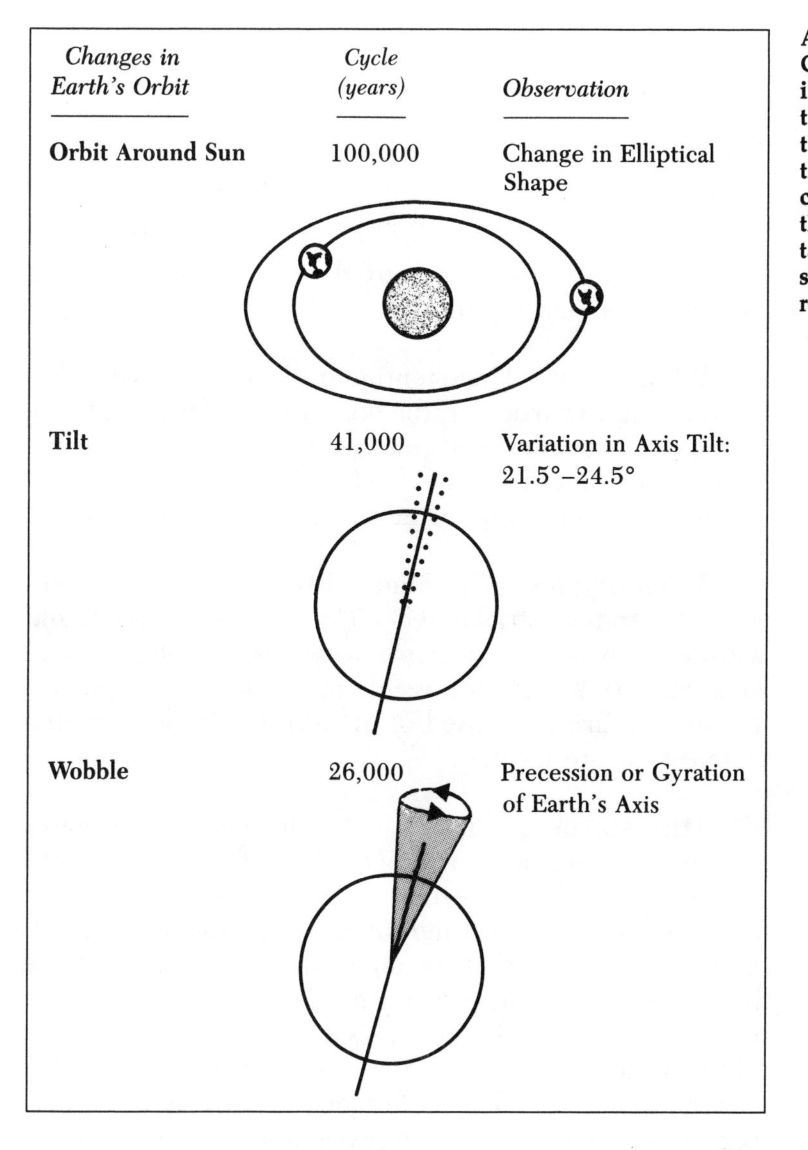

ASTRONOMICAL CYCLE or changes in the Earth's orbit, tilt, and wobble that control the timing of climate changes by altering the angles and distances from which solar energy reaches Earth.

Astronomical Unit (AU) A measure of distance in space, defined as the mean distance between Earth and the Sun, equal to 92.81 million miles. Astronomical distances are so great that special units are required. Were astronomers to use miles or kilometers to measure distances in space they would soon be filling the page with zeros. Our mental image of the universe should include some understanding of the vast distances involved. Estimated mean distances of the closer planets from the Sun in AUs and miles is shown below:

	AUs	Miles
Mercury	0.387	36,000,000
Venus	0.723	67,000,000
Earth	1.000	93,000,000
Mars	1.524	141,000,000
Jupiter	5.202	480,000,000
Saturn	9.539	900,000,000

Two other measures of distance in space are LIGHT-YEARS, and PARSECS.

Astronomy The scientific study of the universe beyond Earth. In particular, the observation of the positions, motions, and evolution of celestial bodies and phenomena. Astronomy should not be confused with ASTROLOGY, a pseudoscience not supported by any creditable evidence.

Astrophysics The study of the physics and chemistry of extraterrestrial objects. The alliance of physics and astronomy was made feasible when the development of SPECTROSCOPY made it possible to determine what celestial objects are composed of, in addition to determining where they are located.

Atmosphere, Earth's A thin envelope of gases held in place around Earth by the gravitational pull of the planet's mass. Earth's original atmosphere was probably composed of ammonia and methane. It was not until 20 million years ago that an atmosphere approaching the modern composition evolved because of changing physical conditions on the Earth's surface and the evolution of plant life. Today's atmosphere on Earth, apart from its water vapor and various pollutants, is made up of 78 percent nitrogen and 20 percent oxygen, with argon and carbon dioxide each making up less than 1 percent. Trace amounts of helium, hydrogen, krypton, methane, neon, ozone, and xenon also exist.

Oxygen is part of Earth's atmosphere today because of life and the process of PHOTOSYNTHESIS—the ability of early plant life to thrive by harnessing the energy of sunlight. A waste product of photosynthesis is gaseous oxy-

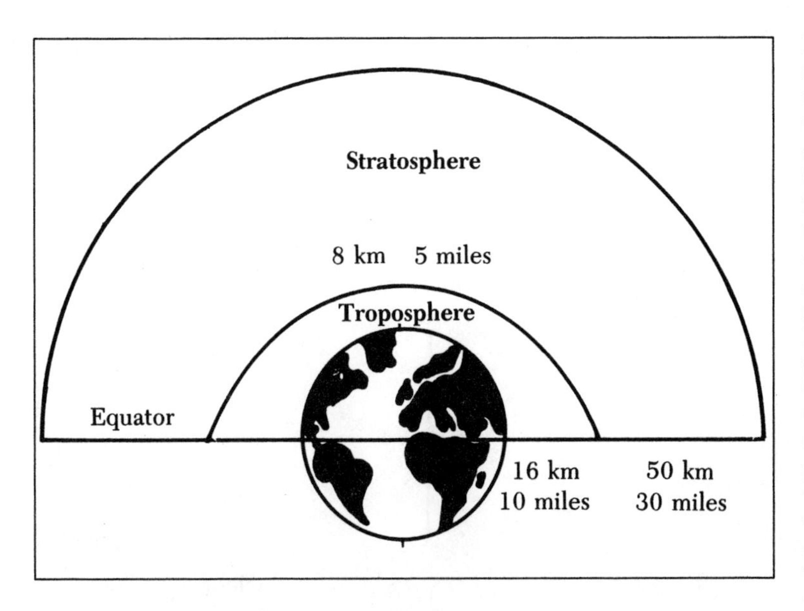

EARTH'S ATMOSPHERE consists of two main layers: the TROPOSPHERE, or lowest layer, extending from 8 to 16 kilometers and the STRATOSPHERE, extending up to an altitude of 50 kilometers. In the lower layer, temperatures decrease with increase in altitude, whereas in the upper layer temperatures rise with altitude and warmer air sits above cooler air. The troposphere is the region that contains winds, rain, storms, and other weather events.

gen—photosynthetic organisms take in carbon dioxide and release oxygen. Our atmosphere is an integral part of our global ecosystem. Altering the concentration of the natural component gases of the atmosphere, or adding man-made new ones, can have serious consequences for the Earth's life system. See **BIOSPHERE**, **ECOSYSTEM**, **STRATOSPHERE**, and **TROPOSPHERE**.

Atoms Every form of matter in the world is made up of atoms, which are the fundamental units of a chemical element. Each atom is composed of a central, positively charged nucleus—only a small fraction of the atom's volume, but containing most of its mass—surrounded by a cloud of much lighter, negatively charged **ELECTRONS**. The number of electrons in an atom matches the number of charged particles, or **PROTONS**, in the nucleus and determines how the atom will link to other atoms to form molecules. Electrically neutral particles (**NEUTRONS**) in the nucleus add to its mass but do not affect the number of electrons and therefore have almost no affect on the atom's link to other atoms.

Combinations of atoms make a **MOLECULE**. For instance, a molecule of water is made of two atoms of hydrogen bound to one atom of oxygen. Other kinds of molecules are much more complex than water. A single protein molecule, as an example, may be composed of tens of thousands of atoms. See **MATTER**, and **SUBATOMIC STRUCTURE**.

ATOMS consist of an electron or clouds of electrons orbiting an atomic nuclei. The nuclei is composed of nucleons—protons and neutrons. The protons and neutrons are in turn composed of trios of quarks. Quarks account for most of the matter in the universe. Shown is a hydrogen atom consisting of one neutron and one proton.

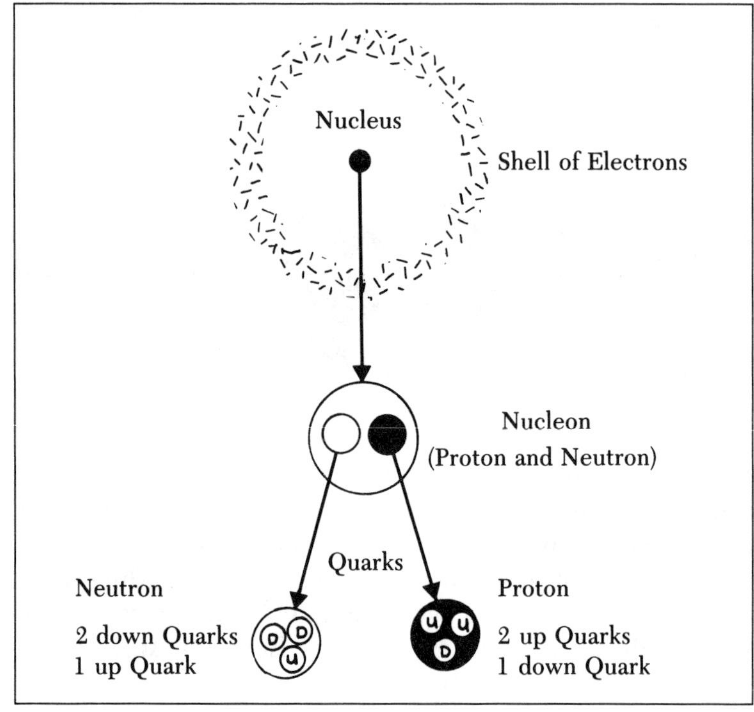

Atomic Number See **PERIODIC TABLE OF THE ELEMENTS**.

Atomic Weight See **PERIODIC TABLE OF THE ELEMENTS**.

Australopithecus Our common ancestor, this is the name given to any of several extinct humanlike primates judged to be the earliest known human ancestors after the human lineage evolved away from the ape lineage. The literal meaning of the term is *southern ape,* because the first fossil remains of this type were found in southern and eastern Africa.

Australopithecus was a short, small-brained creature with a snoutlike face, who walked erect, used tools, and probably used a primitive form of speech. In 1977, a **HOMINID** fossil perhaps four million years old was discovered by American archaeologists Donald Johanson and Tom Gray. Enough bones were found to make up 40 percent of a complete individual, about three and a half feet tall. The scientific name is *Australopithecus afarensis,* but she is better known today as Lucy, now only one of many examples of this species, the earliest undisputed hominid.

AZT (Zidovudine) An antiviral drug used in the treatment of AIDS. AZT is not a cure for AIDS, but the drug has prolonged the lives of patients who can tolerate its strong side effects. In some cases, AZT blocks the replication of HIV, the retrovirus that causes AIDS. See **AIDS**, and **HIV**.

Bacteria The smallest and simplest living organisms on Earth. (Viruses are smaller but they exist at the borderline between living matter and nonliving matter and are thus not considered an organism.) Bacteria, sometimes called germs, are single-cell microorganisms without organized nuclei. The term *microorganism* refers to any form of microscopic life. Bacteria are neither plants nor animals but fall into a different classification called MONERA. They differ from plants and animals in their much simpler internal anatomies. Bacteria DNA is scattered within the cell rather than being contained in a discrete nuclei as it is in animal and plant cells.

Most bacteria species are harmless to humans, existing as free-living organisms that are the chief agents of organic decay. A few are parasites and carry disease. Bacteriological diseases are treatable with antibiotics whereas virus-based illnesses are not. Other bacteria, such as *Escherichia coli,* inhabit the human intestine and are needed to help digest food. Because of their simplicity, bacteria have proven valuable in the study of genetics. Bacteria are used in RECOMBINANT DNA technology to synthesize proteins; animal genes, including human genes, spliced into bacterial genes work perfectly. See CELL, DNA, GENETIC ENGINEERING, TAXONOMY, and VIRUS, BIOLOGICAL.

Balloon Therapy See ANGIOPLASTY.

Beta Particles A high-speed ELECTRON or POSITRON, especially one emitted in radioactive decay. A beta ray is a stream of beta particles and one of three main types of electromagnetic RADIATION, the other two being ALPHA and GAMMA particles.

Big Bang The accepted theory or STANDARD MODEL for the origin of the universe. This theory holds that the universe began when a single point of infinitely dense and infinitely hot matter exploded spontaneously. It is believed that this point, smaller than the head of a pin, exploded with an almost unimaginable burst of energy. Out of the debris of this explosion, all of the galaxies, stars, and planets were formed. This event, sometimes called a **SINGULARITY** by astrophysicists, is believed to have occurred between 10 and 20 billion years ago. Time begins at the Big Bang.

The discovery in 1929, by the American astronomer **EDWIN P. HUBBLE**, that the universe was expanding led logically to the conclusion that if one could reverse the outward-expanding motion of the galaxies, they would all eventually come back to one point. In short, the universe would have an origin. As experimental and theoretical evidence mounted, it became clear that the universe indeed must have had a beginning in time. In 1970 this was finally proven mathematically by Roger Penrose and Stephen Hawking, on the basis of **EINSTEIN'S GENERAL THEORY OF RELATIVITY**.

Big Crunch The Big Bang in reverse. Although the Big Bang is the accepted theory for the beginning of the universe, there is no generally accepted theory for the end. The universe may go on expanding forever, getting colder, emptier, and deader. This is called the **OPEN UNIVERSE** theory. On the other hand, it may someday collapse back into a cataclysmic implosion called the Big Crunch, breaking down the **GALAXIES**, **STARS**, **ATOMS**, and atomic nuclei into their original constituents. This latter belief is called the **CLOSED UNIVERSE** theory. If the Big Crunch is the fate of the universe and the end of time, it will be billions of years in the future, because the universe should take as long to collapse as it did to expand.

Binary In our everyday life, we represent numbers using ten symbols: 0,1,2,3,4,5,6,7,8, and 9. We call this method of representation the base-10 or decimal system. A computer can use only two symbols, 0 and 1, because it depends on two-state internal components: A transistor is either conducting or nonconducting, or a switch is either open or closed, or a memory unit is either magnetized or not magnetized. Systems using the two symbols 0 and 1 are called base-2 or binary number systems. The two symbols are known as binary digits or **BITS**. To permit computers to process information in character form (letters of the alphabet, numbers, punctuation, and other signs), some system for representing each character had to be devised. This character code, analogous to the Morse code in transmitting telegraph messages, is called the **AMERICAN STANDARD CODE FOR INFORMATION INTERCHANGE** or **ASCII**. See also **BYTE**.

Binary Star A double star system in which two **STARS**, or a star and a **BLACK HOLE**, are bound together by their mutual gravitation. See **GRAVITY**.

Biodegradable Capable of decaying through the action of living organisms. The ability of a compound to decompose, or be absorbed by the environment, by natural or biochemical means. An example would be biodegradable paper that, when discarded to a landfill, would eventually decompose as opposed to plastic containers, which would never decompose naturally.

Bioelectrodes Tiny electronic components that can be implanted in the body to release small amounts of drugs or other biologically active materials in response to a pulse of electric current. A likely application of this device would be in the treatment of **PARKINSON'S DISEASE**, a degenerative neurological disorder. The bioelectrode could be implanted in the brain of a patient and, at a signal, release small amounts of *dopamine*, a chemical that Parkinson's patients lack. Because the device would be implanted directly in the affected organ, far smaller amounts of drugs would be effective. Bioelectrodes could also be designed to deliver insulin to diabetics on preset schedules.

Bioelectrodes are based on the discovery of conductive polymers—chemicals that intrinsically conduct electric current—that can be chemically bound with drugs. When a pulse of current passes through the bioelectrode, a chemical reaction releases the drugs from the polymer. See also **DRUGS, DESIGNER**.

Bioethics The field of study concerned with the ethical problems inherent in some biological and medical procedures, technologies, or treatments. Examples include the philosophical implications of ORGAN TRANSPLANTS, GENETIC ENGINEERING, or prolonging life for the terminally ill.

The determination of who is to receive available organs for transplantation is an example of a bioethical problem. Research involving human fetal tissue transplants is another example of a controversial technology and ethical or moral issue. By transplanting human fetal brain tissue into patients suffering from neurological problems, researchers have, in some cases, been able to relieve the symptoms. Currently, the Bush administration's view of the moral status of the human embryo has limited federally funded research on fetuses. The controversy centers on the abortion issue. Antiabortionists are concerned that the medical use of human fetal tissue may in some way encourage abortion.

Biological Clock The intrinsic biological mechanism responsible for the periodicity or time-dependent aspects of behavior in living organisms.

In humans it is the biological clock that controls the natural rhythms of life. Women, for instance, answer to their internal biological timer when they ovulate or otherwise respond to their menstrual cycle. Plants are responding to their internal clocks when they open or close their petals. Virtually all species of life have clocks that regulate their metabolisms in a 24-hour, day-night cycle called a CIRCADIAN rhythm. When we travel long distances rapidly, we upset our internal biological clocks because we have difficulty matching our activities with those of the locals: Our internal timer tells our bodies that it is time to be sleeping while everybody else is working. When we try to ignore this, our hormonal secretions do not match our activities, and we end up feeling tired and inefficient. When this occurs, we say we are suffering from JET LAG. See CHRONOBIOLOGY.

Biomass The weight or volume of all living things in a given geographical area. It is a measure of the abundance of life, both plant and animal, that a particular area can support. It is often used in reference to organic plant matter that is available for conversion to fuel and, therefore, is considered a measure of potential energy source.

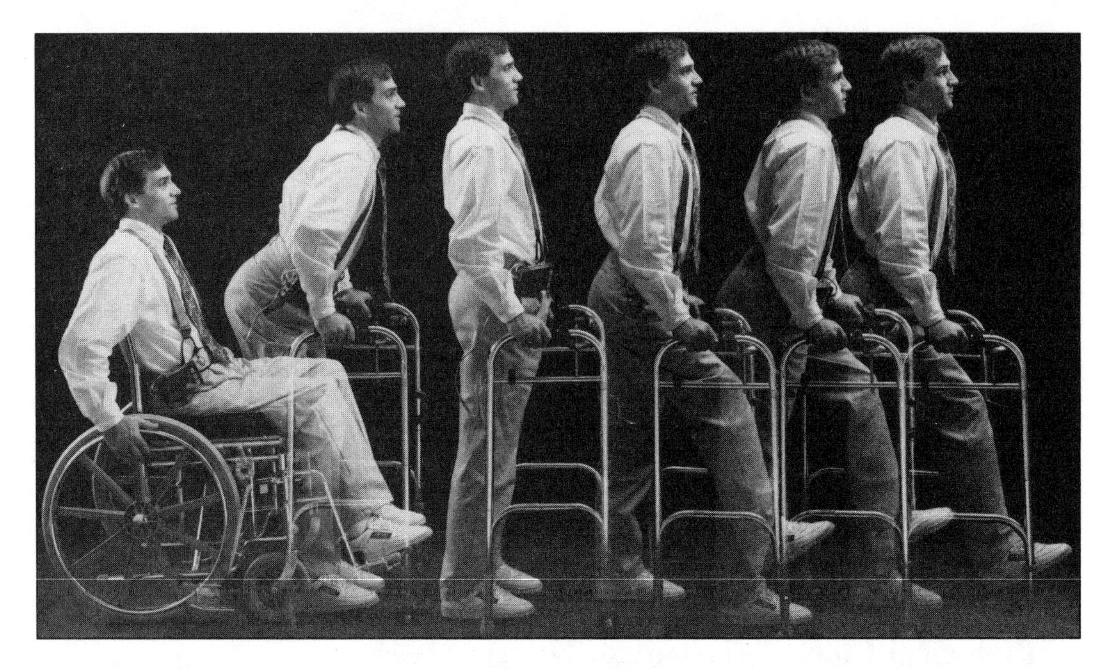

BIONICS. A new beeper-sized device has been developed by Sigmedics, Inc., of Northfield, Illinois, that may someday help some of the 250,000 paraplegics in the United States move about without a wheelchair. The Parastep System™ works by electrically stimulating specific leg muscles. The muscles then contract in much the same way they do when a nonparalyzed person walks.
Source: Sigmedics, Inc. Photo.

Bionics In the medical sense, this term refers to the development of **ARTIFICIAL ORGANS**. A remarkable variety of spare parts have been derived from this new science including artificial ears, synthetic skin, artificial blood vessels, a pocket-sized artificial kidney, and artificial joints and limbs. The term also refers to any electromechanical device that is made to emulate the behavior of a living organism. Robots of various kinds that move about on jointed limbs are called bionic machines. Chess-playing computers that imitate the thought processes of master chess players by the use of **ARTIFICIAL INTELLIGENCE** may also be thought of as bionic.

The most common use today is the medical one, as once portrayed in the popular TV series "The Bionic Man." Today's bionic technology has almost turned TV science fiction into medical reality. Artificial joints (hip, shoulder, elbow, knee, and ankle) can now be made to order. Bioengineers use a reference bank of implant designs, special software, and a computer to design joints to meet a patient's specific needs. New generation artificial legs and arms are controlled electromechanically by the will of the wearer.

Biophysics The study of the physical forces and phenomena involved in living processes. Biophysics is one of the many branches of physics, which includes astrophysics, geophysics, atmospheric physics, and acoustical physics, to name a few. All of these branches study and analyze the physical properties, interactions, and processes of matter and energy. In biophysics, the methods of physics are applied to the study of biological structures and processes. Today, the complex molecules that form the key units of living tissue, the PROTEINS and NUCLEIC ACIDS, are being studied using the techniques made possible by modern advancements in chemistry and physics. The new science of molecular biology represents a merger of biochemistry and biophysics.

Bioremediation The use of bacteria and other microbial forms of life to help clean up hazardous waste in soil or water. BACTERIA that have an appetite for oil have been used on the beaches of Alaska to assist in the Exxon Valdez cleanup. The Mega Borg accident in the Gulf of Mexico in 1990 marked the first time that laboratory-grown bacteria were applied to an oil spill in the open seas. Bacteria are also used in oil wells to break up paraffins and other materials that clog production. Others have been used in restaurants to consume grease and substances that cause odors. Researchers are currently at work on developing species that will attack metals, inorganic chemicals, and COMPOUNDS containing chlorine.

Bacteria have been used for many years to process sewage in municipal plants. Industrial plants and food processors also have used bioremediation to reduce the amount of organic pollutants they dump into rivers or sewers.

Biosphere The totality of regions of the planet that support self-sustaining and self-regulating ecological systems—in short, the totality of life on the planet. This life-supporting region of the planet is a combination of part of the atmosphere, the HYDROSPHERE (the waters of the Earth) and the LITHOSPHERE (the solid part of Earth).

The biosphere can be thought of as thin film on the surface of the planet. From the deepest ocean depths to the upper limit of the TROPOSPHERE, this life-supporting layer is no more than 20 miles thick. The biosphere supports life and is affected by the presence of life. Plant life, for example, removes carbon dioxide from the air, uses the carbon for synthesizing sugars, and releases oxygen. This process is responsible for the oxygen in the atmosphere today. In his book *Gaia, a New Look at Life on Earth,* scientist J. E. Lovelock suggests that life orchestrates the state of the atmosphere to its own advantage—that, in effect, the biosphere itself is alive. See **GAIA**.

Bit A binary digit or single character of a language having only two characters, 0 and 1. A unit of information (either the 1 or the 0) reflected by electrical or mechanical devices in computers that have only two states such as open or closed, charged or uncharged, or magnetized or unmagnetized. Information is manipulated, processed, or stored in computers using combinations of eight bits called **BYTES**. See **AMERICAN STANDARD CODE FOR INFORMATION INTERCHANGE (ASCII)**.

Black Holes Regions in space into which matter is sucked by extremely intense gravitational attraction and from which nothing, not even light, can escape. The concept of black holes is attractive to most astronomers and physicists because it is a logical extension of the **GENERAL THEORY OF RELATIVITY**, and because black holes explain a lot of otherwise unexplainable phenomena; however, not all scientists are convinced that black holes exist. MIT's noted physicist, Philip Morrison, has said, "I'll believe in black holes when I see one." (Because a black hole will not let light escape, there is no observational evidence for their existence and Professor Morrison was making his little joke.)

Do massive **STARS** end their lives as black holes? One approach to this problem is to use a mathematical model of a massive star and then project its evolution and end. According to models of this type, a massive star will form an iron core at its center as it burns out its fuel. The iron core will then collapse, and the rest of the star will follow, forming a black hole. Computer models indicate that stars starting with 20 solar masses (i.e., 20 times the size of our **SUN**) will indeed form black holes.

Boolean Logic Named after the English mathematician George Boole, who developed symbolic logic that reduced logical relationships to simple expressions—And, Or, and Not. Boole demonstrated how types of statements employed in deductive logic could be represented by symbols and manipulated by fixed rules to yield appropriate conclusions. Boolean algebra, as it is sometimes called, permitted the automation of logic and the formulation of logic problems for application on computers.

Bosons One of the many postulated subatomic particles that physicists use to describe the nature of MATTER and ENERGY as simply as possible. This description, or set of theories, known as the STANDARD MODEL, postulates that all matter in the universe, from your pet cat to the most distant GALAXIES, is composed of just four elementary particles (two types of QUARKS, ELECTRONS, and NEUTRINOS) and an entirely separate set of particles called *bosons* that transmit forces back and forth between the elementary particles. See FERMIONS, and SUBATOMIC STRUCTURE.

Brahe, Tycho Danish astronomer (1546–1601), founder of Western observational astronomy, and patron of JOHANNES KEPLER. Brahe's observations over a 20-year period added more to the stock of astronomical facts than anyone before him. See also COPERNICUS, and GALILEI, GALILEO.

Breeder Reactor See REACTOR, BREEDER.

British Thermal Unit (Btu) The common unit used to measure energy. A Btu is defined as the amount of energy required to raise 1 pound of water by 1 degree FAHRENHEIT (specifically from 39.2°F to 40.2°F).

Because ENERGY comes in many forms, with different physical qualities and different capacities, it is sometimes difficult to convert one measure to another or to compare one form to another. The often-used common unit is the Btu. A barrel of crude oil, for example, contains 5.8 million Btus. When large amounts of energy are discussed, it is convenient to use a unit called a *quad*, defined as a quadrillion (10^{15}) Btus.

Brown Dwarfs Celestial objects that are bigger than planets but smaller than STARS. Brown dwarfs form in the same way as a star but have too little MASS to become hot enough inside to trigger hydrogen fusion. Sometimes called "failed stars," brown dwarfs are notoriously difficult to detect, and astronomers have so far come up with only a few candidates. They have discovered a number of objects in the MILKY WAY that appear to have surprisingly low masses and appear alone rather than as a satellite to a larger star. See **DWARF STARS**, **FUSION, NUCLEAR**, and **HYDROGEN**.

Brownian Motion The irregular movement of minute PARTICLES of MATTER when suspended in a liquid; named for botanist Robert Brown, who first noted this phenomena while watching microscopic pollen grains float around in water. Brown noted that although the movement of any particular grain was unpredictable, all the grains moved faster when the water got hotter and slower as it cooled. EINSTEIN later wrote a paper on Brownian motion in which he theorized that the grains were being battered around by water MOLECULES and the hotter the water, the faster the water molecules moved. Brownian motion is now an important substantiating factor of the kinetic molecular theory of matter, which states that matter is composed of tiny particles (molecules) that are constantly in motion. See **HEAT**, and **ATOMS**.

Buckyballs Name given to a newly discovered material—actually symmetric MOLECULES of carbon found in soot. This new class of material is an extremely stable form of 60-carbon called a Buckminsterfullerene or "Buckyball" because its shape resembles that of Buckminster Fuller's geodesic domes. This new class of materials joins diamonds and graphite as the third known form of carbon. Like graphite, the new materials might be used as lubricants or as a superhard coating.

Bugs Computer jargon for errors in the SOFTWARE code that cause the program to do strange things like hang up, crash, dump, or otherwise ruin your data. Bugs are design flaws, sometimes called "undocumented features," by the software companies that are responsible. Because software programs are complicated, they are routinely subjected to debugging procedures during their design phase. Nevertheless, design flaws do sometimes slip through. Bugs are unintended errors not to be confused with a computer VIRUS, which is designed to cause trouble.

Bus In computerese, the term refers to the system of wires that carry data throughout the computer, to and from the microprocessor (CPU). The bus also determines the rules for hooking up memory units, disk drives, and other add-ons. The industry standard in the personal computer world is called the *AT Bus*, the bus used by IBM in their Advanced Technology (AT) models. IBM clones—machines like IBMs—also use the AT bus. Apple (and Macintosh) machines use a different system called the *Apple Bus*.

Byte In computer language, a collection of eight binary digits, or BITS. A binary digit can be either 1 or 0. A typical byte would look like this: 11010011 00110110 11110001 11101100.

In the ASCII code used by most computers, each byte represents an upper- or lowercase letter of the alphabet or a digit, punctuation marks, mathematical and other signs, and a set of control characters used to facilitate data handling. See **AMERICAN STANDARD CODE FOR INFORMATION INTERCHANGE (ASCII)**, and **BINARY NUMBERS**.

CAD/CAM Computer jargon for Computer-Aided Design/Computer-Aided Manufacturing. SOFTWARE programs intended to provide a basic data base for designers to use in the creation of new products, buildings, equipment, or whatever. CAD programs also permit designers to test new concepts and ideas without actually constructing or building the design. CAM software provides control by computers of highly complex manufacturing systems. Increasing automation of production lines requires less direct labor and fewer skilled crafts, but more engineering and computer programming.

CAD/CAM, computerjargon for computer-aided design/computer-aided manufacturing. CAD/CAM provides engineers and designers with an invaluable tool with which they can create new products or equipment or test new concepts.
Source: IBM.

Calorie A unit of HEAT energy—the amount of heat energy required to raise 1 gram of water 1 degree Celsius at sea level. Also used to measure the inherent heat energy in food that the body burns when at work or exercise, or stores as fat when at rest. The Calorie content of a particular food is measured by actually burning a specified amount of the food and measuring the amount of heat generated.

When this unit is used by physicists in their research, a lowercase c is used. This unit is too small for practical use by food nutritionists, who use the term with a capital C, which is equal to 1,000 calories. It's a little confusing, but when we see the units listed on a food package of some sort, the reference is to Calories (uppercase C). See **ENERGY**.

Cambrian A period of the **GEOLOGIC TIME SCALE** corresponding to 500 million years ago. The first or oldest period of time in the **PALEOZOIC ERA**, characterized by warm seas and desert land areas.

The Cambrian period has long intrigued evolutionary biologists because it was the focus of an unprecedented sweep of events that changed the face of nature. For more than 3.5 billion years, Earth's oceans had been filled with bacteria, algae, and a host of other single-celled microorganisms. Over a few million years, those microorganisms gave way to the complex plants and animals that are the ancestors of today's life. See **EVOLUTION**.

Carbon Cycle Circulation of carbon from plants, which take in **CARBON DIOXIDE** (CO_2) from the air and convert it into carbohydrates (any compound made of only carbon, hydrogen, and oxygen) by means of **PHOTOSYNTHESIS**.

Carbon Dating A method of estimating the age of any organic material. Because every living thing on Earth contains carbon, and because the **HALF-LIFE** of carbon-14, for example, is known to be 5,570 years, this radioactive substance is particularly useful in establishing the age of artifacts or **FOSSILS** found by archaeologists. Radiometric age-dating has a wide range of applications in fields extending from geology and geophysics to astrophysics and **COSMOLOGY**.

After 5,570 years, half of the carbon-14 atoms in any given sample will have decayed into atoms of nitrogen-14. By comparing the amount of the carbon-14 to nitrogen-14, it is possible to age-date the sample. For instance, if three-quarters of the carbon-14 has turned to **NITROGEN**, then the material under examination is 11,140 years old. Carbon dating is useful for objects up to 80,000 years old. Older material does not have enough carbon left to accurately measure. Other radiometric age-dating methods are available using radioactive **ATOMS** that decay at a slower rate than carbon. See **ISOTOPE**, and **RADIATION**.

Carbon Dioxide (CO_2) A colorless, odorless gas produced when we breathe or when organic material decays or is burned. Animal life takes in **OXYGEN** and exhales a mixture of oxygen and carbon dioxide, whereas plant life takes in carbon dioxide and produces oxygen. Although carbon dioxide is a natural constituent of our atmosphere, the amount of carbon dioxide in the atmosphere has increased by half again its original amount since 1900, as a result of the burning of oil and coal. The increase in CO_2 creates no problem as far as breathing is concerned. It does, however, add to the **GREENHOUSE EFFECT**, which could lead to **GLOBAL WARMING**.

Most energy used in the world today is derived from the burning of fossil fuels—coal, oil, and natural gas. All burning of fossil fuels dumps into the atmosphere waste

CARBON DIOXIDE (CO_2) has increased in the atmosphere by a significant amount in the past 30 years. CO_2 is thought to be the major contributor to the greenhouse effect and possible cause of global warming. *Source: Maunaloa Observatory.*

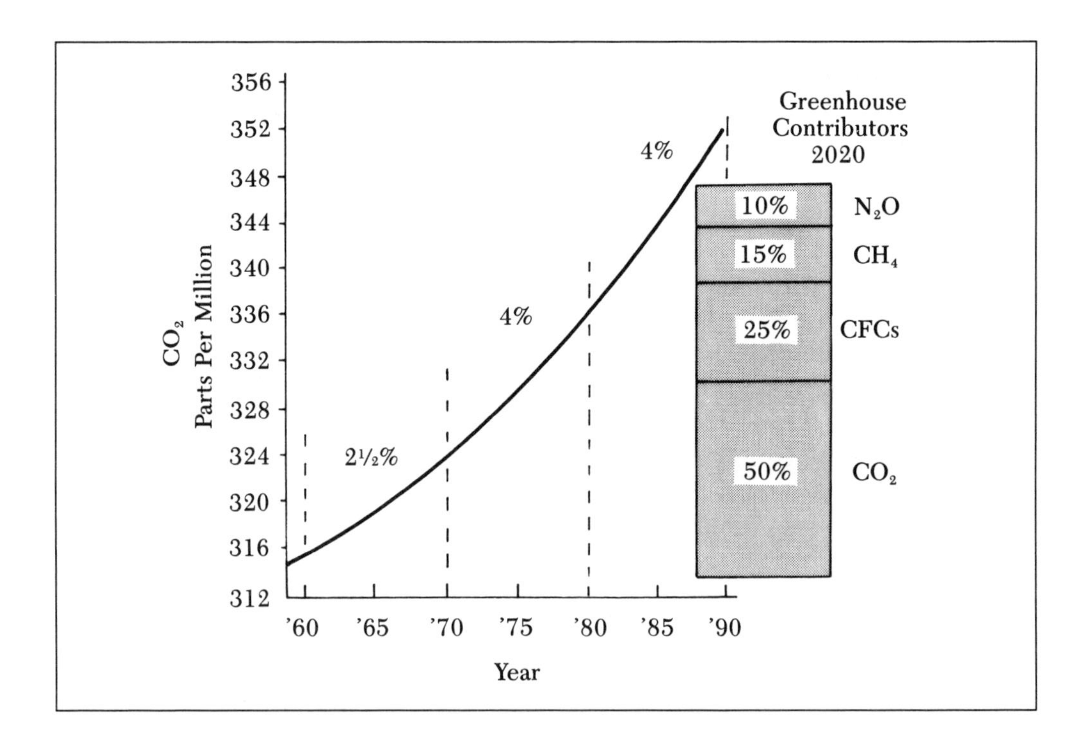

products, primarily CO_2, that may threaten health and life. A buildup of CO_2 can trap heat and raise the planet's average temperature. Climatologists argue about whether the greenhouse effect is changing our climate now, or if global warming will not occur for several decades. There is general agreement, however, that the continued burning of fossil fuel and the continued emission of CO_2 will eventually lead to a warmer Earth. Adding to the concern were reports by climatologists in January 1991 that global average surface temperature during 1990 was the highest in more than a century of weather measurements. See CLIMATE.

Carbon Monoxide (CO) A colorless, odorless, poisonous gas produced when carbon burns with insufficient air. Most of the carbon monoxide in our air comes from automobile engines; emission standards for cars as well as clean air requirements for cities are set by the federal government. The U.S. Environmental Protection Agency (EPA) has set allowable levels of exposure to carbon monoxide at 35 parts per million over an hour, or nine parts per million for eight hours, but many cities in the United States exceed these standards.

Carcinogen Any cancer-causing substance. Biologists believe that humankind has introduced many carcinogenic (cancer-producing) substances into our environment within the last two or three centuries. The increased use of coal and the burning of oil on a large scale, particularly in gasoline engines, are two examples of the introduction of carcinogens into the atmosphere. Cigarette smoking is, of course, the most obvious cancer-causing activity. The growing use of synthetic chemicals in food and cosmetics are two more common sources of carcinogens.

The OZONE layer in the upper atmosphere shields the surface of Earth from carcinogenic ULTRAVIOLET (UV) RADIATION. Decreases in the amount of ozone in this shield can lead to increased UV and increases in SKIN CANCER. A dangerous source of skin cancer is overexposure to sunlight. Health officials have warned Americans to minimize sunbathing, saying the death rate from the most deadly form of skin cancer linked to sunlight exposure jumped more than 25 percent in the past decade.

The number of malignant MELANOMA deaths increased 26 percent from 1973 to 1985, the U.S. Centers for Disease Control in Atlanta reported in 1990. A direct cause-and-effect relationship between ozone reduction and increased skin cancer has been speculated. See **CHLOROFLUOROCARBONS (CFCs)**, and **CLIMATE**.

CASSINI spacecraft shown arriving at Titan, Saturn's largest moon, in this artist's drawing. *Source: NASA/JPL.*

Cassini Planned as a joint mission with the European Space Agency, the NASA-built *Cassini* spacecraft will be launched in 1996, fly by JUPITER in 1999, and arrive in the orbit of the planet SATURN in 2002. Once there, the spacecraft will launch the European-built Huygens probe, a saucerlike object, to the surface of TITAN, one of Saturn's known 14 moons. Scientists are interested in Titan because it is encased in a dense atmosphere that possibly could be hiding an ocean. It is believed that the organic-rich nitrogen atmosphere on Titan may nurture chemical processes similar to those that were at work on Earth before life developed. The *Cassini* spacecraft, named after the French-Italian astronomer who discovered several of Saturn's moons in the 17th century, will be launched from the U.S. space shuttle. See **SPACE EXPLORATION**.

Catalyst A substance, usually present in relatively small amounts, that modifies and increases the rate of chemical changes without being itself affected. A good example of a catalytic effect is the process by which small amounts of synthetic chemicals called **CHLOROFLUOROCARBONS** (CFCs) can destroy significant amounts of the Earth's protective ozone layer. In this example, a catalytic chain reaction causes relatively small amounts of CFCs to destroy large amounts of ozone in a process that is repeated over and over, each **MOLECULE** of the catalyst destroying thousands of ozone molecules.

Catastrophism The belief that a great and sudden calamity occurred sometime in the past that brought about violent changes in the Earth's surface and environment. It was once believed that the Earth's geological features, such as mountains, canyons, and oceans, were caused by sudden cataclysmic events, such as Noah's flood or an encounter with a huge comet. We now know that this concept is erroneous. Geological evidence shows that the Earth's major features were caused by gradual events continuing more or less uniformly over long periods of time. See **PLATE TECTONICS**, **CONTINENTAL DRIFT**, and **UNIFORMITARIANISM**.

Despite the considerable evidence to the contrary, catastrophism still has an appeal to many. In 1950, a psychoanalyst named Emmanuel Velikovsky wrote a book called *Worlds in Collision* in which he contended that in 1500 B.C., an encounter with a giant **COMET** caused the Earth to slow down and possibly stop spinning altogether. This in turn, Velikovsky contended, led to floods, fires, the collapse and rise of mountains, and, among other events, the biblical parting of the Red Sea. The book, and subsequent sequels, garnered a cult following that still believes that Velikovsky was not given a fair hearing. Cultish fads like Velikovskianism (or *Dianetics*, now called

Scientology, to name another example) have lasted and thrived for decades.

The belief that a large body of some sort collided with Earth sometime in the past is in fact supported by considerable scientific evidence (although not precisely in 1500 B.C.). However, it was Velikovsky's mixture of literary and biblical evidence on the one hand and sweeping claims about astronomical events—claims that violated laws of celestial mechanics—on the other hand that led the scientific community to disregard his theories. See **ASTEROIDS**, and **DINOSAURS, EXTINCTION OF**.

CAT Scan Computerized axial tomography (CAT), also called computerized tomography (CT), machines represent a major improvement over conventional X ray radiography. CAT scanners use narrow, focused X rays. The source of the X rays rotates rapidly (almost once per second) around the body, and a detector measures how much radiation passes through the patient. A computer reassembles the data produced by the scanner into thin cross-sectional slices of the human body.

CAT scanners are used not only for detecting and diagnosing numerous conditions but also for observing the effects of therapy and following up on the results of sur-

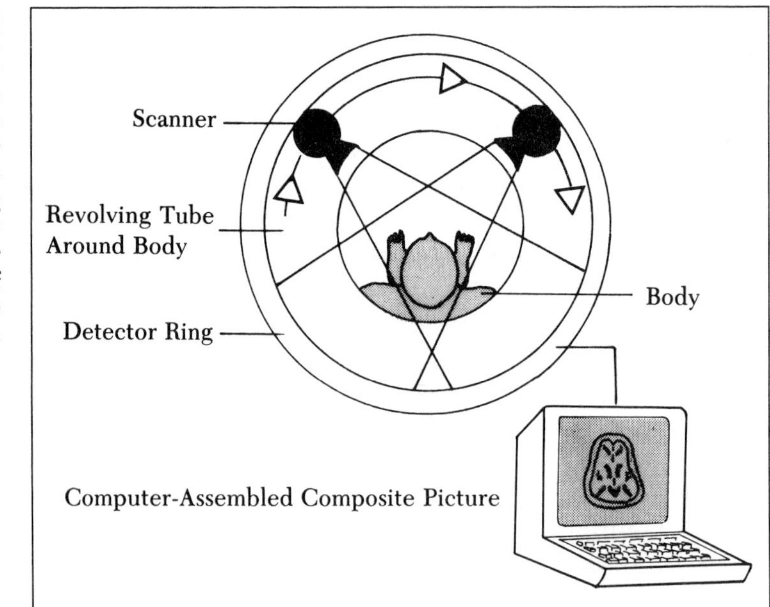

CAT SCAN uses narrow, focused X rays from a scanner that is rotated rapidly around the body. A detector measures the radiation passing through the body and a computer builds an image of a thin slice of the body.

gery. Three-dimensional CAT scanners are under development that can aid physicians in planning surgical procedures. See also **Magnetic Resonance Imaging (MRI)**, and **Positron Emission Tomography (PET)** for descriptions of other medical imaging devices.

CD-ROM Computer jargon for Compact Disc-Read Only Memory. ROM is used in personal **computers** to permanently store programs that are frequently used. Data stored in ROM memory cannot be changed, but they are also not erased when the computer is turned off. CD-ROM devices are similar to CD audio players. Information is retrieved by a **laser** beam that scans tracks of microscopic holes in a rotating disk. A single five-inch CD-ROM disk can store the equivalent of 1,500 floppy diskettes worth of data—the entire contents of an encyclopedia, with articles, pictures, and sound effects. The drawback is that CD-ROM disk players are relatively expensive and new information cannot be added to them. See **RAM**, and **ROM**.

Cells All self-replicating forms of life are composed of cells—from simple, single-celled **bacteria** to whales, with their trillions of cells. Most cells are microscopic in size, although a few, such as hen's eggs, are large. It is at the cell level that many of the basic functions of a living organism are carried out: **protein** synthesis, extraction of energy nutrients, and replication. The common elements of most cells include an extremely thin *plasma membrane* surrounding the cell, a **nucleus**, containing genetic material, and a viscous, transparent material called *cytoplasm* that fills the space between the nucleus and the membrane.

The genetic information encoded in DNA molecules in the nucleus of a cell provide instructions for assembling protein **molecules**. The work of a cell is carried out for the most part by the many different types of protein molecules. Some molecules assist in replicating genetic information, carrying out cell division, changing cell shape, repairing cell structures, and regulating molecular interactions. Other molecules are exported from the cell: hormones; antibodies; digestive enzymes; carriers for oxygen and other molecules in the blood; and material for hair, nails, and other body structures.

CELLS, the elementary particles of life, the smallest components of an organism that may be said to be alive. Cells differ in size, shape, and function and there is no such thing as a "typical" cell. Most cells, however, have many things in common and it is helpful to imagine a "typical" animal cell such as this one.

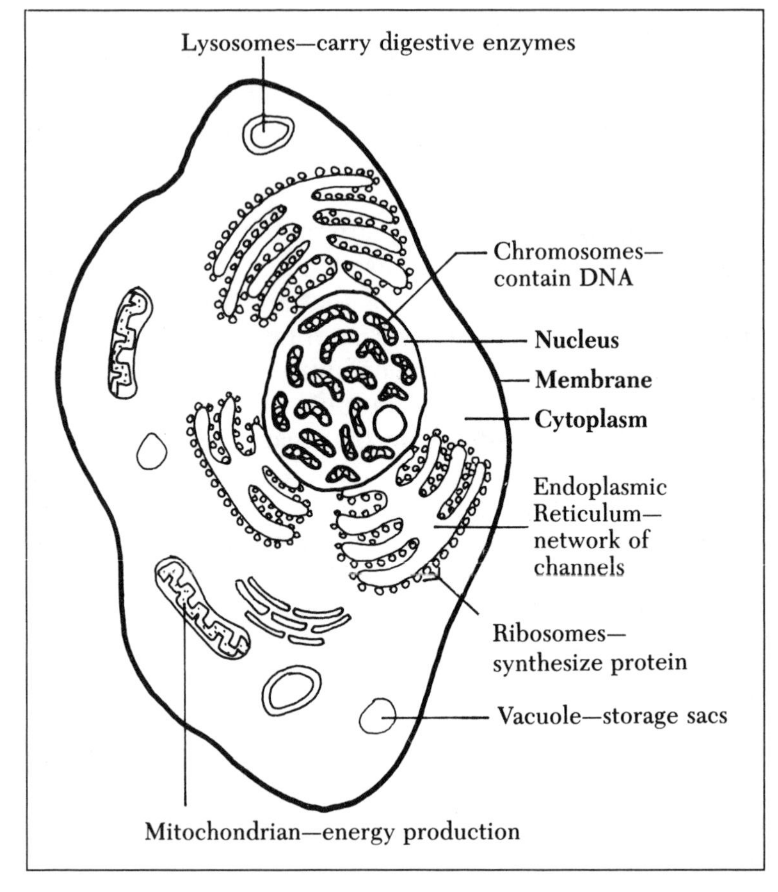

In addition to the basic cellular functions common to all cells, most cells in multicelled organisms are specialized; that is, they perform some special functions that others do not. For example, gland cells secrete hormones, muscle cells contract, and nerve cells conduct electrical signals.

Plant cells differ from animal cells in having a rigid cell wall made up mainly of *cellulose,* a tough material that gives shape to the plant cell. Because of this rigid wall, plants tend to retain their forms and shapes to a much greater degree than animals. The relatively soft plasma membrane of animal cells makes animal tissues soft and pliable. See **DNA**, **CHROMOSOMES**, and **GENES**.

Cellulite There is no such thing as cellulite. The term is used to describe a supposedly unique kind of tenacious fat that causes bumps and ripples on the thighs and buttocks. The term was popularized by women's magazines and marketeers trying to sell useless cures. As the University of California Wellness Letter put it in 1990, "fat is fat."

Celsius In most countries of the world, **TEMPERATURE** is measured in degrees Celsius or centigrade (°C). The Celsius temperature scale is named after its creator, the Swedish astronomer Anders Celsius. In this scale, the freezing point of water is set at 0 and the boiling point of water is set at 100 so that there could be 100 equal subdivisions (called degrees Celsius).

In the United States we are more accustomed to temperature being measured in **FAHRENHEIT** (°F), but if we travel abroad, we have to be able to convert degrees Celsius to degrees Fahrenheit. A fast approximate way of doing this is to double the Celsius figure and then add 30. There are, of course, more exact conversion methods. See **KELVIN**.

Cenozoic The most recent era of geologic time, characterized by the evolution of mammals, birds, plants, modern continents, and glaciation. This era includes the Tertiary and Quaternary periods—from the present to 65 million years ago. See **GEOLOGIC TIME SCALE**.

Cepheid A pulsating variable **STAR** whose *periodicity*, the time it takes to vary in brightness, is directly related to its absolute magnitude. Astronomers use the term *absolute magnitude* to indicate brightness at a standard distance—a star's intrinsic brightness. How bright a star appears to be is called its *apparent brightness*. If absolute magnitude can be judged, then apparent brightness can be measured and distances can be calculated. This correlation between brightness and period makes Cepheids quite useful in measuring galactic distances.

Thanks to Cepheids, sometimes called Cepheid variables, determining the distance to relatively nearby **GALAXIES** is possible. Attempts to extend the Cepheid method to more distant galaxies have been futile, because Cepheids in these far-off galaxies are too faint to be seen by Earth-based telescopes. The Hubble space-based telescope was intended to extend the range of Cepheid measurement to much more distant galaxies. See **HUBBLE SPACE TELESCOPE**.

CERN The European Organization for Nuclear Research and location of the Large Electron-Positron (LEP) collider. Since its huge collider became operational in 1989 CERN, located in Geneva, has become a major high-energy physics research center for the entire world. More than one-third of the Soviet Union's particle physicists are registered to work there as are a quarter of China's. CERN is planning to build a proton-proton collider in the same underground facility as LEP. Called the Large Hadron Collider (LHC), it is planned to be almost half as powerful as the U.S.'s planned SUPERCONDUCTOR SUPER-COLLIDER (SSC). Because the CERN machine will cost only half as much as the supercollider and will be operational sooner, some critics of the U.S. SSC have called for this country to join with CERN instead of building the costly ($8 billion) SSC. See COLLIDER, PROTON, and PARTICLE PHYSICS.

Chain Reaction See CRITICAL MASS.

Chandrasekhar Limit Named for Indian astronomer Subrahmanyan Chandrasekhar, who calculated that a cold STAR of more than about one and a half times the MASS of the SUN would not be able to support itself against its own GRAVITY. The Chandreskhar limit, then, is the theoretically maximum possible mass of a stable cold star, above which it must collapse and become a BLACK HOLE.

Chaos A new field of research, sometimes called *nonlinear science* or *dynamical systems*; known informally as chaos. Chaos is a theory that is highly mathematical but unusually interdisciplinary. Physicists, chemists, biologists, mathematicians, and engineers are among its students. These researchers specialize in turbulent, hard to describe, apparently random systems, and they deal in disorder. The field, however, is not without its skeptics. Some mathematicians say the methods of chaos theoreticians lack rigor, are based on unreliable models, and threaten traditional ways to test solutions. Nevertheless, chaos theory has gained a following and has advocates at every major university or research center. Chaos theory is providing an approach to study systems that defy description by conventional methods. To many researchers, chaos theory is a fresh way of looking at problems that are very difficult and for which new ideas are welcome.

Since the days of NEWTON, scientists have been trying to explain complex behavior with linear (straight-line) equations, where a given input produces a proportionate output. If one knew all the variables, it was thought, and one had a large enough computer to handle all these uncertainties, it would be possible to model, or describe in mathematical terms, any system no matter how complex. Long-term weather predictions are a case in point. Meteorologists were among those who thought that the new supercomputers would make long-range weather forecasting finally possible, but it didn't work out that way. Working with computer models of the weather, Edward Lorenz, a meteorologist at Massachusetts Institute of Technology (MIT), has demonstrated that models of chaotic systems have an exquisitely sensitive dependence on initial conditions and minute but unpredictable variables—in other words, the weather was inherently chaotic.

In all chaotic systems, from the crashing course of a mountain river, to the population of Midwest locusts each year, a slight disturbance can make a world of difference. "A very small perturbation, in due time, can make things happen quite differently from the way they would have happened if the small disturbance hadn't been there," explains Lorenz. Researchers refer to this phenomenon as the *Butterfly Effect*—a term coined by Lorenz in a lecture given in 1970 when he posed the intriguing question, Could the delicate motions of a butterfly deep in the Amazon forest spawn a tornado over Texas? For more information on this interesting new field see *Chaos: Making a New Science* by James Gleick (Viking Penquin, Inc., N.Y., 1987).

Charm The fourth *flavor* of QUARK. Predicted by theory, charmed quarks were first discovered in 1974. We are here venturing into the esoteric and sometimes whimsical world of subatomic particle physics. We will get into this world a little deeper when defining MATTER and QUANTUM PHYSICS, but for this definition it is enough to understand that as far as is known today, quarks and LEPTONS are the fundamental bedrock PARTICLES of all matter. Quarks are the building blocks of PROTONS and NEUTRONS, which in turn constitute the nuclei of ATOMS. Physicists investigating the quark theory say that quarks have to exist in pairs. Pairs of quarks are called flavors and one pair is thought to consist of an s-quark (s for strange) and a c-quark (c for charm). See also **PARTICLE PHYSICS**.

Charon The planet PLUTO's one known MOON. Pluto and Charon pass in front of each other as they orbit each other. By studying the light as it is blocked, astronomers now have determined accurate sizes for both objects. Pluto is considered only two-thirds the size of our Moon, and Charon is half of Pluto's size. Some astronomers think that both Pluto and Charon were once moons of **NEPTUNE**. See **PLANETS**.

Chemical Bonding The process or mechanism by which unstable atoms (atoms with incomplete outer shells) are joined or bound in a MOLECULE. The ELECTRONS in an atom exist and move about the NUCLEUS in cloudlike shells. If the outermost shell of an atom is not "full" (i.e., contains the maximum number of electrons possible for that type of atom), it is considered unstable. Atoms seek to achieve stability by transferring or donating electrons (*ionic bonding*) to other atoms or by receiving electrons (*covalent bonding*) from other atoms. These two processes, donating or receiving, are the two forms of chemical bonding by which all molecules are made.

When two hydrogen atoms stick to one oxygen atom a molecule of H_2O, or water, is formed. When one sodium atom sticks to one chlorine atom a molecule of sodium chloride, or table salt, is formed. These two examples are simple ones but the same mechanisms operate in the formation of all molecules. See **ATOMS**, and **COMPOUNDS**.

Chernobyl City in the Soviet republic of Byelorussia and site, in April 1986, of the worst nuclear accident in history. The release of damaging RADIATION at Chernobyl was one million times greater than that at THREE MILE ISLAND in the United States. Studies of the consequences of the Chernobyl nuclear power plant disaster show that eventually the death toll from cancer will be in the hundreds of thousands along with an appalling number of serious ailments. Soviet representatives told the U.N. General Assembly that about 2.2 million people, or 20 percent of the republic's population, have become victims of Chernobyl in one form or another. Their figures include 800,000 children with health problems and 150,000 people who live in contaminated areas awaiting resettlement to "clean areas." Thyroid-related ailments among children in southern Byelorussia have doubled. Anemia cases have jumped 700 percent, and birth defects 200 percent. Chronic nasal and throat ailments are up tenfold. Altogether, changes in people's immune, endocrine, blood, and nervous systems have created what the Soviet representatives called a "radiation **AIDS**." See also **CHINA SYNDROME**.

China Syndrome A hypothetical nuclear reactor accident in which the fuel melts through the floor of the containment vessel and burrows into the earth. Nuclear **REACTORS** have many backup safety systems, but none are designed to cope with the failure of the vessel structure itself—the last line of defense. A "worst case" scenario for a reactor accident is that once the fuel melts and operators lose the ability to control the core, a nuclear reaction reestablishes itself and the core burns through the vessel and then the reactor basement, and from there into the earth beneath, and thence, in the engineer's jest, all the way to China. In actuality, if the core reached groundwater, the result would be a steam explosion that would spew forth vast amounts of radioactive fallout. Information now available makes it clear that the Three Mile Island accident of 1979 came closer to a core meltdown than was publicly acknowledged at the time. See **THREE MILE ISLAND**.

Chip Because they are made with thin slices or chips of silicon, a semiconductive material, microelectronic circuits are known generically as chips or *microchips*. Microchips, then, are tiny bits of silicon on which numbers of electronic devices have been etched microscopically to form **INTEGRATED CIRCUITS**. In general, chips are composed of **TRANSISTORS**, which function as switches. Electric current travels from one end of the device to the other, crossing through a gate that sometimes allows the current to pass and sometimes does not, thus providing the on/off switch needed for **BINARY** systems (1s and 0s) common to computers. Since their development in the 1960s microchips, sometimes called solid-state devices, have universally replaced vacuum tubes in all of our electronic equipment. They have made possible everything from small hand-held radios to home computers to satellite communications systems.

Developing new generation microchips that could store ever-increasing amounts of information has been a major goal of the international semiconductor business. The memory chip currently in broad use in most computers can store one million **BITS** of information. In mid-1990, American and Japanese microelectronic companies were manufacturing and shipping 4-megabit chips. By the end of 1991, 16-megabit chips are expected to be in use. In June of 1990, Hitachi Ltd, one of Japan's biggest electronic companies, announced the development of a working prototype of a memory chip that can store 64 million bits of information. The experimental chip comprises 140 million electronic devices, all crammed onto a surface area that measures only 9.74 millimeters by 20.28 millimeters, a little larger than a fingernail. This 64-megabit chip, a long-sought milestone in semiconductor technology—will make the development of advanced supercomputers easier and cheaper. See **SEMICONDUCTOR**.

Chlorofluorocarbons (CFCs) Synthetic chemical compounds used in refrigeration and air conditioning; as aerosol propellants and solvents; to form foams, including those used in fast-food packaging; and as rigid insulation. Scientists now see these synthetic chemicals as the main threat to Earth's protective **OZONE** layer. Because CFCs are immune to destruction in the **TROPOSPHERE** (bottommost layer of the atmosphere), and because they eventually float upward, their manufacture and release have led to the accumulation of large amounts in the **STRATOSPHERE**. In the stratosphere, CFCs are broken down by sunlight into chlorine, which has a catalytic and destructive effect on ozone. The result has been a significant decline in the global ozone shield and an increase in the amount of harmful **ULTRAVIOLET (UV) RADIATION** reaching the surface of Earth.

Environmental officials from the industrialized nations have now agreed to phase out production of these chemicals by the year 2000. However, poor nations are not likely to participate in CFC cutbacks unless rich nations help them develop and use technologies that do not rely on ozone-depleting compounds. If large Third World nations such as China and India do not join in the agreement, ozone destruction will continue. According to a United Nation's study, every 1-percent drop in ozone will lead to a 3-percent increase in non-MELANOMA skin cancers in light-skinned people, as well as dramatic increases in cataracts, lethal melanoma cancers, and damage to the human immune system. Higher levels of ultraviolet light may also worsen ground-level pollution and hurt plants, animals, and especially light-sensitive single-celled aquatic organisms. See ENVIRONMENT.

Cholesterol A steroid chemical present in certain foods, mainly, though not exclusively, fatty foods. High cholesterol levels in the blood may lead to a narrowing of the arteries as a result of the formation of large deposits of *atheroma,* a type of fatty tissue, in arteries. If only small quantities of saturated fats and other high-cholesterol foods such as eggs are included in a diet, blood cholesterol level may be significantly lowered, and the risk of hardening or blockage of the arteries is minimized.

The amounts of the two kinds of cholesterol in the blood have been shown to affect a person's risk of suffering a heart attack. One kind, LDL (low-density lipoprotein) cholesterol, appears to encourage the formation of artery-clogging fatty deposits, whereas the other, HDL high-density lipoprotein), which is the good form of cholesterol, washes excess cholesterol out of the body and reduces the risk of heart disease. Research continues to help identify which foods contain which type of cholesterol. In the U.S., blood cholesterol is usually measured in milligrams per deciliter (one-tenth of a liter) of blood, or mg/dl. The National Cholesterol Education Program has set up the following guidelines:

LDL level (mg/dl)	
below 130	Desirable
130–159	Borderline-high
160 or greater	High risk

HDL level (mg/dl)

below 35 is considered a risk factor for heart attack.

Your total cholesterol should be no more than about four times higher than your HDL. The "ideal" ratio is under

3.5. A ratio of about 4.5 suggests that your risk for coronary heart disease is below average, while a ratio higher than that represents a higher-than-average risk.

Chromosomes Collections of genes (the basic units of heredity). The long string of genes that make up a chromosome has been likened to a string of pearls. Each species of plant or animal life has a fixed and characteristic number of chromosomes. Humans, for instance, have 23 pairs or 46 chromosomes. One member of each pair comes from each parent. Cells of pea plants have seven pairs, those of a fruit fly four pairs, and bacteria have only a

CHROMOSOMES are the small bodies in the nucleus of cells that carry the chemical instructions for reproduction of the cell. They consist of strands of DNA wrapped in a double helix around a core of proteins. Each species of animal has a characteristic number of chromosomes. For humans, it is 23 pairs or 46 chromosomes. Females have two X chromosomes while males have one X and one Y chromosome.

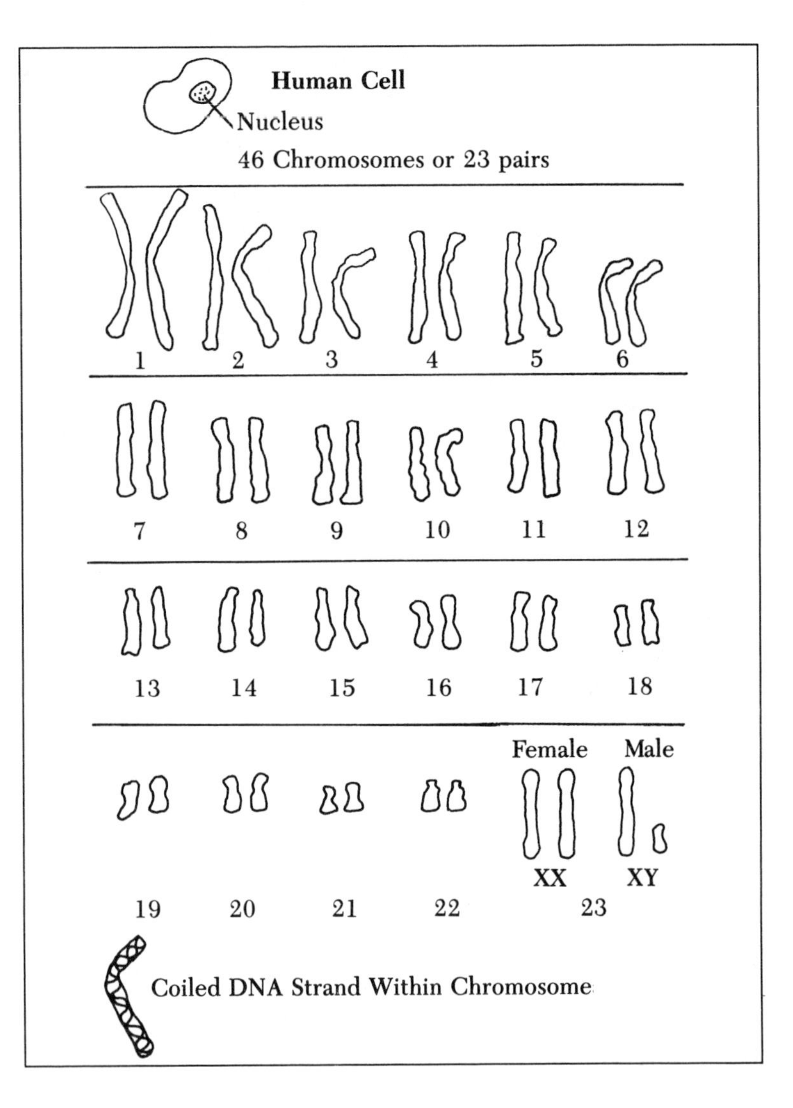

single, unpaired chromosome. The nucleus of a CELL contains genes in the form of DNA coiled into chromosomes. One chromosome contains 3,000 to 4,000 genes composed of 150 million chemical building blocks known as base pairs. See DNA, and GENES.

Chronobiology The study of the daily rhythms of living organisms. Natural rhythms are inherent in all living things—plants or animals. In humans, this research is attempting to explain why the body TEMPERATURE reaches a low point in the morning and why blood pressure peaks in the afternoon. Research in this field may also explain why most babies are born in the early hours of the morning and why so many heart attacks occur between 6 and 9 A.M., or why most Olympic records are set in the late afternoon. Research into humankind's sensitivity to the hours of the day and the seasons of the year may also shed some light on why the hours of 1 to 6 A.M. seem to be the time when many major accidents have taken place—accidents wherein human error played a major role. CHERNOBYL, THREE MILE ISLAND, and Bhopal all occurred in that time period. It is also the time that the Exxon Valdez ran aground. See BIOLOGICAL CLOCK, and CIRCADIAN.

Circadian The daily rhythm, or 24-hour periodicity, of all forms of life on Earth. See BIOLOGICAL CLOCK, and CHRONOBIOLOGY.

Classical Physics The study of physics prior to the introduction of the QUANTUM principle. Classical physics incorporates Newtonian mechanics and cause and effect, and it envisions a knowable, and therefore predictable, universe. The world was simpler and easier to understand prior to the acceptance of quantum physics and HEISENBERG'S UNCERTAINTY PRINCIPLE, but it was not, physicists now tell us, a true picture of the subatomic world. See NEWTON.

Climate Long-term meteorological conditions, such as TEMPERATURE, precipitation, and wind, that prevail for a specified period of time. Climate on Earth is driven by the radiant ENERGY received from the SUN. This energy is received in the form of LIGHT. Earth absorbs some of this energy, which warms the planet's surface, but reradiates most of the heat energy back into the atmosphere as INFRARED (invisible but detectable as HEAT). If Earth only absorbed RADIATION from the Sun without giving back an equal amount of heat, our planet would continue to grow warmer each year until the lakes and oceans would boil. Energy balance, then, is the first crucial factor necessary to a livable climate.

The motion of the Earth and its position relative to the Sun is the second crucial element governing climate on this planet. The Earth's one-year revolution around the Sun, because of the tilt of the Earth's axis, changes how directly sunlight falls on one hemisphere or the other. This difference in heating different parts of Earth's surface produces seasonal variations in climate—when it is summer in the Northern Hemisphere it is winter in the Southern Hemisphere and vice versa. The Earth's climate has changed radically over long periods of time, due mostly to the effects of geological shifts such as the advance or retreat of glaciers over centuries of time, or huge volcanic eruptions, or possibly the impact of large asteroids. Most scientists see our climate as a fragile state where even small changes in atmospheric content or ocean temperature could have widespread effects. Others, in particular followers of the GAIA hypothesis, see Earth's climate as a self-correcting mechanism. See also ASTRONOMICAL CYCLES.

Clone A genetic copy of an individual organism that has come into existence by asexual reproduction in which the NUCLEUS of a CELL from the body of the single parent is stimulated to divide itself. Because it inherits all its GENES from one parent, the clone will be genetically identical to that parent. Horticulturists, by cutting, grafting, and budding, have been cloning various plants for a long time. BACTERIA and one-celled animals clone themselves naturally.

Cloning came into the media spotlight when biologists made a new frog from the DNA in the nucleus of an intestinal cell of an old one. Biotechnology now enables livestock breeders to clone large numbers of identical farm animals from a single embryo. This ability to successfully clone large mammals hints at the possibility in the future that similar techniques might be used on humans. The possibility, however remote at the present stage of knowledge, that a human embryo could be manipulated in the laboratory to produce numerous genetically identical babies carried to term in the wombs of surrogate mothers adds to the controversy about how far we want to go in GENETIC ENGINEERING. See BIOETHICS.

Closed Universe Cosmological theory that envisions the EXPANDING UNIVERSE as "closed" or destined to stop expanding at some future time, to be followed by a collapse of all the galaxies in a sort of reverse BIG BANG, and then a rebound into a new expansion phase. See OPEN UNIVERSE.

Cloud Chamber A glass container filled with moisture-saturated air, or some other liquid, in which PARTICLES can be detected by photographing the tracks of water droplets that the particles leave behind when they pass through the chamber.

Physicists can read the tracks and determine much about the particle that has passed through the chamber. The particle's response to a magnetic field, for instance, will indicate whether it is positively or negatively charged. The curve of the track it leaves behind will indicate its mass and energy while the width of the track will identify the type of particle. Cloud chambers, of whatever type, are an essential tool in the study of PARTICLE PHYSICS.

Clouds Visible bodies of fine droplets of water or particles of ice dispersed in the atmosphere above the Earth's surface at various altitudes ranging up to several miles.

Clouds are the wild card in the climate prediction and modeling game. Clouds both cool and warm the Earth and this paradox has presented scientists with a problem for a long time. A more thorough understanding of the effect clouds have on the climate is necessary if scientists are to make confident forecasts about GLOBAL WARMING and the climatic changes that are expected to result from the GREENHOUSE EFFECT. At the present time, scientists are finding that clouds are more complex than had been realized. For instance, clouds over water seem to differ in character and heating effects than clouds over land. Or clouds in the tropics differ from clouds in the temperate zones in the manner in which they reflect sunlight. Season and time of day also seem to influence cloud behavior. In sum, scientists have not reached a state of knowledge regarding clouds and their behavior that permits climatic modeling with any certainty.

Codon A term used in genetics to indicate a sequence of three adjoining nucleotides that specifies the insertion of an AMINO ACID in a specific structural position during PROTEIN synthesis. The concept of triplet combinations of the four chemical bases—adenine, guanine, cytosine, and thymine—that form DNA was integral to deciphering the GENETIC CODE. When these four chemical bases, usually called A, G, C, and T, are considered as threesomes, there are 64 different ways in which they can be arranged—AAA, AAG, AGA, AAC, GAA, and so on. From these 64 codons (the word comes from a combination of *code* and *on*) the 20 different kinds of amino acids that make up protein are formed. See CHROMOSOMES, DNA, and GENETIC ENGINEERING.

Cold Fusion The original report in 1989 of a successful cold FUSION experiment by two scientists in Utah caused considerable excitement, because the discovery could signal an environmentally safe, inexpensive way to produce ENERGY. The production of energy by causing small atomic nuclei, at room TEMPERATURE, to join together (or "fuse") into larger ones was presumably a major breakthrough. Unfortunately the early excitement turned to doubts when the experiment could not be duplicated. Although most scientists have written off the original cold fusion experiments as wishful thinking or persistent instrumental errors, numerous ongoing experiments in many countries amount to a substantial effort to reproduce the cold fusion phenomena.

Examples of fusion that science knows of today are the processes that take place in the center of stars or in an exploding HYDROGEN bomb. In both of these cases atomic nuclei are forced together under conditions of high energy, high temperature, and extreme pressures. Scientists are hard at work trying to develop controlled fusion in the laboratory because the long-term potential of nuclear fusion as an energy source is enormous. See FISSION.

Collider A type of PARTICLE accelerator in which subatomic particles are first speeded up to velocities close

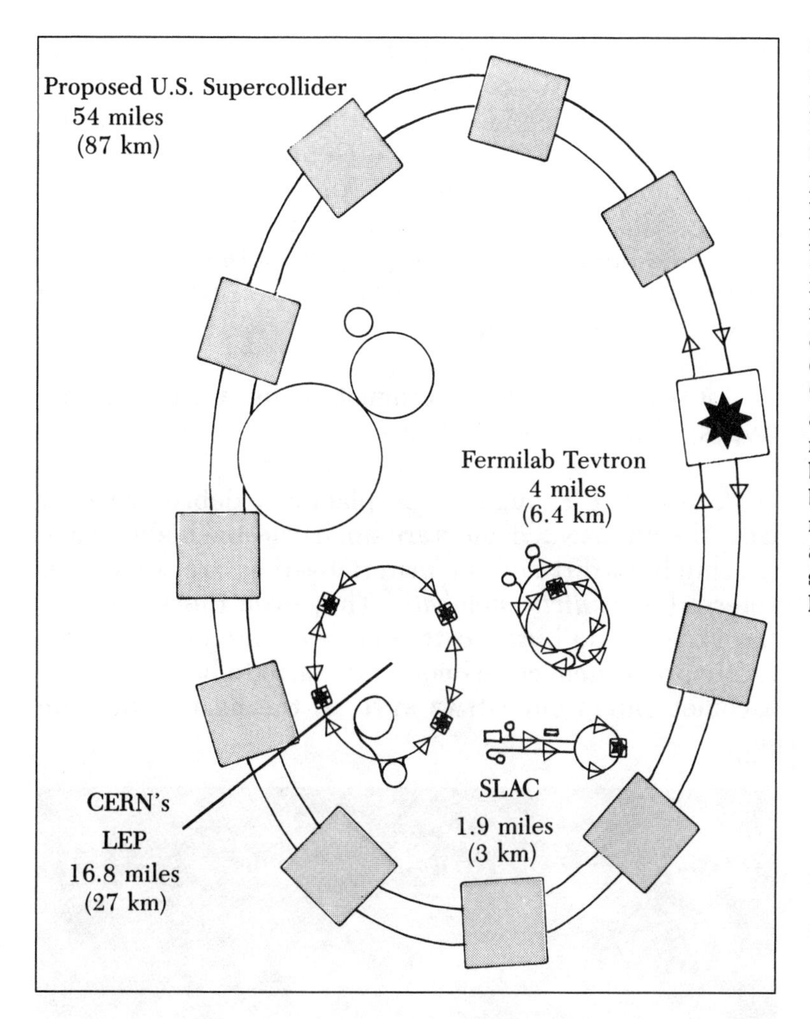

COLLIDERS are machines that enable physicists to see inside the atom. Shown here are the proposed U.S. superconducting supercollider, CERN's Large Electron-Positron Collider, Fermilab's Tevtron, and Stanford's Linear Accelerator Collider. Each time a bigger and better machine has been built, physics has been pushed to a new energy level, and a new understanding of matter has been possible.

to the speed of LIGHT (186,000 miles per second) and then made to collide head-on with other particles traveling in the opposite direction. The resulting explosion produces exotic particles that can then be analyzed.

Colliders are machines that enable physicists to see inside the atom. The collider method of doing this has been likened to the smashing together of two Swiss watches to find out what is inside. As the scientists try to find smaller and smaller particles they need larger and larger colliders. Their goal is to answer the "big question" of subatomic physics: What is the universe made of, and what are the forces that bind its parts together? The three largest colliders in the world are Fermilab's Tevetron, a circular tunnel device that is 6.4 km (4 miles) in circumference; Stanford's Linear Accelerator (SLAC), a machine that shoots electrons and positrons down a two-mile long straightaway and then loops them through two semicircular sections into a collision course; and CERN's large—27 km (16.8 mile) in circumference—electron-positron collider, called LEP, located in Geneva. Dwarfing all of these would be the proposed 87 km (54 miles) in circumference U.S. **SUPERCONDUCTOR SUPERCOLLIDER** scheduled to be built at a site in Texas just south of Dallas. See **ACCELERATOR**, **CERN**, **PARTICLE PHYSICS**, and **SUBATOMIC STRUCTURE**.

Colors A kind of designation for various flavors (types) of quarks. See **QUARKS**.

Comets Thought to be planetary debris left over from the **BIG BANG**, comets are small celestial bodies made up chiefly of dirt and icy materials—they are sometimes referred to as *dirty snowballs.* They orbit the **SUN** just as the **PLANETS** do but in extremely elongated ellipses that rarely bring them close enough to Earth to be seen. Once a comet enters our **SOLAR SYSTEM**, the heat of the Sun

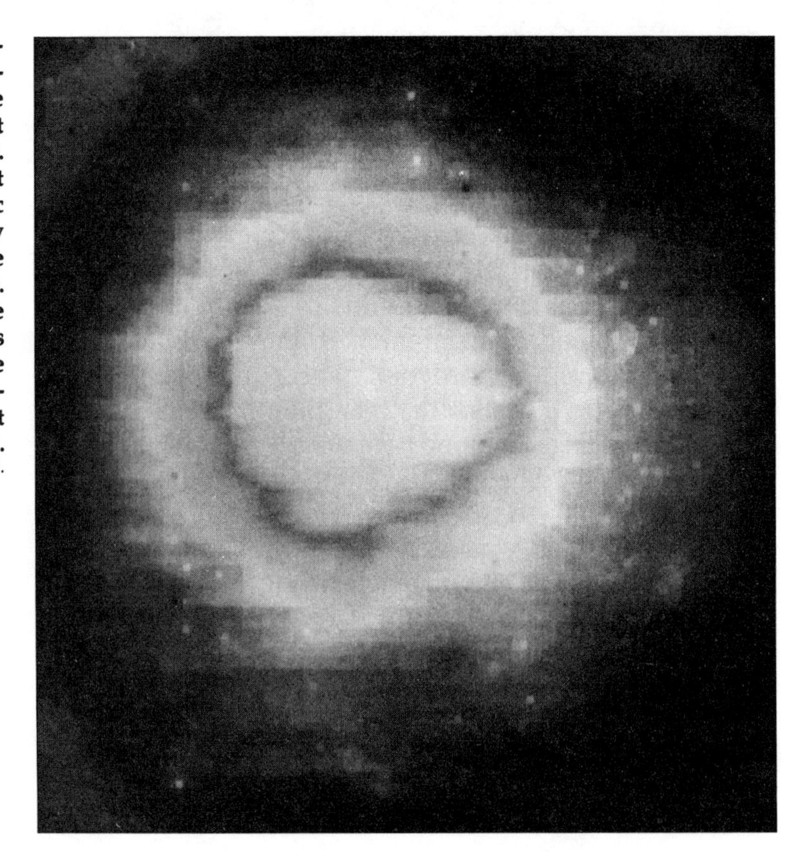

COMETS are relatively small celestial bodies made up chiefly of dirt and icy materials. Halley's comet makes a periodic visit to the vicinity of Earth about once every 77 years. This image of the coma of Halley's comet shows the cloud of gases surrounding the comet nucleus.
Source: NASA.

vaporizes the icy material and the resulting vapor and dust help to form the glowing tails visible in the skies above Earth. Astronomers estimate that there are 100 billion of these objects, ranging in size from 0.5 to 5 miles in diameter residing in a region of outer space beyond PLUTO called the OORT CLOUD.

In 1985–86, an international fleet of five spacecraft flew by and made a close-up examination of Halley's Comet when it made its periodic (about every 77 years) visit. The fleet included two Russian Vegas, two Japanese probes, and one from the European Space Agency. These spacecraft analyzed the cometary grains that they encountered and determined that the basic chemistry of some of these particles are hydrogen, carbon, nitrogen, and oxygen—the same chemicals you and I are made of. It has been postulated that comets that hit Earth when it was formed may have brought these chemicals with them, which helped—or even triggered—the formation of life. To the chagrin of U.S. scientists, NASA, due to budgetary restrictions, did not participate in this important space research.

If budgetary approval is obtained, NASA plans to launch a mission called CRAF (Comet Rendezvous Asteroid Flyby) in 1996. The spacecraft will spend five years in space before meeting with the Kopff comet and dropping a probe to its surface. The probe is designed to penetrate the core of the comet and study the chemistry of its material, which may date from the beginning of the solar system. See **ASTEROIDS**, and **METEOROIDS, METEORS, AND METEORITES**.

Compounds Substances containing ATOMS of two or more different ELEMENTS in definite proportion. (See diagram on page 56.) Chemical bonds hold the elements together. Examples: The compound water (H_2O) contains atoms from HYDROGEN and OXYGEN, and the compound CARBON DIOXIDE (CO_2) is made up of carbon and oxygen. See **CHEMICAL BONDING**.

Computer A machine that manipulates the symbols of information such as numbers and letters. Essentially a collection of on/off switches, a computer can be described as an electronic device designed to accept data, perform prescribed computational and logical operations at high speed, and output the results of these operations.

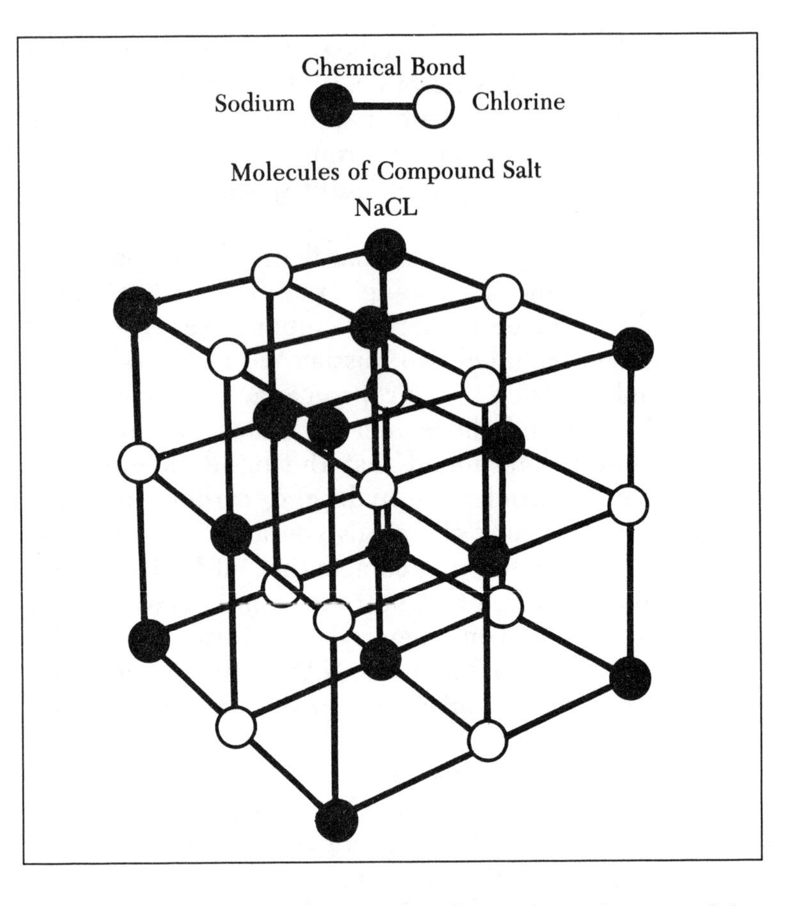

COMPOUNDS are substances containing atoms of two or more different elements in definite proportions. Chemical bonds hold the elements together. Shown is a three-dimensional diagram of the molecules of the compound salt.

Computer **HARDWARE** is the physical machinery of the system; **SOFTWARE** is the program—or set of instructions—that directs the operation of the machine. Computer programs come in many different flavors but all are essentially detailed, step-by-step directions telling a computer how to proceed in order to solve a problem, or process a specific task. See **AMERICAN STANDARD CODE FOR INFORMATION INTERCHANGE (ASCII), BINARY NUMBERS, BIT, BYTE, CPU**, and **DISK, COMPUTER.**

Conservation Laws The laws of science that state that **ENERGY** can neither be created nor destroyed. Energy can be converted from one form to another, say from the chemical energy in gasoline into the mechanical energy necessary to move your automobile down the freeway, but the amount of energy remains unchanged throughout the transformation. See **THERMODYNAMICS, FIRST AND SECOND LAWS OF.**

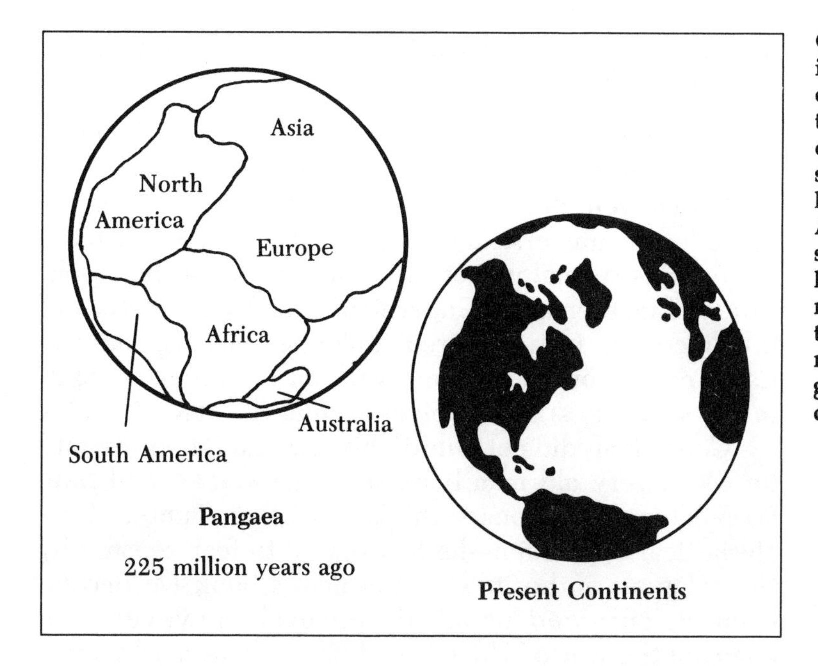

Pangaea

225 million years ago

Present Continents

CONTINENTAL DRIFT is the currently accepted concept that the continents are drifting across the surface of the globe like huge icebergs. At a very early stage of Earth's history, about 225 million years ago, the separate continents formed a single land mass called PANGAEA.

Continental Drift The concept, confirmed by oceanographers and geophysicists, that the continents are drifting across the surface of the globe like so many huge icebergs. The rate of this leisurely movement of continents across the surface of Earth has been estimated by geologists to be about eight-tenths of an inch per year.

When we look at a map or globe of the world we notice that the east coast of North and South America and the west coast of Europe and Africa roughly match like parts of a huge jigsaw puzzle. When the submerged shelves of the continents are also considered, it can be seen that the fit matches quite closely. At a very early stage of Earth's history—about 225 million years ago—the separate continents formed a single land mass called PANGAEA. With time, this single land mass broke up into the continents we know today. See PLATE TECTONICS, and EARTHQUAKES.

Convection The process by which HEAT rises, cools off, and sinks again, producing a continuous circulation of material and transfer of heat. One example of convection is the transfer of heat by massive motions of air within the atmosphere from the warmer equatorial regions to the North Pole and South Pole. Another example is the heat transfer from the interior of the Earth to the surface. See EARTH, COMPOSITION OF; and PLATE TECTONICS.

Copernicus Prior to the revolutionary work of Polish astronomer Nicolaus Copernicus in 1540, the accepted conventional wisdom was that the Earth was the center of the universe and all the heavenly bodies orbited around this special planet. By observation and the application of mathematics, Copernicus established the fact of a heliocentric (sun-centered) universe—that is, that the Earth rotates on its axis and, with all the other PLANETS in the SOLAR SYSTEM, revolves around the SUN.

Copernicus did not publish his "radical" theories until he was a very old man because, like GALILEO and DARWIN—later revolutionary thinkers who encountered ecclesiastical opposition—he had reason to fear censure by the religious authorities. Copernicus's book *De Revolutionibus* survived papal disapproval, however, and changed the world. The work of Copernicus marks a major milestone in humankind's comprehension of the universe. Although these important findings were published in 1543, the word is not getting around as fast as one would expect. In a joint poll conducted by Northern Illinois University and Oxford University in 1989, only one-third of the British adults and one-half of the Americans knew that the Earth revolves around the Sun and takes one year to do so. See SCIENTIFIC LITERACY.

Coriolis Effect The apparent acceleration of a body in motion with respect to the Earth, as seen by an observer on Earth. This effect is a result of the Earth's rotation. The surface of Earth moves faster at the equator, where it must make a circle of 25,000 miles in 24 hours, than it does at the higher or lower latitudes, where a spot on the Earth's surface must make a smaller circle in the same 24 hours. Near either the North or South Pole, the circles are quite small, and at the poles the surface is motionless. The movement of air masses over the surface of the Earth is an example of the Coriolis effect. As an air mass moves northward from the equator, its speed, which matches that of the equator, appears to be faster than the surface over which it is traveling. The result of the Coriolis effect on air masses apparently is to set them turning in a clockwise twist in the Northern Hemisphere and apparently in a counterclockwise twist in the Southern Hemisphere.

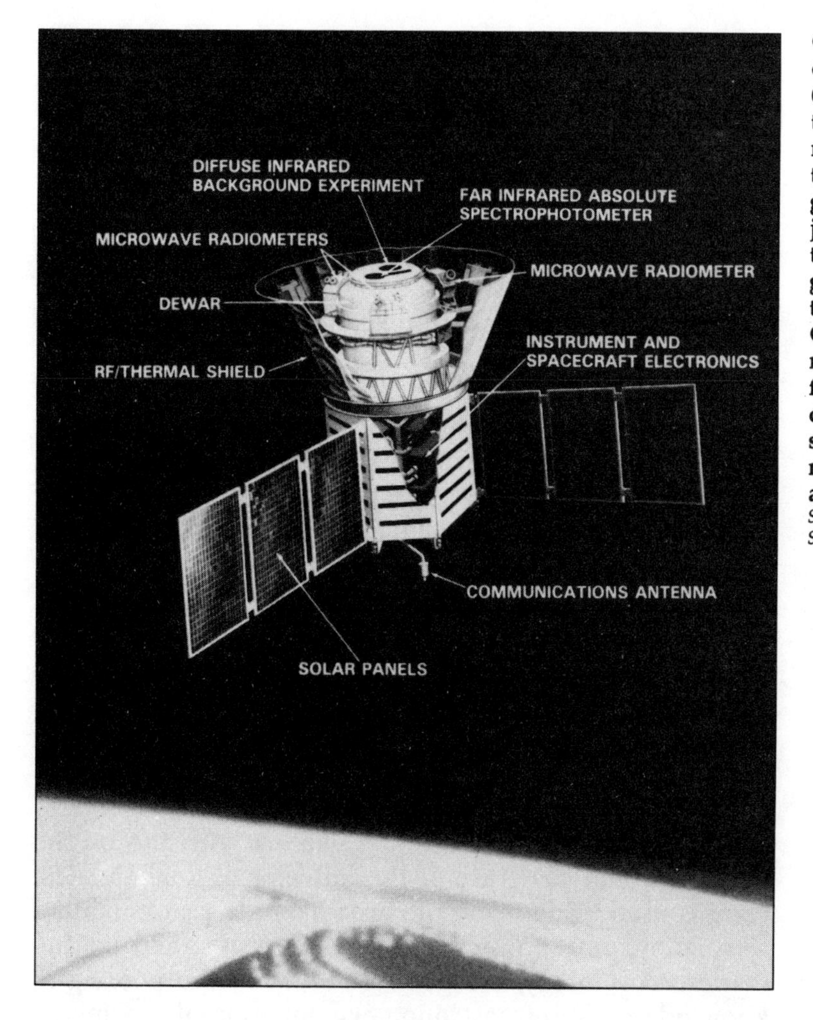

COSMIC BACKGROUND EXPLORER (COBE) is designed to map the sky and measure the radiation emitted by a great variety of objects in addition to the cosmic background radiation of the BIG BANG. COBE is 19 feet (6 meters) long and 29 feet (9 meters) in diameter once the sunshade, solar arrays, and antenna are fully deployed. *Source: NASA Goddard Space Flight Center.*

Cosmic Background Explorer (COBE) Launched in December of 1989, NASA's unmanned COBE satellite is designed to monitor the INFRARED, radio, and optical emissions that many astronomers regard as the radioactive relics of the Big Bang. The COBE program is intended primarily to provide a detailed infrared map of the sky. By looking for regions in the sky with the smallest possible infrared signal, investigators hope to find the fossil residue of light given off by the first luminous objects created after matter started to collapse into lumps early in the universe's history. See **BIG BANG**.

Cosmic Rays Radiation of extraterrestrial origin, chiefly of PROTONS, ALPHA PARTICLES, and other atomic nuclei. Cosmic rays include some high-energy electrons and PHOTONS that enter the atmosphere and cause secondary RADIATION. It is known that our planet is being constantly bombarded with radiation from outer space. From this knowledge, it has been speculated that the entire universe is suffused by a cosmic background radiation composed of primordial photons left over from the BIG BANG.

The theory of cosmic background radiation originated with the brilliant physicist George Gamow. He speculated that if the universe was still expanding, and the evidence supported that concept, then there must be some residual heat left over from the Big Bang. He further theorized that the photons carrying the energy of the Big Bang would have originated in the wavelengths of light and, with the expansion of the universe, would have by now shifted to the lower frequencies of electromagnetic energy, specifically microwave radiation. In late 1989, NASA launched its first satellite dedicated to the study of the phenomena related to the origins of the universe. The satellite, named COSMIC BACKGROUND EXPLORER (COBE), will measure and analyze what is called the remnant radiation left over from the dawn of the universe. See ELECTROMAGNETIC SPECTRUM.

Cosmology The science dealing with the origin, evolution, and structure of the universe as a whole. The term is used to describe the area of study that combines astronomy, astrophysics, PARTICLE PHYSICS, and mathematics. Cosmology is a good example of the pooling of knowledge wherein astronomers, for example, combine their specialized information with that of classical physicists, or some other technical discipline, to gain a better understanding of our world.

CPR (Cardiopulmonary Resuscitation) The standard method used by emergency medical personnel (or other trained individuals) to revive heart attack victims. Rhythmic pressure is applied to the chest to decrease chest volume, which increases pressure in the thoracic cavity, and forces blood to the heart and brain during the critical minutes before spontaneous respiration and heartbeats can be restored.

The problem with CPR is that the intense local pressure near the sternum can cause serious injury to the rib cage and lungs. This danger has led a pair of Johns Hopkins colleagues, Joshua Tsitlik and Henry Halperin, to de-

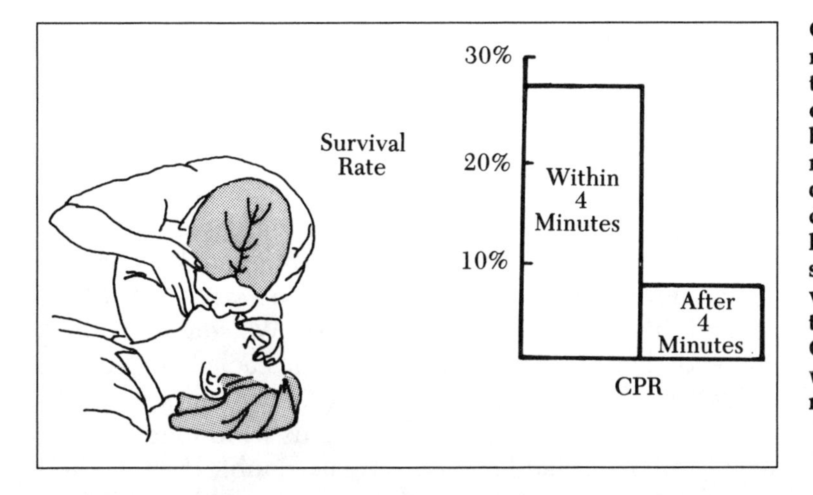

CPR (Cardiopulmonary Resuscitation) is the standard method used by emergency medical personnel or other trained individuals to revive heart attack or shock victims. Survival rate is close to 30 percent when CPR is applied within four minutes.

vise a computer-controlled, inflatable vest that distributes bursts of pressure equally around the chest and back. In experiments with human patients who had failed to respond to standard CPR, the vest increased blood pressure to nearly double the levels achieved using the traditional chest compressions. The vest will have to be approved by the FDA before general use is permitted, but the testing process is underway.

CPU (Central Processing Unit) The brain of a COMPUTER, the CPU (consisting of one or more microchips) controls all the functions of the machine and makes all the calculations. The CPU coordinates all of the computer's activities from interpreting commands from the keyboard or MOUSE, retrieving data from a memory bank, or sending data to the display screen or printer. The CPU acts like a central clearinghouse—all commands are received by the CPU and all instructions to the various subdivisions of the machine are issued from the CPU. The CPU also controls calculations by determining which instructions to obtain from the machine's main memory and the sequence of operation of these instructions. See **CHIP**, **RAM**, and **ROM**.

CRAF (Comet Rendezvous Asteroid Flyby) Proposed NASA mission planned for launch in 1996. (See photo on page 62.) See **COMETS**.

Cretaceous A period in the GEOLOGIC TIME SCALE corresponding to about 65,000,000 to 140,000,000 years ago. The third and last period of the MESOZOIC era, noted for the disappearance of the dinosaurs and the first appearance of flowering plants. See **EARTH, AGE OF**; and **DINOSAURS, EXTINCTION OF.**

Crick, Francis Codiscoverer (along with James D. WATSON) of the structure of DNA, and cowinner of the 1962 Nobel Prize. Prior to the findings of Watson and Crick, nobody knew exactly what GENES were, what they looked like, or how they worked. Watson and Crick showed that the DNA MOLECULE in each living cell is made up of a pair of strands, one of which is a special kind of copy of the other. The strands wind together like a twisted rope ladder (the famous double helix), their rungs composed of chemical subunits called nucleotides

CRAF (Comet Rendezvous Asteroid Flyby) spacecraft shown in this artist's drawing ejecting a penetrator probe toward the nucleus of a comet. *Source: NASA/JPL [Jet Propulsion Laboratory].*

that are bonded together. It is the sequence of these subunits that spells out the genes and that gives the instructions a cell needs to assemble another cell.

Along with EINSTEIN's theory of relativity, the Watson and Crick breakthrough may go down in history as the major intellectual triumph of the 20th century. Schoolchildren in the future will know about NEWTON, they will know about DARWIN, and they will know about Watson and Crick. See **DNA**, **GENETIC ENGINEERING**, and **GENOME PROJECT**.

Critical Mass In physics this term is used to define the amount of a given fissionable material necessary to sustain a chain reaction. A chain reaction is a self-sustaining phenomenon in which the FISSION of nuclei of one generation of nuclei produces PARTICLES that cause the fission of at least an equal number of nuclei of the succeeding generation.

Cro-Magnon Regarded by most anthropologists as the prototype of modern *Homo sapien*. Named after caves in southwest France where the first remains were discovered. Oldest remains date to about 40,000 years ago. See **HOMO SAPIENS**, and **EVOLUTION**.

Cryogenics The study of the behavior of substances, materials, or gases at extremely low temperatures. The odd behavior of substances at TEMPERATURE near ABSOLUTE ZERO has been the object of considerable research. The phenomenon of SUPERCONDUCTIVITY, wherein a material loses all resistance to the flow of electricity, is an example of this field of science.

Curie Named for Pierre Curie, the codiscoverer of radium, the term is used as a measurement of RADIATION. A curie is defined as the number of disintegrations per second in one gram of radium. Smaller subunits are millicurie (one thousands of a curie) and microcurie (one millionth of a curie). It is these smaller units that we read about in newspaper accounts of nuclear accidents or radioactive fallout. The curie or its smaller subunits are only useful as an indicator of what is going on in the radioactive material itself and not as a measure of possible biological harm to living organisms. Other measurements of radiation that take into account the different kinds of radiation and their effects on living tissue are **RADs**, **REMs**, and **ROENTGENS**.

Curved Space The concept, described by **EINSTEIN** in his **GENERAL THEORY OF RELATIVITY**, in which the four dimensions of space-time are curved. The essence of Einstein's theory is that the presence of matter distorts space and makes it curve. Experiments carried out in 1919 proved Einstein's theory, which had predicted that light waves would bend when they passed close to a large gravitation field.

One popular analogy used to help explain curved space is to picture a large rubber sheet held taut at the edges. If a heavy weight such as a bowling ball is placed on the sheet, it forms a depression. If we now propel a marble across the sheet it will tend to curve toward the depression in the sheet made by the large weight. If the depression is deep enough (i.e., if the gravity is strong enough) the marble will be captured by the depression and will circle around and around the bowling ball. According to Einstein, a large **MASS** in space such as our **SUN** distorts the fabric of space just like the bowling ball on the rubber sheet. See **SPECIAL THEORY OF RELATIVITY**, and **SPACE-TIME**.

Cybernetics A branch of robotics, cybernetics (from the Greek word for helmsman) is the term originated by MIT mathematician Norbert Wiener to cover the study of the fundamental mathematical relationships governing the handling of feedback control systems. The theory of automated computer control is based on obtaining information from the output of a system and then feeding back this data to the input in order to control the overall system without human intervention. This field of technology is now more commonly called feedback control engineering. A simple example of a feedback device is the thermostat that controls room temperature. A more complex example is the antilock braking system on new cars. Here the braking of a specific wheel is controlled by sensors attached to each wheel and a computer-controlled central unit that overrides the driver and prevents wheels from locking up. See **FEEDBACK CONTROL**, and **ROBOTS AND ROBOTICS**.

Cyberspace A field of advanced computer technology wherein the user enters a *virtual reality* that exists as pure computer data. Using devices such as fiber-optic gloves and data helmets or goggles displaying the data in three-dimensional form, the user can interact with all forms of computer data, as if he or she were physically in another world (see ARTIFICIAL REALITY).

Researchers are exploring a wide range of technologies to find new ways to communicate with computers. Computers are being designed to recognize human speech, or a particular handwriting, or track eye movement, or even recognize gestures. Cyberspace research is part of this quest to design computers that do not depend on keyboards and graphical interfaces.

Cyclotron An early type of ACCELERATOR—a machine designed to speed up subatomic PARTICLES to very high speeds and high energies useful in research. In a cyclotron, charged particles generated at a central source are accelerated spirally outward in a plane at right angles to a fixed magnet field by an alternating electric field, in other words in a circular motion. See COLLIDER.

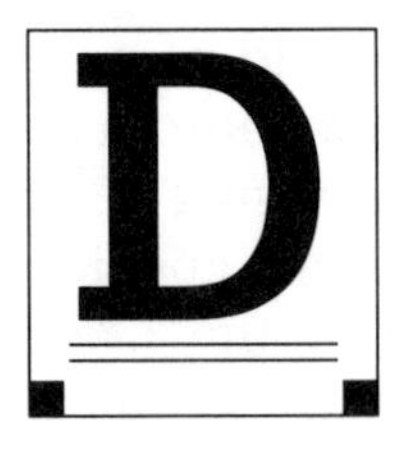

Dark Matter Undetected material in the universe that has puzzled scientists for a long time. The MASS of a galaxy can be deduced by measuring the velocity at which STARS orbit the center of the galaxy. In the same way, the mass of a cluster of GALAXIES can be deduced by measuring the velocity at which the separate galaxies orbit the center of the cluster. Whenever this is calculated, it adds up to roughly five or ten times the mass of all the visible stars in the universe. The implication, then, is that every thing we see and photograph in the sky amounts to only a fraction of the gravitational matter that should be there.

Astrophysicists speculate that this unseen matter may be BLACK HOLES or small BROWN DWARF STARS, but it might also consist of subatomic PARTICLES so far unidentified.

Darwin, Charles British naturalist, originator of the theory of EVOLUTION in his book *On the Origin of Species* (1859). The theory of evolution developed by Charles Darwin states that species of plants and animals develop through NATURAL SELECTION of variations that increase the organism's ability to survive and reproduce. At the time of its publication, the book created a storm of controversy and debate. The book, and the theory, survived because Darwin's carefully documented facts could not be ignored.

We know that three and a half billion years ago life on Earth consisted of simple forms such as BACTERIA and today consists of complex forms such as humans. It seems apparent then that the simpler forms led to the more complex forms. All fossil data collected since Darwin's time support and demonstrate the fact of evolution beyond any rational question. The word *theory* in theory of evolution is where the trouble lies because it implies scientific doubt where in fact there is none. The effort by creationists to have the Biblical story of the origin of life on Earth taught as a science in school has led to controversy. To scientists, demanding equal time for creation "science" in a biology class makes as much sense as demanding equal time for the flat Earth theory in a geography course, or the stork theory in a sex education class. See **FOSSILS**.

Death Star See **NEMESIS**.

Decay, Radioactive The progressive decrease in the number of radioactive atoms in a substance by spontaneous nuclear disintegration. An atom decays when it disintegrates, when it changes from instability to stability. Half of the atoms have decayed in a half-life. The process of **RADIOACTIVITY** involves the emission of **RADIATION** (in the form of **ALPHA PARTICLES**, **NUCLEONS**, **ELECTRONS**, or **GAMMA RAYS**) either directly from unstable atomic nuclei or as a consequence of a nuclear reaction.

Radioactive decay is a natural process going on all the time all around us. It is the radioactive decay of such elements as **URANIUM**, thorium, and potassium that heats the Earth today. The internal heat of Earth's core is also generated by radioactive decay of elements originating in the **STARS** and incorporated into primitive Earth at the time of the **BIG BANG**. This internal heat is, in turn, what energizes the tectonic activity of Earth. See **ATOMS**, and **HALF-LIFE**.

Decibel Unit used to measure the intensity of **SOUND**. Sound is a series of air pressure waves, or alternate peaks of high pressure and troughs of low pressure, traveling through the atmosphere. The loudness of sound or noise is measured in units called decibels. (See diagram on page 68.) Sounds quieter than 10 decibels are difficult for the human ear to detect, whereas sounds of 120 decibels or greater are usually painful. Normal conversation has a sound intensity of about 60 decibels while a rock concert will have a loudness of 100 decibels.

Deforestation The cutting down and clearing away of trees from the forest, a process that has been going on since humankind discovered wood as a source of energy—usually with ecologically disastrous results. By the beginning of the Christian era, Greece, northern Af-

DECIBELS are units of sound as measured by a decibel meter. Sounds quieter than 10 decibels are difficult for the human ear to hear, and sounds that are louder than 120 decibels are painful to the human ear.

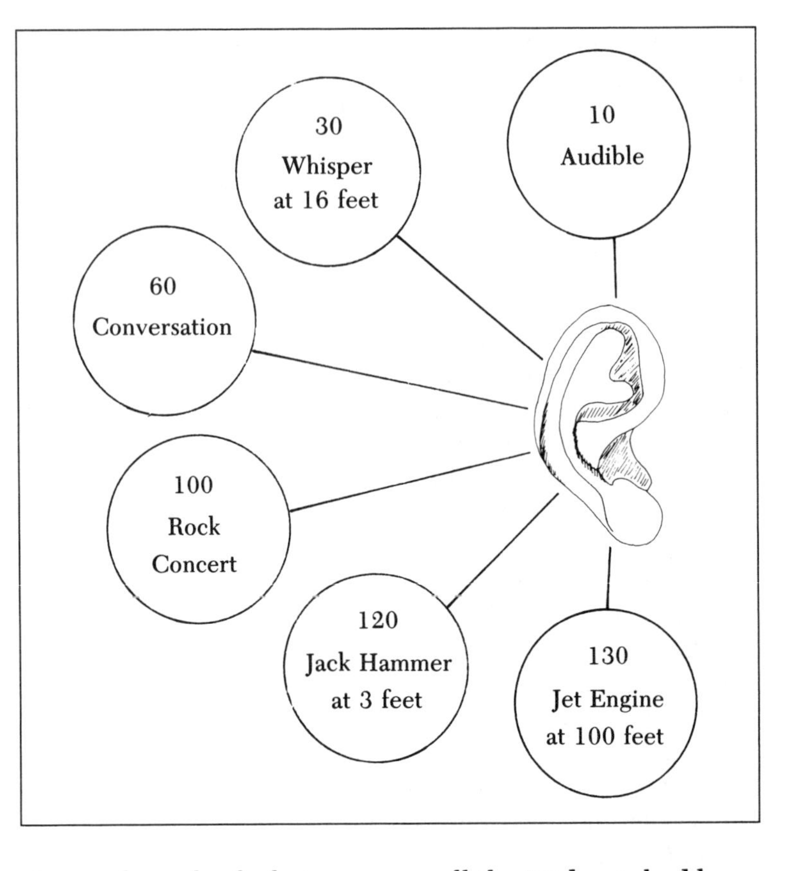

rica, and much of what we now call the Mideast, had been deforested, partly to gather fuel and partly to clear the land for agriculture. The gradual deforestation of western Europe took place in the Middle Ages and more modern times have seen the rapid deforestation of North America. The result is that, except in Canada and Siberia, no great forests remain in the world's temperate zones. The current rapid destruction of tropical rain forests is a subject of serious ecological concern for three main reasons: (1) Live trees absorb CARBON DIOXIDE—the principal GREENHOUSE gas—whereas dead trees, whether burned or left to rot, add carbon dioxide to the atmosphere, thus increasing the danger of GLOBAL WARMING; (2) the drastic denuding of an area means destruction of fertility and eventually a sterile unproductive land; and (3) destruction of the tropical rain forest will mean extinction for a large number of species now dependent on this wildlife habitat.

According to a World Resources Institute report, a total of 40 million to 50 million acres of forest disappear yearly. This is about the equivalent of cutting and burning all the trees in the state of Washington each year. The lesser developed countries are the ones most involved in deforestation, including mainly Brazil, India, Indonesia, Myanmar (Burma), Thailand, Vietnam, Philippines, and Costa Rica. The reasons for deforestation are, of course, economic: sale of timber and clearing land for agriculture and cattle raising. See ENVIRONMENT.

Dendrochronology The science of tree-ring dating. Developed from the work of astronomer A. E. Douglas in 1904. He was looking for the effects of SUN spots on climate and in the process worked out a method for developing good climatological maps as well as archaeological dating methods from tree-ring data.

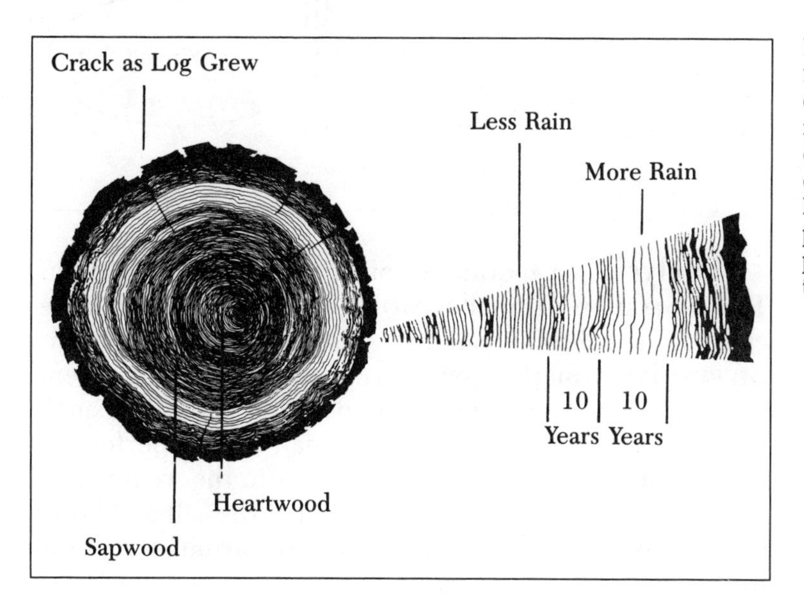

DENDROCHRONOLOGY is the science of the study of annual tree rings to determine the dates and chronological order of past events, most particularly climatological changes.

Desalination The process of changing sea water to fresh water by distilling, evaporating, and then condensing the water, leaving the dissolved material behind. This is an expensive process even if SOLAR ENERGY is used, but there are situations where it is appropriate. Large ocean-going ships supply themselves with fresh water by using oil to both power the engines as well as run the desalination equipment.

Fresh water increasingly is becoming a scarce commodity and in those parts of the world where sea water is available, desalination techniques of various types will be tried.

DESALINATION by means of reverse OSMOSIS, the most widely used process for desalting water from the ocean. Heavy duty pumps build up pressure in a large tank of ocean water and this pressure eventually squeezes water through a membrane that removes the salt particles and creates pure water in an adjacent holding tank.

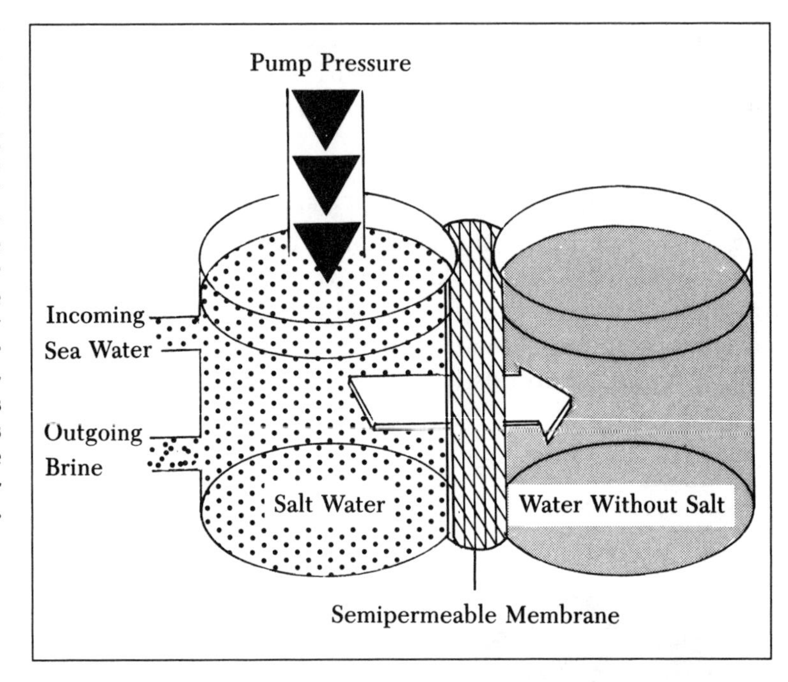

Dialysis A method of purifying **PROTEIN** solutions. The separation of smaller **MOLECULES** from larger molecules in a solution can be accomplished by selective **DIFFUSION** through a semipermeable membrane (membranes with pores large enough to permit passage of small, but not large, molecules). The human kidney performs a somewhat similar filtering function in the removal of waste products and in the regulation of the salt and liquid contents of the body. Kidney failure of various sorts leads to the necessity for periodic treatment by a dialysis machine to adjust the blood's composition. See **OSMOSIS**.

Diastolic One of two components of blood pressure measurement—the second and lowest number in the usual reading. Blood pressure measured when the heart is resting between beats as opposed to **SYSTOLIC** pressure, which is measured when the heart is contracting to pump blood. Readings are in millimeters of mercury as it rises in a testing device; the higher the pressure the higher the number. A diastolic pressure of 80 is normal for a 25-year-old, whereas diastolic pressure of 90 to 104 would be an indicator of mild hypertension. Blood pressure rises with age and a modest rise is better than a steep one.

Diesel A type of INTERNAL-COMBUSTION ENGINE that uses the heat of highly compressed air, instead of a spark plug, to ignite a spray of fuel. The diesel engine's main advantage is its superior fuel economy; however, there are drawbacks. The high pressure required in diesels is hard on engine components and generates more noise than conventional internal-combustion engines. Diesels must be built heavier and sturdier than their gasoline counterparts to permit them to stand the harder wear. Another drawback is the significant exhaust problem. Although considerable research is underway to minimize these exhaust problems, diesels do produce soot and other undesirable emissions.

Diffusion The manner in which the MOLECULES of two substances brought into contact will intermingle; the intermingling of molecules, IONS, and so on resulting from changes in TEMPERATURE as in the dispersion of a vapor in air. The term is also used in connection with reflection or refraction of LIGHT or other electromagnetic radiation when striking an irregular surface—the light is then diffused or unfocused. See ELECTROMAGNETIC SPECTRUM.

Digital Relating to the use of numbers. It is likely that the first counting tools humans ever used were fingers; the word *digit* derives from the words for fingers or toes and is used for a numerical integer. A digital COMPUTER performs operations with quantities represented electronically as digits, usually in a BINARY system. Any device or system that incorporates information in a numerical form is called a digital system. Space exploration SATELLITES, for example, convert information obtained photographically into digital form for transmission to Earth where it is then converted back to a visual image for analyses by astronomers.

Digital recording equipment converts sounds into a numerical equivalent for high-quality and high-fidelity reproduction. Digital audiotape (DAT) decks and cassettes went on sale in mid-1990. DAT decks offer the sound

DIGITAL transmission of radio or TV signals involves the conversion of images and sound into the ones and zeros that can be processed by a computer. Currently, analog signals are broadcast as waveforms. Experts say that digital broadcasting will result in higher quality pictures and sound.

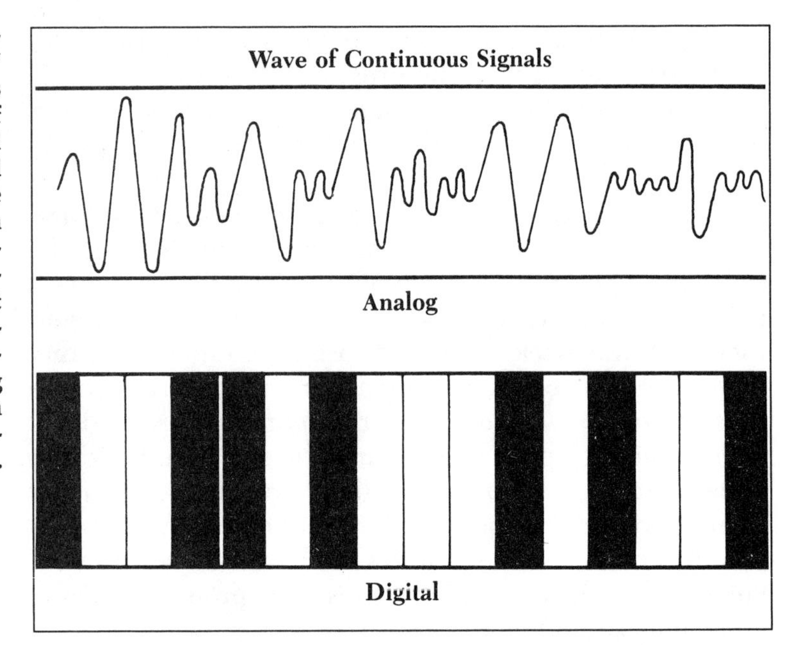

fidelity of compact discs. Digital TV broadcasting systems are also in development and may eventually replace the current analog broadcast systems. Currently, analog signals are broadcast as waveforms. In a digital system, images and sound are converted into the 1's and 0's that can be processed by a computer. Once material, visual or audio, is in digital form, it can be rapidly stored and manipulated in ways not possible with conventional TV. Digital radio broadcast systems are also in the planning stages. Again, higher-fidelity sound is the goal. When a radio station plays a compact disc today, the higher pitched sounds are lost because FM radio can reproduce frequencies only up to 15,000 cycles a second. AM radio is even more limited. Another advantage of digital broadcasting is the elimination of "noise" or distortion. Digital receivers can discriminate between the intended signal and noise and filter out the distortions. Digital radio broadcasting will, however, need frequencies on the broadcast spectrum now allocated to other uses. See ANALOG, and ELECTROMAGNETIC SPECTRUM.

Digital Subtraction Angiography (DSA) A medical diagnostic technology that produces clear views of flowing blood or, more important, its blockage in constricted blood vessels. The term *digital* means numerical and *angiography* means blood vessel imaging. Subtraction comes about in the process itself. First, a fluoroscopic image is made of the body area being examined and that image is stored in a computer. Second, a contrasting dye containing iodine that is opaque to X rays is injected into the patient's vein. The shadow that this opacity creates enables physicians to see the flow of blood. A second contrasting X-ray image is then made, highlighting the flowing blood revealed by the dye. The COMPUTER then subtracts image one from image two, leaving a sharp picture of blood vessels.

DSA is an important diagnostic tool enabling physicians to see blood vessels as small as 1 millimeter (0.039 inch) in diameter. DSA measures the rate at which blood diffuses into the heart muscles, giving the physician a good indication of whether a heart attack is likely to occur. See ANGIOPLASTY.

Dinosaurs, Extinction of Scientists have long speculated about why dinosaurs became extinct at the end of the CRETACEOUS geologic period, 65 million years ago. The various theories are all controversial and none is considered accepted by the scientific community. Dinosaurs are classified as reptiles although some may have been warm-blooded. Fossil records indicate that they lived about 200 million years ago and died out about 135 million years later. After the extinction of the dinosaurs, smaller reptiles, birds, and mammals came to dominate Earth. It is important to understand that Cro-Magnons, early humanlike creatures, did not appear until some 50 million years after the disappearance of the dinosaurs. Some scientists hold that the demise of the dinosaurs—and 70 percent of all other living creatures on Earth—was gradual and took many millions of years. Others believe the animals starved to death from overpopulation in drying swamps. Some scientists see the extinction as sudden and attribute the disappearance to surges of cosmic rays, or on immense volcanoes throwing up enormous amounts of debris that darkened the skies and brought about a continuous winter.

The theory that a comet or meteorite impact caused the mass extinction has been a controversial one since it was first postulated by a research team led by the late University of California-Berkeley physicist and Nobel laureate, Luis Alvarez, and his geologist son Walter in 1980. Considerable evidence has subsequently been found to support this theory. The Alvarez team found unusually high concentrations of the rare metallic element iridium lying in layers within the Earth's crust. Because the iridium was too abundant to have originated on Earth, Alvarez and his team thought that it must have come from extraterrestrial objects that crashed into Earth at various times in the past. They dated the deposits they found to two distinct geologic periods, one about 65 million years ago and the other at about 230 million years ago. Scientists now believe they have found the impact site of the huge meteorite that struck Earth 65 million years ago. The discovery in May of 1990 of thick layers of mud and debris on Cuba and Haiti indicates that a six-mile-wide comet struck somewhere in the area between North and

DINOSAURS is the term used for two different orders of now extinct reptiles that were the dominant life-form on Earth for many millions of years: the bird-hipped group and the lizard-hipped group. Both groups of animals disappeared about 65 million years ago at the end of the CRETACEOUS period.

South America. Firm evidence of an impact site is considered the smoking gun supporting the Alvarez theory. Alvarez supporters postulate that an iridium-containing comet or meteorite must have struck Earth, injecting large amounts of matter into the atmosphere. This matter would have blocked sunlight and created an *impact winter* that would have directly or indirectly led to the death of the dinosaurs, from cold or lack of food resulting from the death of plant life. See COMETS; FOSSILS; METEROIDS, METEORS, AND METEORITES; and NEMESIS.

Dioxins Environmental contaminants formed in minute amounts as unwanted by-products of various industrial and combustion processes, including papermaking, garbage incineration, and automobiles. Virtually everyone carries traces of dioxin in their fatty tissues, at a level of about three parts per trillion. Many environmentalists are concerned that the chemical is exceedingly persistent, breaking down slowly in the ENVIRONMENT, if at all. Few pollutants have generated as much alarm over their potentially damaging health effects as dioxins.

Researchers have established that dioxins can affect animals and humans by mimicking steroid hormones, which are themselves extremely potent chemicals. Toxic agents such as polychlorinated biphenyls (PCBs) and some chlorinated pesticides such as DDT cause hormonal disruptions that result in an array of symptoms, including the growth of tumors. Environmentalists have argued for government regulation of discharges from paper mills and have opposed the construction of some municipal garbage incinerators because of known discharges of traces of dioxin. See TOXIC WASTE, and WASTE MANAGEMENT.

Direct Broadcast Satellite System designed to deliver TV via high-powered SATELLITES to homes equipped with receivers small enough to fit in the windowsill. (See diagram on page 76.) When operational, this service, also known as DBS, promises to greatly expand viewing choices—more than 100 channels will be available. An alternative to cable TV, direct broadcast satellite systems will offer many of the programs now offered on cable plus pay-per-view services and specialized channels that appeal to narrow audiences such as opera lovers, chess players, or ethnic subgroups.

DIRECT BROADCAST SATELLITE system enables satellites to beam up to 108 channels to a 12-inch-by-18-inch receiver that can be mounted in a windowsill.

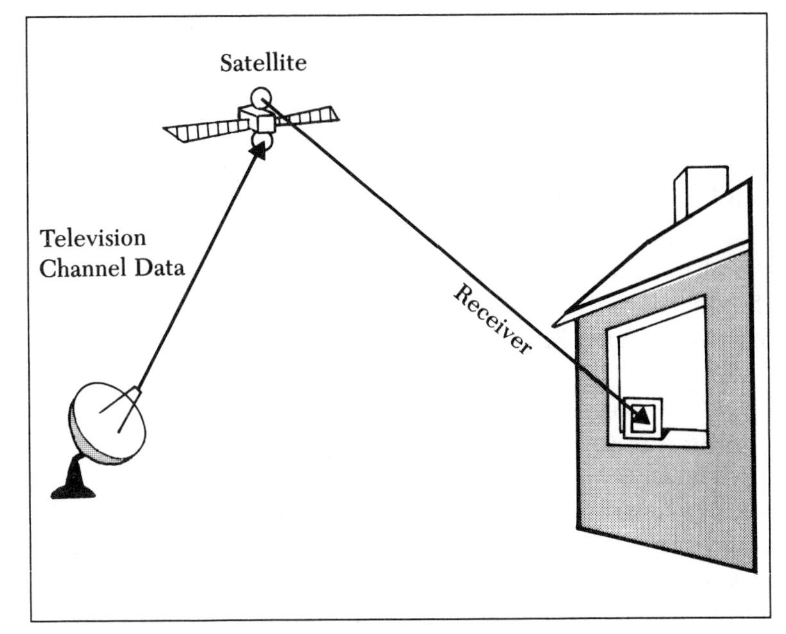

Directed Energy Weapons An integral part of the original Star Wars (or Strategic Defense Initiative) concept. Two types of directed energy weapons were talked about in the early days of SDI: LASERS and *particle beams*, both intended to destroy enemy missiles as they left their launch pads. The beams produced by lasers travel at the speed of LIGHT and, in theory, could strike a target almost instantaneously. Questions arose, however, as to the ability of these weapons to deliver enough ENERGY to destroy a target. The X-ray laser was the key element in the Star Wars concept but numerous technical failures led to rising opposition to the weapon system and a decline in funding.

A particle beam is produced by first accelerating a charged beam of negative IONS (ATOMS that have an additional ELECTRON). The extra electron is then stripped away in a gas cell, leaving a neutral beam. Again, however, serious questions arose as to the practicality of such theoretical weapons. A panel of scientists from the American Physical Society concluded that the power requirements for any of these SDI weapons exceeded any current capability by a factor of 20. Directed weapons were the heart of the original Star Wars concept but when technological reality set in they were quickly deemphasized. See SDI (STAR WARS).

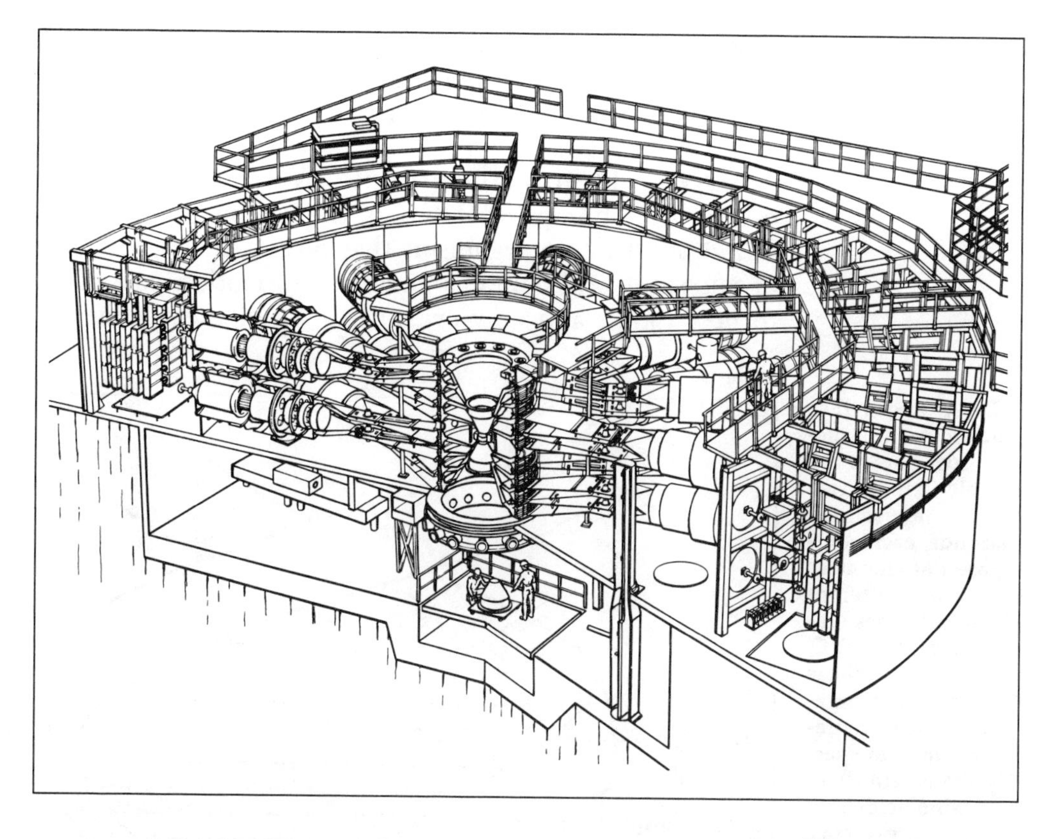

DIRECTED ENERGY WEAPONS under development as part of the Star Wars program include a particle beam accelerator. The particle beam fusion accelerator is designed to produce a 30 million-watt, 5-million-ampere beam of lithium ions and to deliver at least 100 trillion watts per square centimeter for target experiments. The machine consists of 36 pulsed power generators arranged in spoke fashion around a central hub.
Source: U.S. Department of Energy.

Disk, Computer Information storage devices that look something like phonograph records but store information magnetically rather than being engraved in grooves. Disks vary in physical form from lightweight, flexible floppy disks, or diskettes, that can be removed from the COMPUTER and transported, to hard disks, which are usually built in to a computer and can store a great deal more information.

The floppy disk is so named because it is made of thin, flexible plastic coated with a magnetic film. Both floppy and hard disks allow rapid access to stored data. Information is stored on and retrieved from an access arm at the end of which is a read-write head. Just as a phonograph needle can be put down on any part of a spinning record, so too can the access arm quickly move the read-write head to any part of the disk. A *double-density* floppy disk can store up to 400 single-spaced pages of text. A *high-density* disk can store about three times that amount. A hard disk can store hundreds of times more data than a floppy and is therefore used to store application programs or software used by the computer to accomplish specific tasks such as bookkeeping or word processing. See **SOFTWARE**, and **COMPUTER**.

DNA (Deoxyribonucleic Acid) Sometimes called life's blueprint, DNA is the chemical substance found in all living organisms that directs the production of **PROTEINS** and contains genetic information passed on to new **CELLS** and new organisms. All of the recent ad-

DNA (Deoxyribonucleic Acid) structure shows the double helix of two intertwined strands, each composed of chains of four different chemical bases (abbreviated A, T, C, and G). These bases are so organized as to represent a code to specify the assembly of amino acids into proteins.

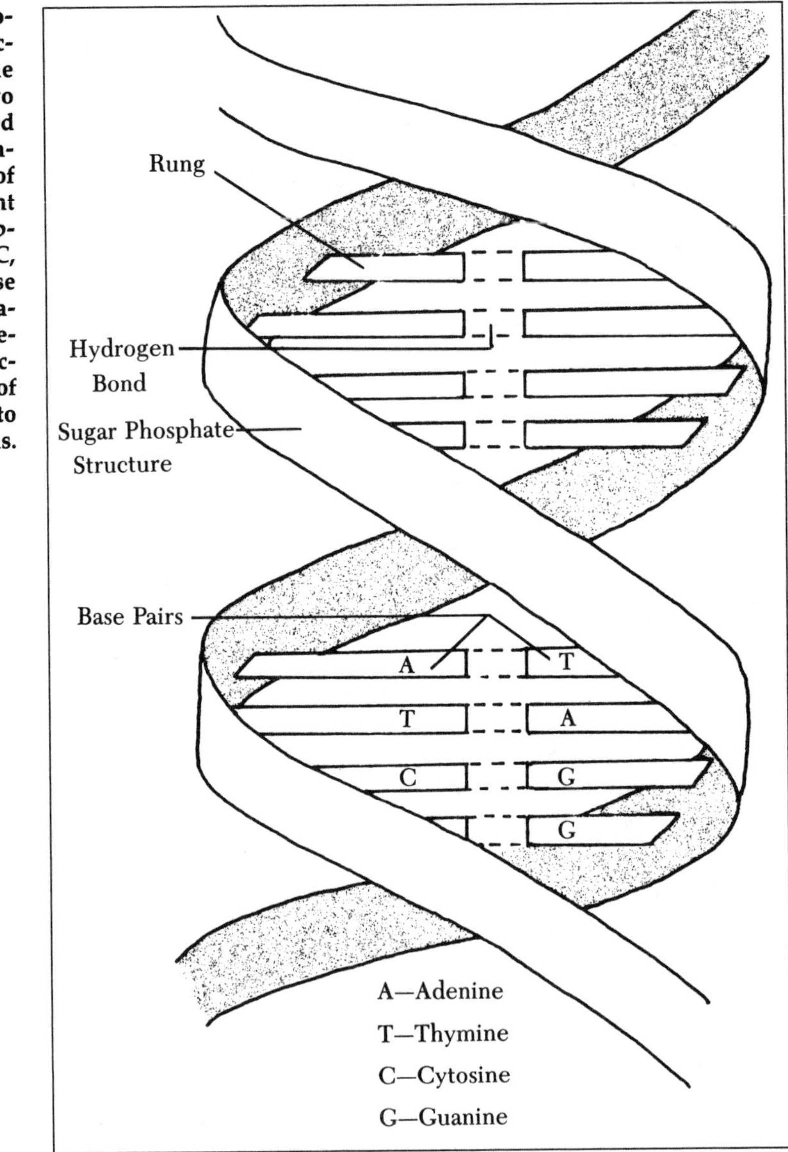

vancements in genetic medicine and genetic engineering are due in a large part to the discovery of the structure of DNA in 1953 by James D. WATSON and Francis H. C. CRICK. They showed that the DNA MOLECULE in each living cell is made up of a pair of strands, one of which is a copy of the other. The strands wind together like a twisted rope ladder, their rungs composed of chemical subunits called nucleotides that are bonded together. Each of the two vertical parts of the ladder has a shape known as a helix, therefore the two together form a *double helix*. It is the sequence of the nucleotide subunits—the rungs of the ladder—that spells out the GENES and gives instructions a cell needs to assemble another cell.

Each of the 100,000-plus genes in the human body consists of the same four bases (any two of which make up a rung of the ladder): (A) adenine, (C) cytosine, (G) guanine, and (T) thymine in a specific sequence or pattern that determines a gene's function. Within a strand the bases can be arranged in any order, but the "rungs" between the strands can be made only between two specific pairs (A to T and G to C). These bases or rungs are organized so as to represent a code to specify the assembly of AMINO ACIDS into proteins. The code for all 20 amino acids used to make protein is found to be different groups of three bases in sequence. These threesomes are called CODONS. A stretch of codons together forms the instructions for the building of protein. Each stretch is in effect a gene. DNA replicates by coming apart, or "unzipping" down the middle. This exposes the bases on each strand of the helix to a group of loose bases. Because they can bind only in complementary pairs, each helix directs the synthesis of a companion helix that is exactly like its original. See GENETIC ENGINEERING, RECOMBINANT DNA, and DRUGS, DESIGNER.

Doppler Effect The apparent change in wavelength of radiation—whether sound or light—emitted by a moving body. This effect is noticeable when the source of SOUND or LIGHT is moving toward or away from an observer. If the source of the waves (light or sound) is moving toward the observer, the frequency of the wavelengths increases and the wavelength is shorter, producing high-pitched sounds and bluish light (called *blue shift*). If the source of waves is moving away from the observer,

DOPPLER EFFECT is the apparent change in pitch as the source of sound approaches and then recedes.

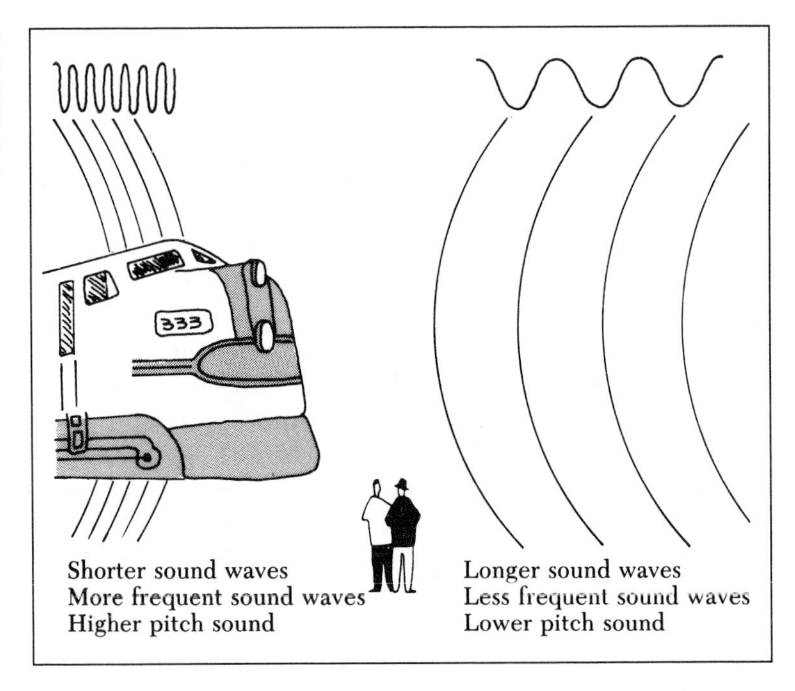

the frequency of the wavelengths decreases, the sound is pitched lower, and light appears reddish (called REDSHIFT).

An often-cited example of the Doppler effect is the apparent change in pitch of a siren as a police car or fire engine approaches and then moves away from a listener. Astronomers use the Doppler phenomenon to estimate the motion and direction of celestial bodies by measuring the color shift (frequency of light waves) given off by an object in space. See EXPANDING UNIVERSE, and HUBBLE.

DOS (Disk Operating System) The SOFTWARE that a COMPUTER must have before it can do anything else worthwhile is called its *operating system*. DOS is one of the most common operating systems in use today. It coordinates activity between the user and the computer. DOS is the master program, the program that makes it easier to run other programs. Operating systems are usually built in to the computer at the factory, and they do most of their work without the user being involved or aware. DOS is the operating system used in most IBM computers as well as in the IBM clones (computers similar to IBMs). DOS was created for IBM by Microsoft, a leading software company. Microsoft also provides DOS for many other computers; these other versions are usually called MS-DOS (for Microsoft DOS). The IBM version is often called PC-DOS but the differences between DOS for one system or another are quite minor. See **OPERATING SYSTEMS**.

	Period	C_D (approx.)
	Late 1950s Early 1960s	0.50
	1970s	0.47
	1980s	.25–.40
	Future	.15–.25

DRAG COEFFICIENT (C_D) of typical cars over the past few years as compared with today's average and tomorrow's potential.

Drag Coefficient (C_D) A numerical measure of resistance to the movement of a body through a fluid medium. Air, for instance, resists the movement of a vehicle passing through it and these resisting or *drag* forces increase with the square of the vehicle speed: Twice the speed produces four times the force.

In automobiles, the engine power required to overcome the drag force increases with the cube of the vehicle's speed: Twice the speed requires eight times the power. As may be seen, even small reductions in drag can result in significant decreases in power requirements at high speeds. Small reductions in drag also result in significant improvements in fuel efficiency. Because of improved aerodynamics, the average automobile in 1990 has a C_D of about 0.4—an improvement over the 1960s when the average was 0.5. Researchers have concluded that a drag coefficient of 0.15 is about the lowest level achievable in a wheeled vehicle. See **AERODYNAMICS**.

Drake's Equation A mathematical formula for estimating the probability of **EXTRATERRESTRIAL INTELLIGENCE** developed by, among others, the American astronomer Frank Drake. In 1960, Drake conducted the first modern search for artificial radio signals from another civilization. He began a search that has been continued off and on by scientists from all over the world.

Drake, and others who believe that there is a good probability of extraterrestrial intelligence (called ETI), start with the assumption that Earth is not unique—that is, that the conditions that led to the evolution of intelligent life on this planet could have occurred elsewhere. Starting with this basic assumption, a number is then assigned to a set of variables that are considered important to the development of ETI, among which are:

- the probable rate at which **STARS** are formed in our galaxy each year
- the probable fraction of those stars that would have planets similar to Earth
- the probability of life of some kind forming on a **PLANET**
- the probability of intelligent life evolving
- the probability of intelligent life trying to communicate with other worlds.

As may be seen, the equation is subjective, and the answer will depend on whether optimistic or skeptical numbers are assigned to the variables. The universe is so large, however, and contains so many **GALAXIES** that even if pessimistic numbers are assigned to Drake's variables, the probability of intelligent life somewhere in the universe must be considered. See **SETI**.

DRAM (Dynamic Random Access Memories) Sometimes called D-RAMs, these **SEMICONDUCTOR** memories store information as a series of 1's and 0's. DRAMs are the highest capacity **CHIPS** in production and are used in the most advanced mainframe **COMPUTERS** and **WORKSTATIONS**. They can store four million **BITS** of information, the equivalent of more than 300 pages of typed text. In 1990, Hitachi Ltd. of Japan demonstrated a prototype 64-million bit DRAM that, although not a complete working model, did demonstrate what is possible in the future.

Drugs, Designer Understanding how the brain's neurotransmitters, the chemical messengers, work is leading the way toward the development of specific drugs that can be applied to intended receptors in the brain. By designing drugs for a specific brain receptor only, undesirable side effects can be avoided. Nonaddictive pain killers are one example of this new field. Other designer drugs are experimental compounds that increase attention spans, improve memory, and prevent suicidal depressions.

The long-range potential for designer drugs is both hopeful and frightening. Drugs to curb anger and limit aggressive tendencies, dispel fear, produce calm, or affect just about any other emotion are now possible. If the biochemical correlates of each emotion can be controlled, as the neurologists tell us, the possibilities for both social good and evil is tremendous. We may have to go back and read Huxley's *Brave New World* to see if this is the future we really want. See NEUROTRANSMITTERS.

Dwarf Stars Stars with masses equal to or less than the Sun. When the surface temperature of various stars is plotted against their absolute magnitude, most stars fall within a narrow band, increasing from dim coolness to bright hotness. This band is called the *main sequence* and our SUN, being a medium-sized star, lies about in the center of this band. Not all stars belong in this main sequence. Massive cooler stars called RED GIANTS burn their fuel at a fast rate and are thus relatively short-lived. Also falling outside the main sequence are the less massive, slower burning, longer-lived WHITE DWARFS. The least massive stars in the white dwarf group have about 1 percent of the MASS of our Sun. If they had less mass than that, they would not be able to generate interior heat for FUSION to take place, and they would instead be PLANETS.

E $= mc^2$ See EQUIVALENCE OF MASS AND ENERGY.

Earth, Age of The Earth is now known to be 4.6 billion years old. It was the geologists of the late 18th and early 19th centuries who first perceived that the Earth could not be as young as the theologians said it was. By examining rock formations, the geologists could see that the emplacement of the rocks called for a lot more than a few thousand years. Also, some of the rocks contained the remains of plants and animals unfamiliar to the scientists and, therefore, evidence that the living world of the past had been a lot different than the present.

EARTH as viewed from *Apollo 17* on its mission to the Moon. This outstanding photograph extends from the Mediterranean Sea area to Antarctica. Almost the entire coastline of Africa is clearly visible and the Arabian Peninsula can be seen at the northeastern edge of Africa. The large island off the coast of Africa is the Malagasy Republic. Astronaut Ronald E. Evans took this picture. Also aboard the *Apollo 17* were astronauts Eugene A. Cernan and Harrison B. Schmitt.
Source: NASA/Johnson Space Center.

Starting with the BIG BANG between 10 and 20 billion years ago, the fundamental forces of nature evolved, the fundamental PARTICLES of MATTER evolved, and GALAXIES and STARS came into being. Then 4.6 billion years ago the SUN and its planets evolved, and 3.5 billion years ago life—in the form of single-celled, blue-green algae—developed on Earth. See EVOLUTION, and GEOLOGIC TIME SCALE.

Earth, Composition of Like all the PLANETS and STARS, the Earth's shape is approximately spherical, the result of mutual gravitational attraction pulling its material toward a common center. Unlike the much larger outer planets, which are mostly gas, the Earth is mostly rock. The distance from Earth's center to its surface is 3,960 miles. The first 2,160 miles is called the core. The inner part of this core is solid, and the outer part is fluid. The iron of the core is thought to be alloyed with other elements such as nickel, sulfur, and OXYGEN. The outer part of the core to the surface is a rocky mantle, except

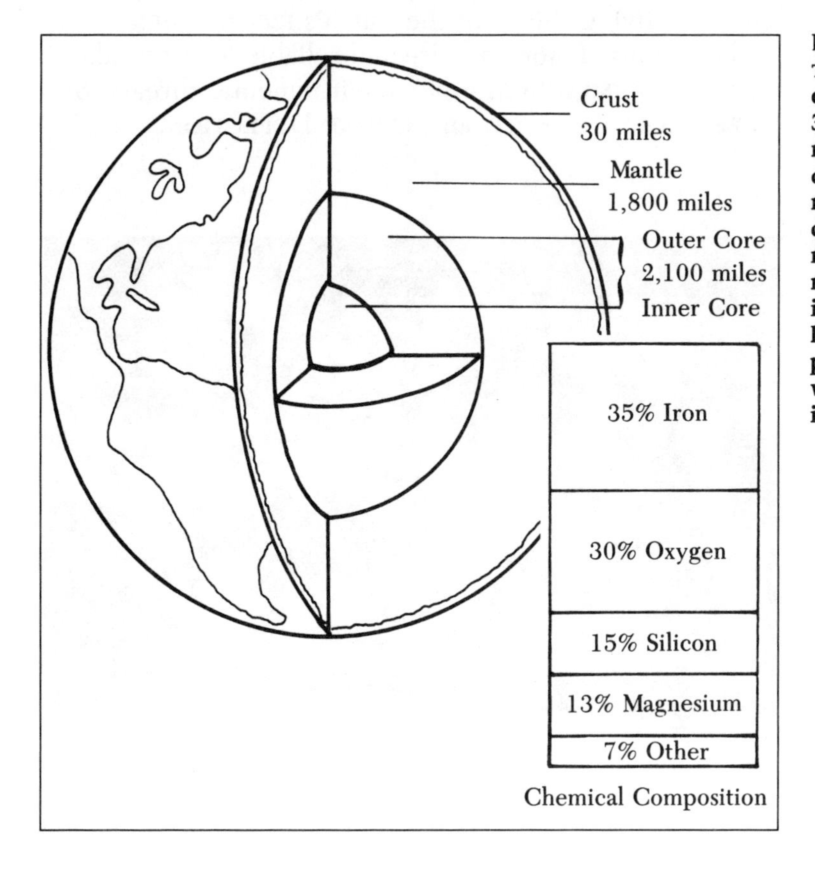

EARTH'S COMPOSITION includes an outer crust about 30 miles thick, a rocky mantle to a depth of 1,800 miles, an outer core of fluid magma, and an inner solid core. The interior of Earth is hot, under high pressure from the weight of overlying layers.

part of the core to the surface is a rocky mantle, except for the thin outer layer that includes the Earth's crust, the oceans, and the ATMOSPHERE. The interior of Earth is hot, under high pressure from the weight of overlying layers. Forces within the Earth cause continual changes on its surface. See **EARTHQUAKES**, and **PLATE TECTONICS**.

Earth Observing System (Eos) An elaborate array of six large (15-ton) SATELLITES, each carrying remote sensing instruments to be launched into low Earth orbit starting in 1998. The purpose of this proposed $30-billion, 30-year NASA enterprise is to help learn how the global environment functions and how it is being altered by human activity.

The unmanned Eos array is planned to be part of NASA's ambitious Mission to Planet Earth, a program to study Earth much as NASA studies other planets in the solar system. Eos will enable scientists to monitor simultaneously many of the complex interactions of air, sea, land, and living things in the Earth's ENVIRONMENT. The goal is to obtain more precise and reliable forecasts about GLOBAL WARMING and other environmental threats. See **GREEN-HOUSE EFFECT**, and **OZONE DEPLETION**.

EARTH OBSERVING SYSTEM (Eos), shown in this artist's concept, is part of a series of missions to help understand the physical, chemical, and biological processes shaping the global environment. Eos will be placed in polar orbit and will carry 20 scientific instruments. Eos is an international cooperative effort involving the European Space Agency and the Japanese Space Development Agency. *Source: NASA/JPL.*

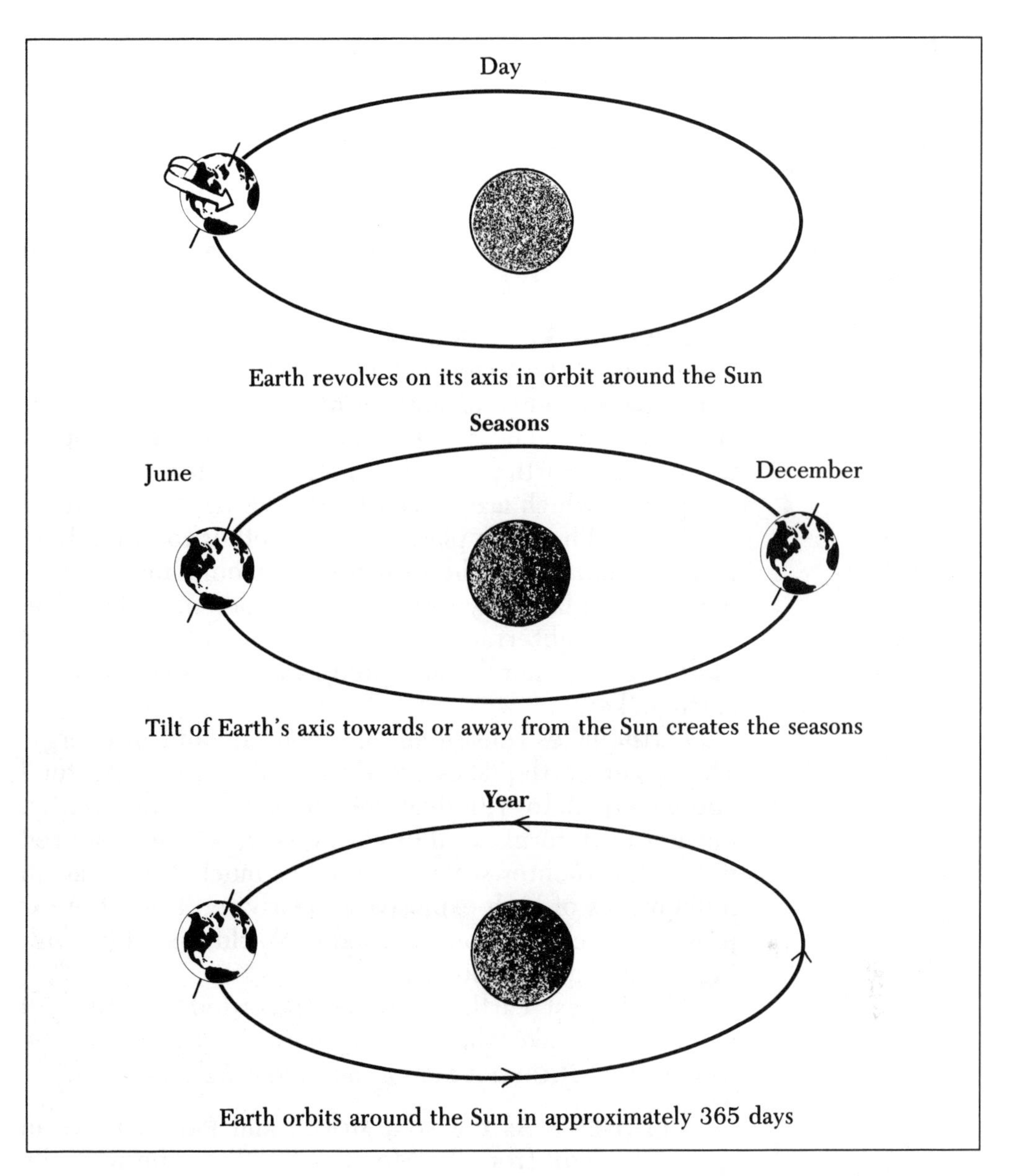

Earth, Orbit of Earth is in orbit around the SUN at a MEAN distance of 93 million miles. Planetary scientists tell us that had Earth been placed in orbit 10,000 miles farther away from the Sun, all of the water on Earth would be frozen. If Earth had been placed in orbit 10,000 miles closer to the Sun, all of the water on Earth would have boiled away. Either way, life as we know it would not exist.

Earth's one-year revolution around the Sun, because of the tilt of the Earth's axis, changes how directly sunlight

EARTH, ORBIT OF. It is the tilt in the Earth's axis toward or away from the Sun that accounts for the amount of time regions of the Earth receive direct sunshine and therefore the seasons. When the **(continued on page 88)**

Northern Hemisphere receives the most sunlight it is our summer (winter in the Southern Hemisphere) and when the Southern Hemisphere receives the most hours of direct sunlight, it is their summer (our winter).

falls on one hemisphere and produces seasonal variations in CLIMATE (summer in the Northern Hemisphere and winter in the Southern Hemisphere, followed a half-year later by a reversal in climates). The rotation of the planet on its axis every 24 hours produces the planet's night-and-day cycle. The combination of the Earth's motion and the MOON's orbit around Earth, once in about every 28 days, results in the phases of the Moon.

Earthquakes Natural disasters resulting from the movement of the *tectonic plates* under the Earth's mantle. Earthquakes (and volcanic eruptions) commonly occur along the boundaries of the tectonic plates. The hot interior of the Earth produces a layer of molten rock under the plates, which are moved by CONVECTION currents in the layer. When two plates come together too rapidly to allow buckling (and the formation of mountains) the surface of one plate may force its way under the other. The subsequent subterranean slippage along what are called *fault lines* is generally accepted as the cause of tectonic earthquakes.

Earthquakes release an enormous amount of energy. The larger earthquakes are estimated to release a total energy equal to 1,000 atomic bombs. Put differently, northern California's 1989 earthquake, which measured 6.9 on the Richter scale, released as much energy as 30 million tons of high explosives—nearly ten times the explosive power of all bombs used in World War II (including the two atomic bombs).

The largest earthquake ever recorded was the 8.9 magnitude quake that hit Japan in 1933. See **PLATE TECTONICS**, **RICHTER SCALE**, and **SEISMOLOGY**.

Eclipses As the SUN, MOON, and EARTH move in space, the light from the Moon or the Sun is periodically obscured by: (1) the intervention of Earth between the Sun and the Moon (**LUNAR ECLIPSE**), or (2) the intervention of the Moon between the Sun and the Earth (**SOLAR ECLIPSE**). A total solar eclipse is a spectacular astronomical sight, so people travel to those regions of Earth where the phenomenon can best be observed. The Moon is 400 times closer to us than the Sun and also 400 times smaller than the Sun. Thus the Sun and the Moon take up almost exactly the same angle in the sky, and the Moon can block the Sun exactly.

Ecology The science and study of the relationships between organisms and their ENVIRONMENT. The study of the interconnectability of life-forms has demonstrated that living species are largely interdependent. Obvious examples include bees and plants. Bees pollinate the plants and are in turn fed by the plants. The two organisms are thus interdependent. The more humankind learns of the sometimes unforeseen consequences of its actions, the more necessary becomes the study of ecology. See ECOSYSTEM.

Ecosystem The entire ecological community—all living organisms—together with its physical ENVIRONMENT, considered as one system. The term is generally used to describe the interaction of living organisms as well as the nonliving environment of land and water, solar radiation, rainfall, mineral concentration, TEMPERATURE, and topography.

The linked and fluctuating interactions of life-forms and environment compose a total ecosystem. The interdependence of organisms in an ecosystem often results in an approximate stability over long periods of time. Ecosystems change when drastic CLIMATE changes occur, when new species are introduced, or when humankind deliberately or inadvertently causes a modification of some sort. See ECOLOGY.

Einstein, Albert When he was only in his twenties, German-born Albert Einstein published theoretical ideas that made revolutionary contributions to humankind's understanding of nature—the SPECIAL THEORY OF RELATIVITY and the GENERAL THEORY OF RELATIVITY. The special theory built on and modified the Newtonian laws of motion—the scientific and philosophical views of the world that had prevailed for 200 years. General relativity is a fundamental theory of the nature and relationship of space, time, and gravitation and has profoundly influenced how humankind views the universe. Einstein's theories have been tested and retested by checking predictions based upon them, and they have never failed. Nor has any new theory of the architecture of the universe replaced them. See ENERGY; EQUIVALENCE PRINCIPLE; GRAVITY; LIGHT, SPEED OF; MASS; and SPACE-TIME.

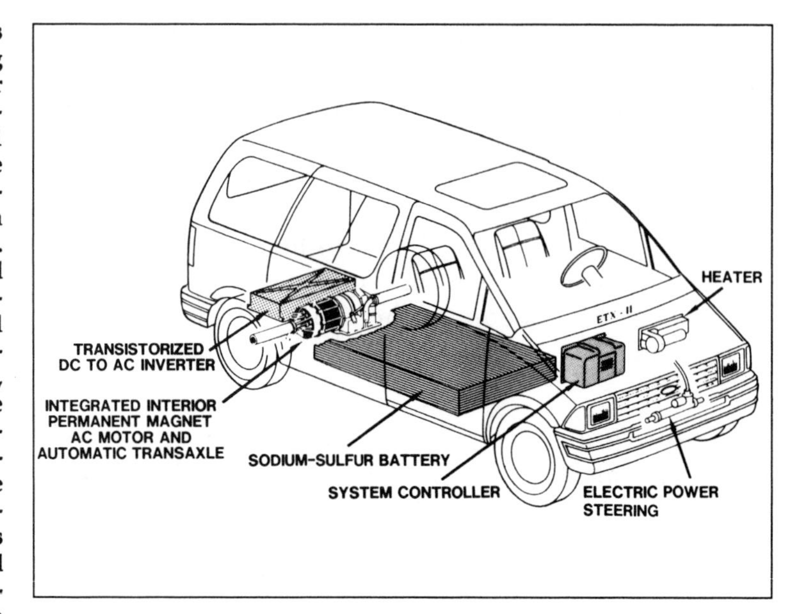

ELECTRIC CAR. This schematic drawing of the Ford Motor Company's proposed ETX-11 Aerostar shows the location of the major components in an electric vehicle. The 1,200 pound battery pack supplies a nominal 200-volt direct current to an inverter, which converts the energy to three-phase, variable-frequency, variable amplitude alternating current. This current is supplied to the traction motor through a power module that can regulate motor speed from zero to 11,000 rpm. *Source: Ford Motor Company.*

Electric Car Replacing gasoline-powered vehicles with electrically powered cars has been talked about and tinkered with since the early days of the automobile. The oil embargo of 1973 and the current concern about curbing pollution from autos has intensified this interest. Clean air is suddenly a national priority. Nearly 60 percent of the population of the United States lives in areas that do not meet federal clean air standards. Poor performance in electric cars, specifically speed and range, has always been the major drawback in the past. However, in mid-1990 General Motors announced what they called a "full-performance" experimental electric car called the Impact, which can reach 75 miles per hour and travel as far as 120 miles per charge. They were careful not to claim that the Impact could do both at the same time. GM said that at current prices, it would cost about $40 per month to fuel and service an electric car that is driven 10,000 miles a year.

Battery technology remains the major stumbling block to the wide use of electric cars. Batteries store energy only one-twentieth as well as gasoline, on a pound for pound basis, and recharging is slow. More important, standard lead-acid batteries used in the Impact and other electric cars have to be replaced every two years at a cost of $1,500, bringing the operating cost of using electrics to about twice that of gasoline-powered vehicles. Until major breakthroughs in battery technology are achieved, electric cars will occupy a modest niche in the market for the foreseeable future and mainly will be used as delivery trucks, service vans, and golf carts that travel short distances each day. See **Electromagnetic Induction**, **Environment**, and **Fuels, Alternate.**

Electromagnetic Field Emissions Electric and magnetic fields abound in nature. They also emanate from the flow of electricity through everything from transmission lines to video display monitors and household appliances. A 1990 study by the Environmental Protection Agency (EPA) has shown a possibly statistically significant link between cancer and exposure to extremely low frequency (ELF) electromagnetic fields. The first warning of the possible health effects of ELF came in 1979 in a study by Nancy Wertheimer and Edward Leeper that suggested a link between electric power lines and childhood leukemia. Since that time, numerous epidemiologic investigations have focused on the connection between electromagnetic fields and human cancer. Biological studies, however, have yet to prove a conclusive cause-and-effect relationship. The strongest epidemiologic studies found elevated risks of cancer among children, pregnant women, and workers in occupations that exposed them to higher than average levels of electromagnetic radiation.

Utility company employees who work near power lines or power line transformers and telephone workers who work on lines seem to be the two groups most at risk, according to the studies that have been made at the end of 1990. Electric blankets are another focus of concern, as one study revealed a quadrupling in risk of brain tumors among children whose mothers slept under electric blankets during the first trimester of pregnancy. So far the potential risks of ELF are controversial and the subject is being handled with extreme caution by the responsible government agencies such as EPA and the congressional Office of Technology Assessment. The possibility of eventual government regulation as well as the threat of legal action by employees have caused the Electric Power Research Institute, and other affected organizations, to view developments in ELF exposure technology with serious concern. See **Carcinogen**, and **Radiation.**

ELECTROMAGNETIC INDUCTION is to be used to transfer electric power from a roadway to cars and buses on the surface in an experimental program underway. This Southern California Edison plan involves a retractable unit that hangs from a car's chassis and picks up electrical energy from cables buried in the roadway.

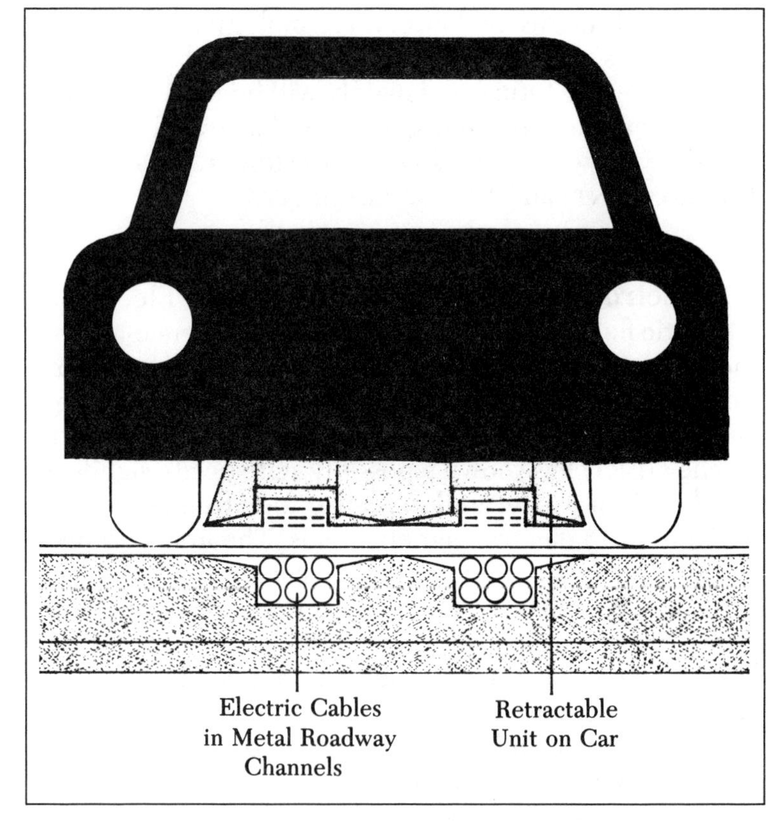

Electromagnetic Induction Production of an electric current by changing the magnetic field enclosed by an electric circuit, as in a generator. This is the technology that Southern California Edison Company proposes to utilize in a radically different approach to commercializing electric cars. The California utility has announced plans to build a road that transfers electric power from underground cables to cars and buses on the surface, without physical contact between the vehicles and the highway. Cables buried in the concrete create a magnetic field at the surface while the vehicles have a metal plate descending to within a couple of inches of the road surface to pick up energy. The magnetic induction process delivers over 90 percent of the energy across the gap between the pavement and the car. The Southern California Edison plan envisions commuters traveling a few miles from their residence to the powered highway using battery power. Once on the freeway, the cables would recharge the batteries and provide motive power.

Because the Los Angeles basin has more than seven million gasoline-powered vehicles and the worst air pollution in the nation, electric vehicles are of special interest for this area. Recent studies have raised questions about whether the exposure to electric and magnetic fields have adverse health effects, and this concern would have to be resolved before the L.A. project can proceed beyond the experimental two miles of roadway planned for 1991. See **ELECTROMAGNETIC FIELD EMISSIONS.**

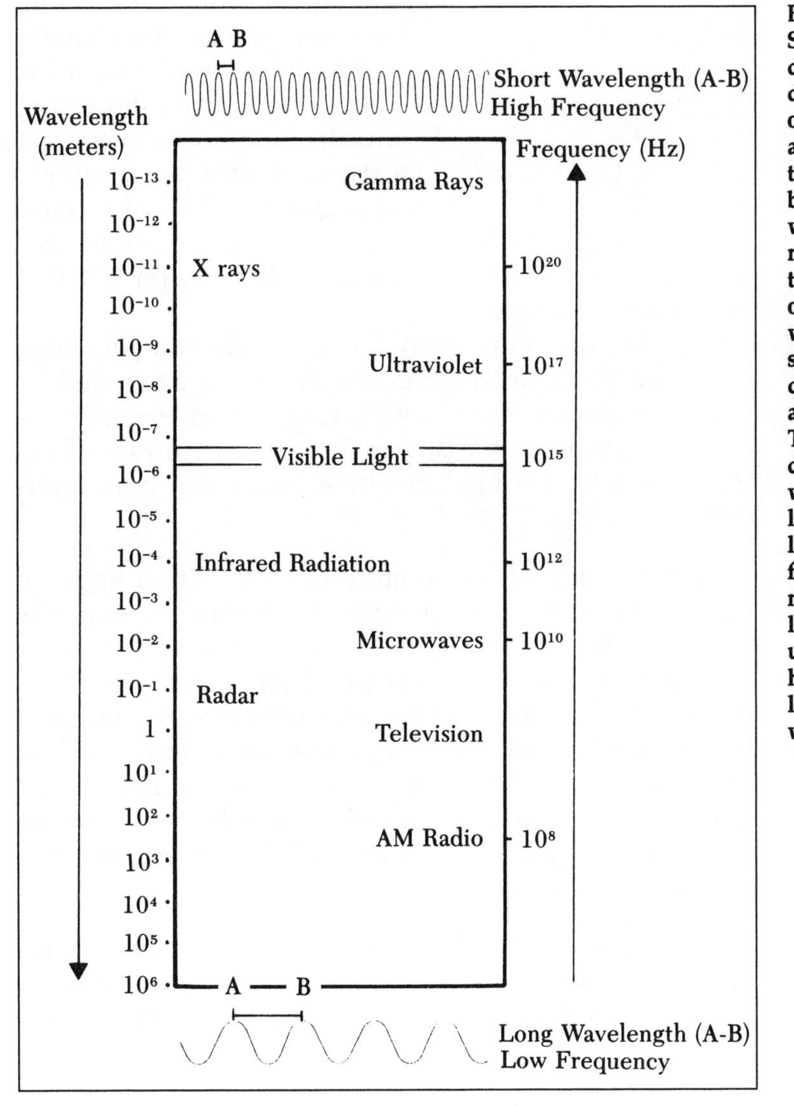

ELECTROMAGNETIC SPECTRUM. Radiated energy is described in terms of its wavelengths and its frequency; that is, the distance between successive wave crests and the number of crests that arrive per second. When the wavelength is short, the frequency is high, and vice versa. These forms of radiation range in wavelength from less than a billionth of a micron for gamma rays (a micron is a millionth of a meter) up to tens and hundreds of miles long for radio waves.

Electromagnetic Spectrum A log of the distribution of ENERGY by wavelength. Energy appears in many forms, including RADIATION. Radiation is described in terms of its wavelengths and frequency, that is, the distance between successive wave crests and the number of crests that arrive per second. This family of electromagnetic waves from the short wavelength, high-frequency GAMMA RAYS to the long wavelength, low-frequency radio waves form what is called the electromagnetic spectrum. (See diagram on page 93.) These forms of radiation range in wavelength from less than a billionth of a micron for gamma rays (a *micron* is a millionth of a meter) up to tens and hundreds of miles for long radio waves. Wavelengths are measured in HERTZ (Hz), the international unit of frequency equal to one cycle per second. From shortest to longest, the spectrum of waves includes: GAMMA RAYS, X RAYS, ultra-violet, visible LIGHT, INFRARED (heat), MICROWAVES, VHF, television, and ordinary radio. All electromagnetic waves have the same speed in a vacuum; that is, the speed of light, equal to 186,000 miles (or 300,000 kilometers) per second.

In addition to the many forms of radiation generated by humankind, electromagnetic energy is generated by natural processes across a wide range of wavelengths, including light from the SUN and STARS, microwaves from the cosmic background radiation, and radio from interstellar clouds.

Electron An elementary particle with a negative charge and a very small MASS. Electrons are normally found in orbit around the nucleus of an atom. The number of electrons in an atom—ranging from 1 to about 100—matches the number of charged PARTICLES, or PROTONS, in the nucleus, and determines how the atom will link to other ATOMS to form the chemical bonds called MOLECULES. The movement of large numbers of electrons through conductors constitutes an electric current. See CHEMICAL BONDING.

Electron-Beam Lithography Most microchips today are made by photolithographic systems based on optical technology. However, as COMPUTER CHIPS shrink and grow more sophisticated, they are nearing the point at which beams of visible LIGHT are not sharp enough to

Wavelength used to etch	Chip linewidth (microns)	Transistor Equivalent (million)	Memory dbl. spaced typed pages	Stage
Visible Light	1.2	1	100	Current Use
	.8	4	400	
	.5	16	1600	
Ultraviolet rays	.35	64	6400	Development
X rays	.25	256		Research
	.17 to .10	1–4 billion		

Data: IBM.

etch circuitry more than 150 times thinner than a human hair. Electron-beam lithography involves steering a powerful beam of ELECTRONS to directly "write" circuit patterns on silicone wafers. Electron-beam lithography permits remarkably fine resolution, giving chip makers the ability to create microscopic electronic circuits denser than is possible with optical methods. The electron-beam method is, however, slower than optical processes and is generally used today to manufacture the *masks,* or master patterns, used in conventional optical systems and in making custom chips, for which cost and speed matters less. See **DRAM**.

Element In chemistry and physics this term refers to a substance composed of ATOMS having an identical number of PROTONS in each nucleus. These materials, such as carbon, HYDROGEN, and OXYGEN, cannot be broken down into more fundamental substances. There are more than 100 known chemical elements. Chemical COMPOUNDS are formed when atoms of different elements are bound together into MOLECULES.

A chart of chemical elements that displays them in horizontal rows according to atomic number, and in vertical rows according to similarity of the structure of their atoms, is called a PERIODIC TABLE.

Elementary Particle A subatomic particle regarded as an irreducible constituent of matter, sometimes called a *fundamental particle.* See **PARTICLE PHYSICS**.

ELF (Extremely Low Frequency) See **ELECTROMAGNETIC FIELD EMISSIONS**.

Elliptical Having the shape of an ellipse; that is, a squashed or elongated circle. The orbits of the nine PLANETS revolving around the SUN and of most COMETS are ellipses.

Embryo A developing living organism, plant or animal. A plant embryo is an undeveloped plant inside a seed. In human development, the embryo is carried in the mother's womb. In the first stage, it develops from the single cell of a ZYGOTE. The term *embryo* is used in the first two months after conception, when organ systems are being formed. After that time, and until birth, the term used is FETUS.

Empirical Derived from observation or experiment rather than by theory. Empirical data is information gathered through practical experience rather than through hypothesized doctrine. In medicine, the term refers to diagnosis or treatment guided by experience rather than theoretical conjecture.

Energy In physics, the potential for work. The capacity for work can be represented by energy in different ways: An object can have energy by virtue of its motion (KINETIC ENERGY), by virtue of its position (potential energy), or by virtue of its MASS. As Einstein showed in 1925, mass and energy are equivalent ($E = mc^2$), and a small amount of mass can contain an enormous amount of energy.

The *first law of thermodynamics* states that energy is conserved; that is, it is indestructible—there is always the same amount of energy in the universe. Energy can neither be created nor destroyed; it can, however, change form, such as from chemical energy in fuel to mechanical energy in an automobile or to heat in a furnace. See PLANCK, MAX; SPECIAL THEORY OF RELATIVITY; THERMODYNAMICS, FIRST AND SECOND LAWS OF; and QUANTUM PHYSICS.

Energy, Sources of The primary sources of energy in the United States today (1989 data from U.S. Statistical Abstracts, 1990) and the percentage of our total energy supply represented by each source are:

Petroleum	43%
Coal	24%
Natural Gas	22%
Nuclear	6%
Hydro	4.75%
Geothermal and other	0.25%
	100%

Energy, Uses of A breakdown of how we use energy in the United States today (1989 data from U.S. Statistical Abstracts, 1990) is shown below:

Generation of electricity	36%
Transportation	28%
Industrial	23%
Residential and Commercial	13%
	100%

Energy use facts: Americans make up only 5 percent of the world's population but use 26 percent of the world's oil. Almost half of the oil consumed in the United States is imported and about half of the United States trade deficit is due to these oil imports. Of the 17 million barrels of oil consumed each day in the United States, 43 percent is used by automobiles and trucks. Burning one gallon of gasoline produces almost 20 pounds of CARBON DIOXIDE, the chief cause of global warming. In fact, transportation generates 33 percent of all carbon emissions in the United States.

Entropy A measure of the degree of disorder, or tendency toward the breakdown, of any system. In physics, the term is defined in the *second law of thermodynamics,* which states in part that "the entropy of the universe tends to a maximum." This is another way of saying the overall disorder of an isolated system must increase. The effect of increased entropy is that things evolve from a state of relative order to one of disorder, and with this disorder there is increased complexity. The term is often used informally to refer to the disorganization or breakdown of a social system or function as in, "his clarifying remarks did nothing but add to the entropy."

As all homeowners are aware of, one has only to stop making repairs around the house to see a convincing example of the tendency toward disorder and breakdown in a system. See **THERMODYNAMICS, FIRST AND SECOND LAWS OF.**

Environment Everything that makes up our surroundings. In the physical world, the term means the global or local conditions affecting our health and well-being. Environmental abuse is the subject of much concern and environmental problems such as the **GREENHOUSE EFFECT, OZONE DEPLETION, ACID RAIN**, population control, and **TOXIC WASTE** disposal may come to dominate political debate in the near future.

A survey by the Roper Organization in 1990 indicated that 78 percent of Americans believe that a "major national effort" is necessary to improve the environment. This survey showed that Americans placed the environment on their list of national priorities, ranking it fourth behind concern about crime and drugs, **AIDS**, and the cost of health care.

Enzyme A protein **MOLECULE** that acts as a **CATALYST** in bringing about chemical changes in other molecules but is not changed itself. Enzymes are vital to the body's chemical reactions. There are thousands of kinds of enzymes and each one makes possible, by its presence, a particular chemical reaction. A well-known enzyme is yeast, which, by its presence, makes fermentation occur. Human digestion is another example of an enzymatic process. Here, foodstuffs are decomposed into simple, assimilable substances as a result of the catalytic reaction of enzymes.

Equinox Either of the two times during the year (about March 21 and September 22) when the length of day and night are equal. The word *equinox* means equal night. At these times the relative position of the **EARTH** to the **SUN** is such that the tilt in the Earth's axis is nullified, and the same amount of sunlight falls on both the Northern and Southern Hemispheres—in other words, when the Sun crosses the celestial equator. The two times are called the vernal equinox (March) and the autumnal equinox (September). See **ASTRONOMY.**

Equivalence of Mass and Energy The principle embodied in the most famous equation in the world:

$$E = mc^2$$

In this equation (formulated by Einstein), E represents

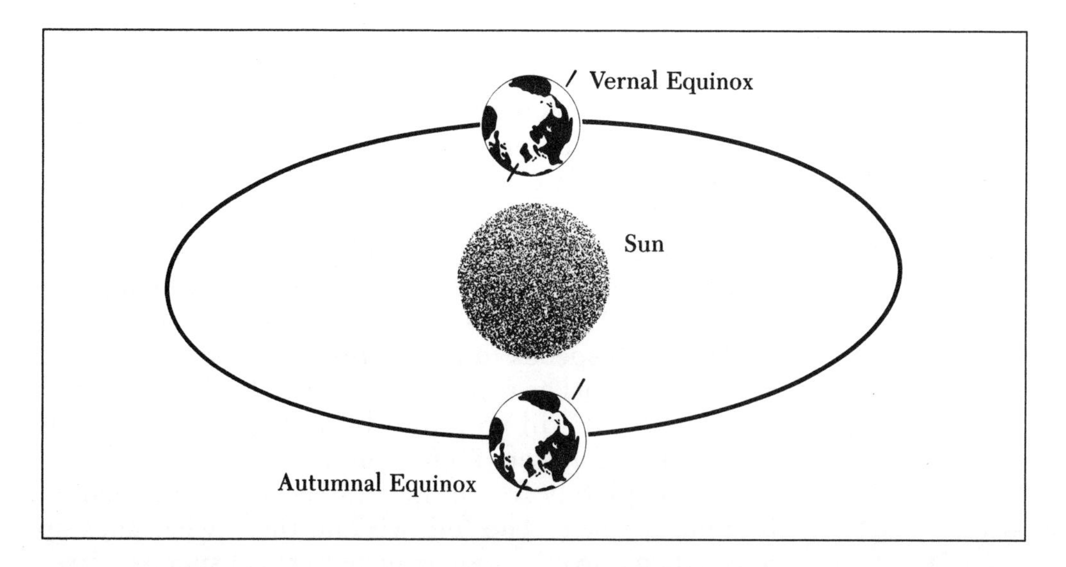

ENERGY in ERGS, m represents MASS in grams, and c represents the speed of LIGHT in centimeters per second. Because light travels at 30 billion centimeters per second, the value of c^2 is 900 billion billion. Conversely, if an object radiates energy E, its mass decreases by E divided by the speed of light squared. It can be seen that mass and energy are equivalent and can be transformed one to the other. Moreover, the transformation of even a tiny amount of mass releases an enormous amount of energy. When this release is accomplished slowly and under control, the energy can be used to generate power, as in a nuclear REACTOR. When this energy is released suddenly, a destructive force is released, as in an atomic bomb. See EINSTEIN.

EQUINOX. Either of the two times during the year when the length of day and night are equal. This occurs in the northern hemispheric winter on about December 21 and in the northern hemispheric summer about March 21.

Equivalence Principle Formulated by EINSTEIN in his GENERAL THEORY OF RELATIVITY, the principle that the weightlessness, or vanishing of GRAVITY, that an observer inside an enclosed, free-falling laboratory would experience was equivalent to life with no gravity. Conversely, a laboratory in space that was being accelerated by rockets was equivalent to one at rest in a gravitational field. From this equivalence, Einstein concluded that SPACE-TIME must be curved, not flat, and that this curvature is produced by the gravitational effects of matter. Einstein's theories have since been confirmed by numerous space experiments.

ERG A very small unit of ENERGY or work used in scientific inquiry and equal to the force necessary to accelerate one gram of MASS over a distance of one centimeter in one second. From EINSTEIN's famous equation ($E = mc^2$), we know that one gram of mass will produce 900 billion billion ergs of energy. See EQUIVALENCE OF MASS AND ENERGY, and JOULE.

Escape Velocity The speed that must be reached at the surface of a planet to escape the pull of the planet's gravity. Rockets and spacecraft launch vehicles must achieve high speeds to break free of the pull of Earth's gravity. If a rocket reaches a speed of 17,500 miles per hour (mph), it will go into orbit around the EARTH. If it travels still faster (25,000 mph, to be exact) it will break completely free of Earth's GRAVITY and head off into outer space. The folk wisdom that "what goes up must come down" is only true on Earth at speeds slower than 25,000 mph.

The velocity of escape from any astronomical body can be calculated from its MASS and size. For instance, the gravity forces on the MOON are much smaller than Earth's and therefore the escape velocity is much less (5,400 mph). JUPITER, on the other hand, is a larger planet than Earth and exerts a much greater gravitational force. Escape velocity on Jupiter is 134,548 mph. To put these speeds in some perspective consider that the speed of the Concord SST is about 1,400 mph whereas the speed of a bullet fired from a .22 caliber rifle is only 886 mph.

ESP (Extrasensory Perception) A pseudoscientific belief that knowledge or perception is possible without the use of the five senses. ESP includes clairvoyance (the supposed power to perceive things out of the natural range of human senses), telepathy (reading another person's thoughts), and precognition (predicting some future event). None of these claimed powers, although quite popular in the supermarket checkout-stand tabloids, has ever been verified by scientific procedure.

The problem for ESP workers, and there are still a few psychologists at work in this field, is the need to prove both that a perception is occurring (as against chance, coincidence, wishful thinking, etc.) and that it is extrasensory. Despite serious attempts to confirm ESP—by Duke University in the 1950s and even by the prestigious SRI International in the 1960s, the belief remains unproven. See PARANORMAL.

Ethanol (See FUELS, ALTERNATE).

Evolution Earth's present-day life-forms have evolved from common ancestors reaching back to the one-cell organisms about three billion years ago. The fact of evolution is confirmed by three main sets of observable data: 1) the enormous number of different life-forms present on Earth, 2) the clear similarities in anatomy and molecular chemistry seen within that diversity of life-forms, and 3) the sequence of changes in FOSSILS found in successive layers of rock and sediment that have formed over more than a billion years.

Central to the concept of evolution is the process of NATURAL SELECTION by which the sequence of evolution over generation after generation favors those life-forms and subspecies best able to cope with their particular environment.

Life on this planet has existed for three billion years. During the first two billion years of life, only microorganisms existed, some quite similar to the BACTERIA and algae that exist today. With the development of CELLS with nuclei about a billion years ago, there was a great increase in the rate of evolution of increasingly complex, multicelled organisms. How evolution works in every detail in still not known, but the concept, first postulated by Charles DARWIN in the 19th century, is now well established and accepted as fact by the scientific community.

Expanding Universe The concept, first postulated by the astronomer Edwin HUBBLE in 1929, that distant GALAXIES were receding from EARTH, and from each other, at a constant rate. Hubble found that the speed of recession was proportional with the distance of the galaxy from Earth.

As with SOUND waves, when the source of LIGHT moves toward an observer, the waves are squeezed together and the colors become bluer (*blueshift*). If the source of the light waves is moving away from an observer, the frequency of light waves decreases and the light appears reddish (*redshift*). Hubble determined from the amount of redshift of distant stars that the universe was expanding. See BIG BANG, and DOPPLER EFFECT.

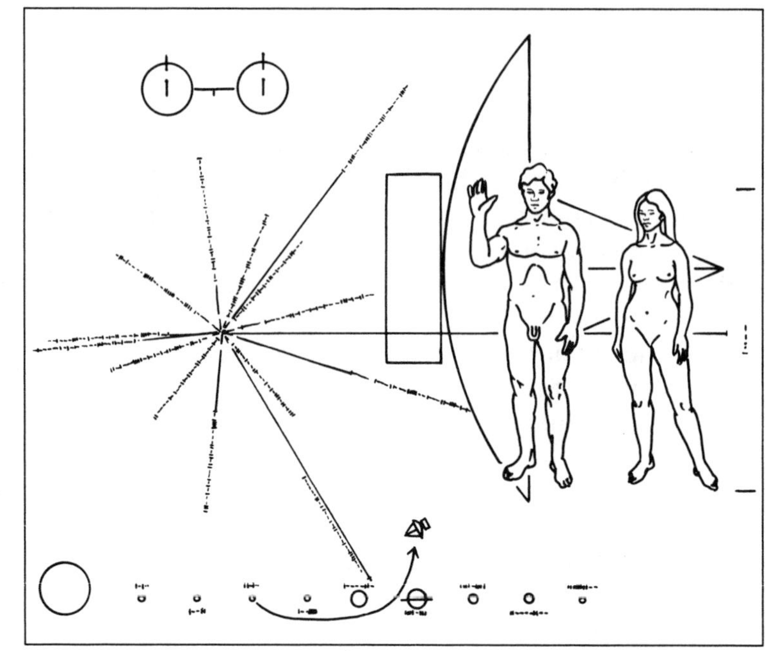

EXTRATERRESTRIAL INTELLIGENCE may exist on some far-off planets in the universe, although there is no evidence to date of any such life. *Pioneer 10* and *11* spacecraft are the first vehicles of humankind to venture into interstellar space. Each carries a 6-by-9-inch plaque intended to convey some information on the locale and the nature of the builders of the spacecraft. The plaque indicates the planet from which the probe was launched, relative size of humans, and the location of the Sun.

Extraterrestrial Intelligence

Is humankind alone in the universe? Or are other intelligent beings out there somewhere in the vastness of space? Some astronomers have long thought that there are probably so many planets in the universe that even if only a small fraction were suitable for life, there should be hundreds of thousands of PLANETS with life. Recent advances in astronomy and physics have strengthened the theory that there could be many planetary systems hospitable to life. Some scientists believe that on planets with an abundant supply of HYDROGEN-rich gases, liquid water, and sources of ENERGY, the probability of life forming is high. These scientists also believe that an alien intelligent life-form that attempts to communicate with us is most likely to be more advanced than we are.

Astronomer Frank D. Drake, a pioneer in the search for some signs of extraterrestrial intelligence in the universe, has quantified the probability than an advanced civilization exists somewhere in the universe in a mathematical formula called DRAKE'S EQUATION. Depending on assumptions about all of the equation's variables, the typical guess is that 10,000 to 100,000 advanced civilizations exist in the MILKY WAY galaxy alone. A scientific search for life on other worlds has been underway, on a modest scale, both here and in the Soviet Union for more than five years. The U.S. project is called SETI, for Search for Extraterrestrial Intelligence. Not all astronomers or astrophysicists agree on the advisability of this search program. Skeptics have long posed the question, If there are extraterrestrial beings, why have they not contacted us by now? The main criticism the opponents have to the search is that no firm evidence of any life elsewhere in the UNIVERSE exists, despite the enormous number of STARs and the possibility that there may be other habitable planets. Opponents maintain that we may be unique, and we may be alone.

Fahrenheit TEMPERATURE in the United States is usually measured in degrees *Fahrenheit* (°F). This scale is named for Gabriel Fahrenheit, a German-born instrument maker, who also improved the thermometer by using mercury as the temperature-indicating substance. Like the other two widely used temperature scales (CELSIUS and KELVIN), the Fahrenheit scale has two fixed points of reference: the freezing point of water (arbitrarily designated as 32) and the boiling point of water (also arbitrarily set at 212). The Fahrenheit scale divides the interval from 32 to 212 into 180 equal parts called degrees (indicated by the symbol °).

Most of us in this country have grown up with the Fahrenheit scale and we are used to it. We know, for instance, that a thermostat set at 69 degrees will provide comfort indoors, and when the weather reporter announces that it will reach 100 degrees that day, it is going to be hot. Most other countries use degrees centigrade (the Celsius scale), and scientists often use the Kelvin scale, which starts at ABSOLUTE ZERO. If you find yourself in parts of the world where the Celsius scale is used, an approximate, and easy to remember, conversion method is to merely double the Celsius figure and then add 30.

Falling Bodies, Law of Of the many fallacies concerning the great Italian astronomer and physicist, Galileo Galilei, the most famous one has him climbing to the top of the Leaning Tower of Pisa in order to drop a 10-pound sphere and a 1-pound sphere simultaneously to show that objects of different weights would fall to Earth at the same velocity. The story is not only apocryphal (it was never mentioned in any of Galileo's notes), but it is also scientifically unsound: The two spheres would fall at the same rate only in a vacuum. See GALILEI.

Feedback Control A process in which a system regulates itself by monitoring its own output and then sending this data back to the input to control the overall operation. A familiar example of a feedback device is the thermostat that controls room TEMPERATURE. A more complex example is the automatic pilot on an airplane that maintains the craft in straight and level flight at a preset altitude.

In physiology, our brain uses feedback information to control various muscles and joints. Feedback control systems are necessary in automation or in robotics. See CYBERNETICS, and ROBOTS AND ROBOTICS.

Fermions Subatomic PARTICLES generally have a property visualized and labeled as *spin*, similar to astronomical bodies rotating on their axes. Those particles which have spins that can be measured in half-numbers are classified and dealt with using a set of rules formulated by the physicists Fermi and Dirac; they are called *fermions*. The proton, the electron, and the neutron are all fermions. Particles whose spin can be expressed as whole numbers are dealt with by a different set of rules (formulated by Indian physicist Satyendranath Bose and Albert EINSTEIN) and are called BOSONS. See ATOMS and SUBATOMIC STRUCTURE.

Fetal Tissue, Medical Use of Tissue taken from an aborted human FETUS and implanted in a patient has proven beneficial in relieving the symptoms of PARKINSON'S DISEASE in some cases. Transplantation of human fetal tissue may also be helpful in treating ALZHEIMER'S DISEASE, leukemia, and quadraplegia. However, research in this field has been the subject of controversy for a number of years. The current administration's view of the moral status of the human EMBRYO has prevented federally funded research on fetuses.

The controversy centers on the abortion issue. Those who oppose abortion object to federal funding of fetal research on the grounds that this tends to "institutionalize" government complicity with the "abortion industry." There is little objection to the use of spontaneously aborted (i.e., miscarried) fetuses in medical research. The outcry has been against electively aborted fetuses. See ORGAN TRANSPLANTS.

Fetus The EMBRYO of an animal that bears its young alive (rather than by laying eggs). In humans, the embryo is called a fetus after all major body structures have formed. This stage is reached about nine weeks after fertilization.

The single cell that results from the fertilization of the female ovum by male sperm is called a **ZYGOTE**. After dividing many times into a cytoplast, it implants in the uterus where it continues to divide, producing more **CELLS** and passing through the stages of embryo and fetus. From about the third month after conception until birth, the unborn is properly called a fetus.

Fiber Optics Hair-thin cables that can carry vast flows of information, including high-quality video and audio signals. The information-carrying capability of fiber optics is considerably greater than the copper wires of telephone or cable television. The tiny cable itself is a channel for pulses of **LASER LIGHT** that carry information as a stream of numbers, giving transmission of a higher quality than ordinary signals. Transmission capacity is so great that two-way, or interactive video, is possible.

Japan and Europe are investing heavily in fiber-optic communication systems, with Nippon Telephone and Telegraph planning to spend more than $200 billion over the next decade. U.S. phone companies will also replace copper wiring with fiber-optic wiring in every home in the country at some unspecified future time. See **DIGITAL**.

Fission, Nuclear A nuclear reaction in which **NUCLEONS** previously united in an atomic nuclei, are split apart or disjoined releasing **ENERGY** in the process. Fission is the process used in atomic bombs and is the basic system used today in nuclear power plants. **URANIUM** is often the fuel used because it splinters readily, releasing two or more neutrons, which in turn strike and splinter other uranium nuclei in a chain reaction. The result of the chain reaction, if it is controlled and gradual, is used to make electricity in a nuclear power plant. More heat comes from splitting atoms than from burning coal or oil. Splitting the atoms in just one pound (0.45 kilogram) of uranium gives off as much **HEAT** as burning three million pounds (1.4 million kilograms) of coal. If the same chain reaction is sudden and uncontrolled, an atomic bomb is the result. For a different type of nuclear interaction, see **FUSION**.

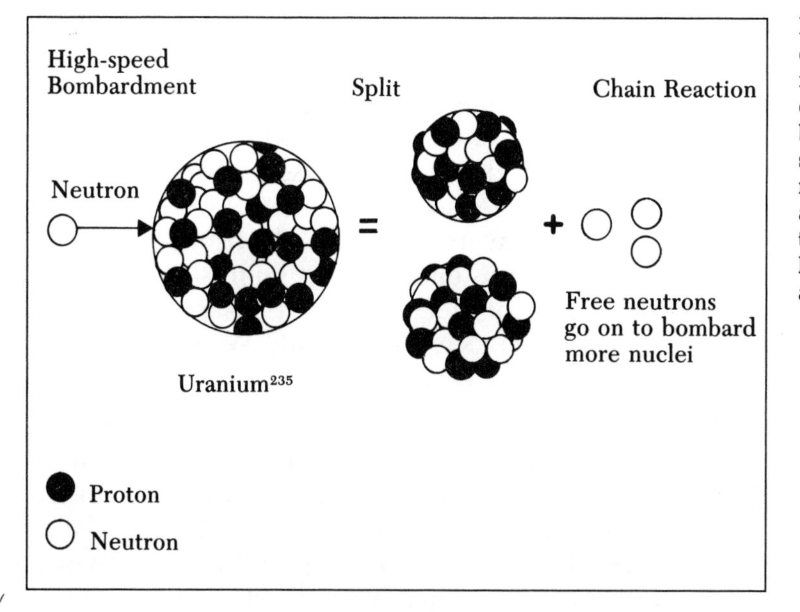

FISSION is the process in which the nuclei of heavy atoms are bombarded by neutrons and split into two nearly equal parts and several additional neutrons, releasing large amounts of energy.

Flavor In physics, the term is used to designate types of QUARKS, the basic building blocks of nature. There are thought to be six flavors—up, down, strange, charmed, top, and bottom. The top quark has yet to be discovered but is believed to exist. Because its discovery would complete the set and thus validate the theory, the top quark is the object of intensive scientific investigation. The top quark is thought to be heaviest of the quarks (its MASS is at least 90 times that of a PROTON), and it will take the energy of the most powerful of the new ACCELERATORS to produce it. See **PARTICLE PHYSICS**, and **STANDARD MODEL**.

Fluorescence The emission of LIGHT from an object because of bombardment by other kinds of electromagnetic RADIATION, such as X RAYS or ULTRAVIOLET radiation. In a fluorescent lamp, a discharge of ELECTRONS from the filament excites the mercury vapor in the tube, producing ultraviolet radiation. The ultraviolet makes the phosphor coating of the tube glow, producing visible light. The color properties of a fluorescent lamp are determined by the chemical elements used in the phosphor coating.

Because it is much more ENERGY efficient, a 40-watt fluorescent lamp produces as much light, and far less HEAT, than a 150-watt incandescent bulb. *Watt* is not a measure of light output but rather a measure of electrical power input. The new compact fluorescent bulbs can save a significant amount of energy used for lighting. Compact fluorescents can cost as much as $17 a bulb, but they may last ten times longer than an incandescent bulb. Also, they use less electricity. For example, nine 60-watt incandescent bulbs cost an average $40 in electricity for 9,000 hours of light, whereas one compact fluorescent will cost an average of $10 for electricity for the same amount of light. See **INCANDESCENT**, **LUMEN**, and **WATT**.

Fluoride A trace mineral that occurs naturally in soil and water and that has been found to prevent tooth decay. About 60 percent of the water supply in the United States is fluoridated, including the water in most large cities. Generally, the mineral is added by local governments. But some areas have naturally fluoridated water, and it was in these communities that scientists first discovered fluoride's properties as a tooth-decay preventive.

The National Institute of Dental Research estimates that schoolchildren in fluoridated areas have about 25 percent less tooth decay than children elsewhere. There is, however, a vocal minority of the American public who remained opposed to fluoridation. Some 1990 reports in the national media raised the possibility that fluoride may cause cancer. These sensationalistic reports were not supported by the available data.

Fluorocarbons See **CHLOROFLUOROCARBONS** (CFCs).

Fly-by-Wire Aircraft control systems that react to movement of pilot-operated controls by sending an electric signal to a computer, which in turn translates the signals into commands to the aircraft's control surfaces—rudder, aileron, flaps, or other airfoil surfaces. In more conventional aircraft control systems, the pilot-operated controls manipulate cables that are attached to motors or hydraulic pumps, which in turn move the airfoil surfaces.

This approach to aircraft control has been used for years in high-performance combat aircraft and more recently is being applied to commercial airplanes (e.g., the French A320). The main advantage of fly-by-wire systems is that the electrical network permits more extensive use of COMPUTERS. Planes under computer control can respond more quickly to turbulence or other changes in flying conditions. See FEEDBACK CONTROL.

FOOD CHAINS are systems of organisms interrelated in their feeding habits, the smallest being fed upon by larger organisms which in turn are fed upon by still larger ones. Humankind depends upon two main food chains, one being sea based and one being land based.

Food Chains Humankind depends on two main food chains, or *webs*, to obtain the ENERGY and materials necessary for life. One web starts with the microscopic ocean plants and seaweed and includes animals that feed on them and animals that feed on those animals. The second food web begins with the land plants and includes the animals that eat them as well as the animals that eat those animals. The complex interdependencies among species serve to stabilize these food webs.

Disruptions or disturbances of living populations or their environments may result in irreversible changes in the food web. Overfishing in a specific part of the ocean or of a particular type of fish could lead to the complete extinction of a SPECIES. The reduction in global OZONE may eventually lead to a significant reduction in the ocean PHYTOPLANKTON and thus, on up through the food chain, to a major disruption to the world's fish harvest. See ENVIRONMENT.

Force There are four types of forces in the universe controlling the ways with which objects interact. The two main forces we are commonly aware of are *gravitational* and *electromagnetic*. Force may be thought of as the agency responsible for change in a system. Gravitational forces in space, for instance, hold PLANETS in orbit and gather cosmic dust together to form STARS. NEWTON's laws of motion define force as an object's mass multiplied by its acceleration. Electromagnetic forces acting within and between ATOMS are much stronger than gravitational forces acting between them. Electric forces between oppositely charged PROTONS and ELECTRONS hold atoms and MOLECULES together. These same electric forces hold solid and liquid materials together. The two other forces in the universe are called the *strong nuclear force* and the *weak nuclear force* and they act only within the atomic nuclei and have no effect on the universe as a whole. See GRAVITY, MASS, and WEIGHT.

Fossil Fuels All material that humankind uses as fuel that was formed from the remains of plants and animals that lived millions of years ago. Coal, petroleum, and natural gas are fossil fuels. All fossil fuels produce CARBON DIOXIDE when burned and thus add to air pollution and to the GREENHOUSE EFFECT.

Fossils Petrified remnants of once living organisms found buried deep in the lower layers (*strata*) of rock. Fossils are formed when minerals in groundwater replace materials in skeletal bones, creating a replica in stone of the original plant or animal. The term *fossil* comes from the Latin word meaning "to dig." Because the lower layers of rock have to be much older than upper strata, remains of plants or animals found in these lower layers have to be from a much earlier geological period than remains of SPECIES found in the upper layers. From this evidence, paleontologists (scientists who study fossil remains) have been able to reconstruct the evolution of life on EARTH from the oldest fossils (blue-green algae), estimated to be 3.5 million years old, up to the species of life on Earth today. See GEOLOGIC TIME SCALE.

Fractals Either a revolutionary approach to mathematics or a fuzzy-headed "pop science," depending on which group of mathematicians you listen to. Fractals are closely associated with CHAOS theory, another controversial notion that has gained wide publicity, thanks in part to a best-selling book, *Chaos*, by James Gleick. Fractals are purported to be able to express in simplified, algebraic form what once seemed inexpressible—the uneven shape of clouds, the topography of sea bottoms, the ruggedness of mountains.

In conventional mathematics, the world is made up of smooth, idealized objects such as curves, cubes, and spheres. But the real world is more uneven or *ragged* than these hypothetical objects. Fractals research is an effort to mathematically express the *raggedness* of the real world. Enthusiasts claim that fractals, along with CHAOS theory, are literally revolutionizing our understanding of the natural world. Critics acknowledge that fractals make pretty pictures on the computer screen and that the concept has boosted public awareness of mathematics but see little practical value.

Freezing Point The temperature at which a substance passes from a liquid to a solid state by loss of heat. Water freezes at 32° FAHRENHEIT or 0° CELSIUS, or 273° KELVIN. See TEMPERATURE.

Freon Trade name for CHLOROFLUORCARBONS (CFCs) produced by E. I. Du Pont de Nemours, the world's largest manufacturer of this environmentally harmful chemical. Freon is widely used as a refrigerant, aerosol spray, cleaning agent, and in the manufacture of foam plastics. See OZONE DEPLETION.

Frequency In physics, the number of repetitions per unit of time of a complete waveform. Radiation of all types is described in terms of its wavelength and frequency, that is, the distance between successive wave crests and the number of crests that arrive per second. When the wavelength is short, the frequency is high, and vice versa. The common unit of frequency measurement is the HERTZ (Hz), corresponding to one wave crest (one cycle) per second. The various forms of RADIATION—from the short wavelength, high-frequency GAMMA RAYS to the long wavelength, low-frequency radio waves—form the ELECTROMAGNETIC SPECTRUM.

Frequency Modulation (FM) In electronics, the encoding of a carrier wave by variation of its frequency in accordance with an input signal. FM radio or TV broadcast signals are modulated, or changed, in frequency in order to carry the required communication, rather than in amplitude as in AM broadcasting. See also DIGITAL broadcasting.

Friction The resistance of an object to the medium, air or water for example, through which it is moving. Also, the rubbing of one object or surface against another as in any mechanical device. Resistance is the major cause for the loss of ENERGY in systems involving the transformation of one type of energy into another. For example, most of the energy stored in a liter of gasoline used during an automobile trip goes by way of friction and exhaust into producing a warmer car, road, and air. From 70 to 80 percent of the energy in gasoline flows out of the automobile in the form of rejected HEAT. See THERMODYNAMICS, FIRST AND SECOND LAWS OF; and ENTROPY.

Fuel Cell An electromechanical device in which the ENERGY of a reaction between a fuel such as liquid HYDROGEN and an oxidant such as liquid OXYGEN is converted directly and continuously into electrical energy.

A fuel cell is a type of storage battery developed for the space program. One extra advantage in the use of fuel cells is that the process also produces water. Fuel cells can be a clean, nonpolluting way of obtaining two essential supplies on a manned spacecraft—electricity and water. Fuel cell technology is not yet efficient enough to be a competitive source of energy on Earth.

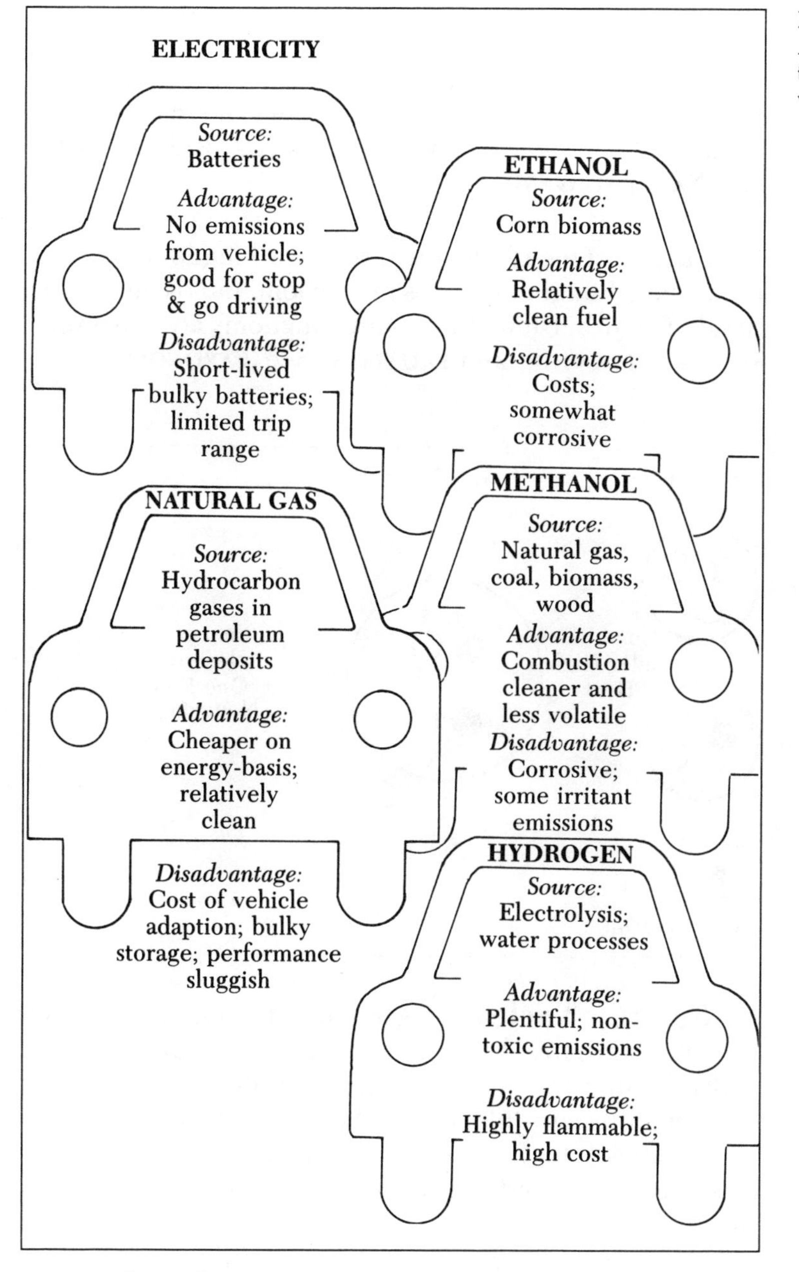

FUELS, ALTERNATE. All have advantages and disadvantages.

Fuels, Alternate Because of air pollution problems common to most U.S. cities, considerable attention is being given to finding an alternative to gasoline as a fuel for our automobiles. So far, a methanol/gasoline mixture called *flexible fuel* is the only serious contender. Equipping cars to operate on this mixture is expected to up the price of a new car by at least $300. The additional costs are because of the need for an expensive fuel sensor and a bigger fuel tank. None of the suggested alternate fuels have as much energy content as gasoline and, therefore, it will take more of it to propel our cars a given distance. See FUEL, ALTERNATES, METHANE, HYDROGEN, and NATURAL GAS.

Fungi One of the five taxonomic kingdoms of living organisms. Fungi comprises mushroom, molds, and other fungi. The other four taxonomic kingdoms are ANIMALIA, MONERA, PLANTAE, and PROTISTA. See TAXONOMY.

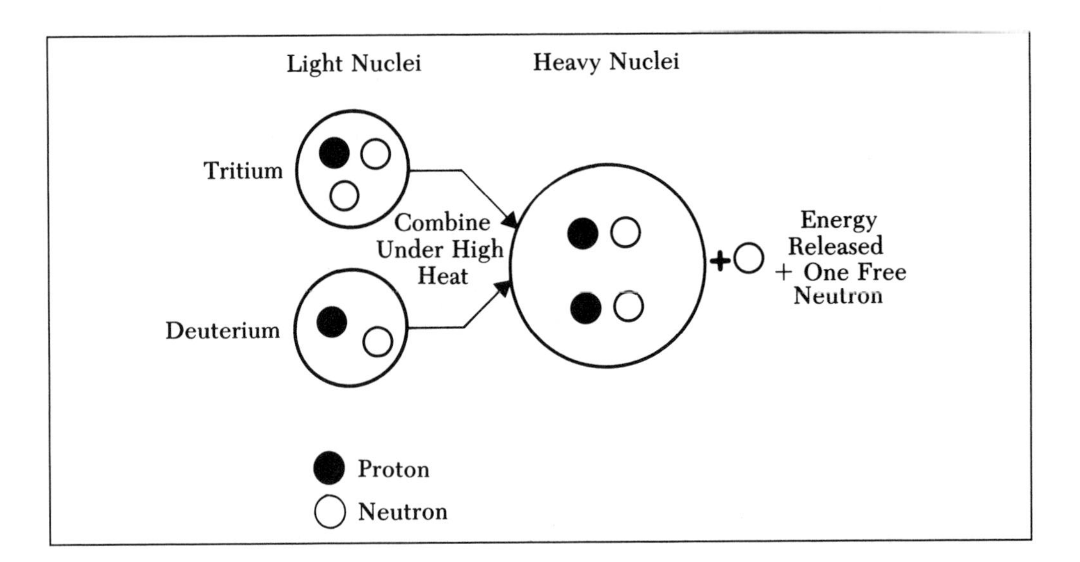

FUSION is the process in which two nuclei of light atoms combine, or fuse together, at high temperature to form a heavier nucleus and an extra neutron, releasing large amounts of energy.

Fusion, Nuclear Atomic interaction in which NUCLEONS are fused together, creating new atomic nuclei and releasing ENERGY. In fusion, the light nuclei of hydrogen atoms are joined together at extraordinarily high temperatures to form a single, heavy helium nucleus, ejecting high-speed neutrons in the process. The atoms resulting from the joining weigh slightly less than the ones that fueled the process, and it is this difference in mass that has been converted to energy (see EQUIVALENCE OF MASS AND ENERGY).

Fusion is the process that makes the SUN and the other STARS burn and powers the HYDROGEN or thermonuclear bomb. Scientists are working to adapt the fusion process to energy production in a power plant. The problem is achieving the high HEAT necessary to initiate the fusion process. Fusion has much to offer as a source of electrical power. In addition to having an almost inexhaustible fuel supply, fusion produces relatively minimal waste. There are, however, numerous technical problems to solve and controlled-fusion power plants cannot be expected before the 21st century. See **COLD FUSION**, and **FISSION, NUCLEAR.**

Gaia Hypothesis The theory that suggests that EARTH is an organism—the sum of all organisms—that can modify and maintain its own optimum ENVIRONMENT. This superorganism is Gaia, mother goddess of Earth. The Gaia theme that the Earth is alive was developed by James E. Lovelock, a British scientist, in his book *The Ages Of Gaia*. Drawing on developments in geology, geochemistry, evolutionary biology, and climatology, Lovelock postulates a new scientific synthesis in harmony with the Greek conception of thc Earth as a living whole. Conventional science has depicted the Earth as an inert rock, upon which plants and animals happen to live. Lovelock's Gaia theory postulates a much different view of the world as one great circuit of life from its core to its outer atmosphere.

The question of how organisms influence, modify, and sometimes control their local environments and their cumulative impact on the global CLIMATE is considered legitimate by geologists, geochemists, biologists, and atmospheric scientists. The Gaia theory, however, is judged controversial at best, and many consider it merely another example of unproven New Age pop science. See also CHAOS.

Galaxies Collections of stars bound together gravitationally. Astronomers classify galaxies as spiral, elliptical, and irregular. Our SUN belongs to a spiral galaxy called the MILKY WAY galaxy. The Milky Way is only one of some hundred billion galaxies that can be seen using modern telescopes, each galaxy itself containing some hundred billion STARS. Our galaxy is about one hundred thousand LIGHT-YEARS across and is slowly rotating. The stars in its spiral arms orbit around its center about once every several hundred million years. Our Sun is an average-sized star located near the inner edge of one of the spiral arms.

In 1929 the American astronomer EDWIN P. HUBBLE discovered that nearly all galaxies in the universe were

Galaxies are giant collections of billions of stars bound together gravitationally. Most galaxies, such as the Cepheus galaxy shown here, are spiral in form.
Source: NASA photo.

moving away from us. Moreover, the farther a galaxy is, the faster it is moving away. This discovery that the universe is expanding is considered one of the most important intellectual achievements of the 20th century. Both the Big Bang concept of the origin of the universe as well as current estimates of the age of the universe resulted from the discovery of an expanding universe. By extrapolating backwards from the current rate of expansion, astronomers were able to infer that the universe started in a burst of energy called the Big Bang and that this event occurred 15 billion years ago, give or take two or three billion years. The present rate of expansion is estimated to be between 5 and 10 percent every thousand million years. See **Big Bang, Doppler Effect, Expanding Universe.**

Galilei, Galileo Seventeenth-century Italian astronomer and logician who is considered the father of modern science. By postulating that bodies fall with a uniformly accelerated motion, Galileo described the first laws of classical dynamics. The story that he did this by climbing to the top of the Leaning Tower of Pisa and dropping two objects of different weights from there is probably not true. The two objects would hit the ground at the same time only if air resistance was not a factor (i.e., in a vacuum). Galileo did not invent the telescope, but he did pioneer its use as an astronomical instrument. He used the telescope to confirm the Copernican concept of a helio-(sun)-centered universe and in so doing aroused the wrath of the Catholic Church, which held to the belief of an Earth-centered universe. Galileo was tried by the Inquisition, but he was never tortured or imprisoned over this issue. He was, however, placed under house arrest and forbidden to disseminate his heretical ideas.

It was Galileo's insistence that humankind could understand how the world works, and that this could be done by observing the real world, that led to the development of modern physics. His methods of observations, experiments, and mathematical analysis in both physics and astronomy had a profound influence on the thinking population of the world and it was upon his basic work that Isaac **NEWTON** developed Newtonian physics.

Galileo See **GALILEI, GALILEO.**

Galileo Spacecraft Launched in 1989, the *Galileo* spacecraft will reach **JUPITER** in 1995. On the way, *Galileo* swung past **VENUS** and **EARTH** in 1990 in order to pick up gravitational assist to increase its speed. This NASA spacecraft will first take a close look at the **ASTEROID** Gaspra in 1991, pass Earth again in 1992 for another slingshot-type boost, fly by the asteroid Ida in 1993, and finally arrive at Jupiter two years later. The spacecraft is expected to provide the first direct sampling of Jupiter's atmosphere and the first extended observation of the fifth planet and its **MOONS**.

Galileo's pictures of Jupiter's four largest moons—Io, Europa, Ganymede, and Callisto—are expected to have 20 to 100 times better resolution than those taken by **VOYAGER** 2 in 1979. After its instruments wear out, the *Galileo* spacecraft will remain in permanent orbit of Jupiter. Because *Galileo's* propulsion system utilizes a plu-

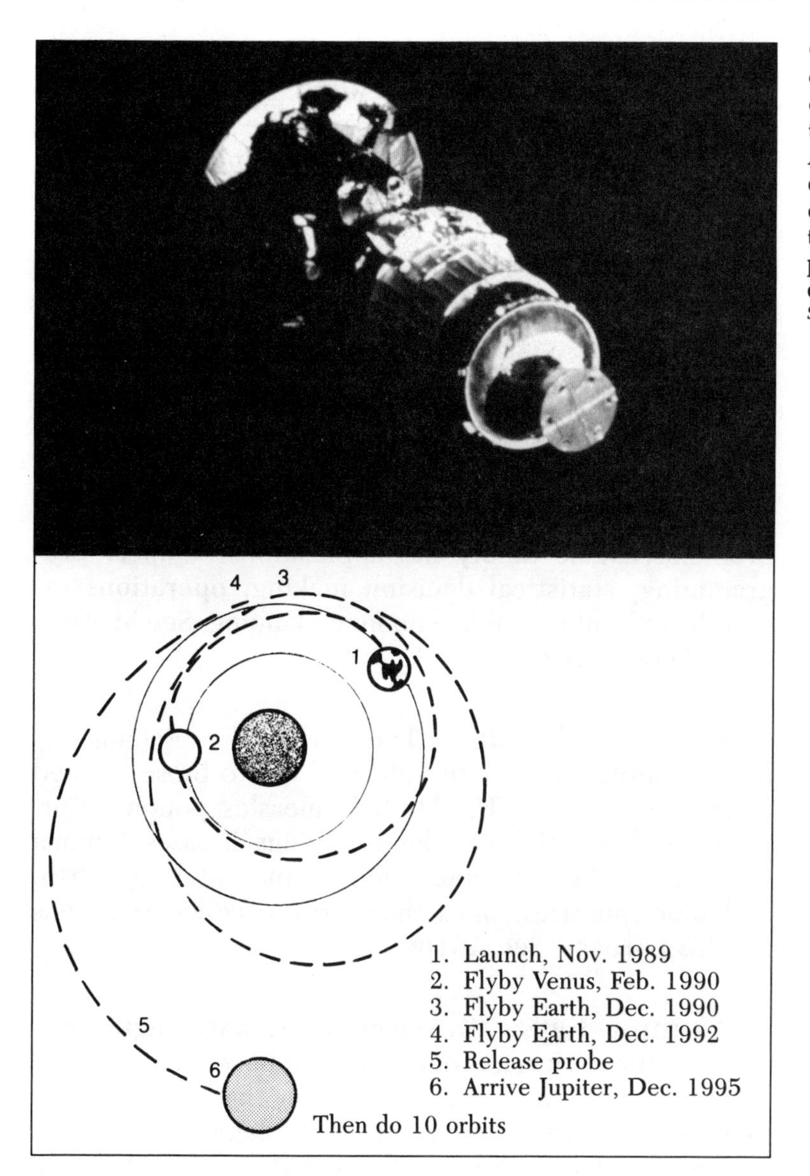

GALILEO SPACECRAFT shortly after deployment from the space shuttle *Atlantis* in October of 1989. The spacecraft will arrive in the vicinity of Jupiter in December of 1995.
Source: Nasa photo.

tonium generator, there was controversy and protest demonstrations at its launch. A subsequent Jet Propulsion Laboratories (JPL) study showed that SOLAR ENERGY might have been a better and safer choice for the mission.

Gallium Arsenide A semiconductive material used in TRANSISTORS, solar cells, and semiconducting LASERS. Like silicon, gallium arsenide has the ability to both conduct and resist the flow of ELECTRONS, depending on how it is chemically treated. The speed of electrons moving through gallium arsenide is faster than that of silicon and future microchips may be made of this material. In addition, at some future time **PHOTONS** (discrete units of light), rather than electrons, may serve as the primary vehicle for transmitting and processing information; gallium arsenide can transmit light whereas silicon cannot. Skeptics make the tongue-in-cheek comment that, "gallium arsenide is the material of the future—always has been and always will be." See **CHIP**, and **SEMICONDUCTOR**.

Game Theory Sometimes called the *theory of games,* the term relates to the mathematical analysis of abstract models of strategic competition, such as in war or business, to determine the best strategy for solving problems. Game theory has applications in linear programming, statistical decision making, operations research, and military and economic planning. See **MODELING, MATHEMATICAL**.

Gamma Globulin The substance containing the blood's antibodies and therefore of use to those in need of greater immunity. Used to treat measles, poliomyelitis, infectious hepatitis, and other infectious diseases. Gamma globulin can be extracted from the placental afterbirth, and some countries, for example France, collect placentas for this purpose. See **ANTIBODY**.

Gamma Rays Extremely short wavelength electromagnetic **RADIATION** emitted by radioactive decay and having energies in a range overlapping that of the highest energy X rays. Radioactivity is the tendency of an atomic nucleus to decay through the emission of **PARTICLES**. Three types of emission occur: **ALPHA PARTICLES** (each one being two protons and two neutrons), **BETA PARTICLES** (each one being an electron or a positron), and gamma radiation.

The study of cosmic background radiation in the gamma-ray region is in its infancy. NASA's 17-ton Gamma Ray Observatory is designed to operate for at least two years and can be reboosted to keep it in orbit for as long as a decade. Among the scientific instruments on board is a gamma-ray telescope developed in West Germany. See also **ELECTROMAGNETIC SPECTRUM**.

Gauss A unit of measure for magnetic-field strength defined by German mathematician Johann K. F. Gauss (1777–1855) and usually expressed as *milligauss*—one-thousandth of a gauss (see **NUMBERS: BIG AND SMALL**). Levels of between two and a half and four and a half milligauss have been associated in several epidemiological studies with the development of cancer in human beings. See **RADIATION**.

General Theory of Relativity **EINSTEIN**'s theories dealing with accelerated motion and gravity. General relativity is a fundamental concept of the nature of space, time, and gravitation and has profoundly influenced how humankind views the universe. Published in 1915, after his special theory of relativity (1905), Einstein's theory views **GRAVITY** as a property of space rather than a force between bodies. As a result of the presence of **MATTER**, space becomes curved, and bodies follow the line of least resistance among curves. Gravity, then, is viewed as the consequence of the curvature of space induced by the presence of a massive object. This concept introduced the principle that gravitational and inertial forces were equivalent. See **SPECIAL THEORY OF RELATIVITY**.

Genes Units of heredity, portions of **DNA** that direct the production of a specific protein or the expression of a specific trait. Genes, which are the blueprints for cell construction, exist in tightly organized packages called **CHROMOSOMES**. Genes are different segments of DNA within a chromosome which have a specific function.

The relationship of the various elements of a cell may be thought of as follows: The nucleus of a cell is a library containing life's instructions. The **CHROMOSOMES** would be the bookshelves inside the library, the DNA would be individual books on each shelf, genes would be the chapters in each book, and the nucleotide bases making up the strands of DNA would be the words on the pages of the individual books. See **CELLS**, **GENETIC CODE**, and **GENOME PROJECT**.

Genes, Transfer of Experiments conducted in mid-1990 have provided strong indications that the transfer of altered or modified genes into human beings offers a safe and effective means of fighting cancer and other diseases. Genetically altered CELLS injected into cancer patients were found in their blood up to two months later, showing that such cells could last long enough to fight tumors or correct disease-causing genetic defects. The transferred genes had no medical treatment value but were only used to help measure the effectiveness of a therapy that employs cells called tumor-infiltrating lymphocytes, or TILs. The 1990 work also confirmed earlier findings that gene transfer causes no side effects and poses no public health threat.

Genetic Code The way in which the sequences of DNA base pairs serve as instructions for making PROTEINS and defining biochemical functions. Each of the 100,000 or so GENES in the human body consists of the same four bases—(A) adenine, (C) cytosine, (G) guanine, and (T) thymine—in a specific sequence or pattern that determines a gene's function. The messages of heredity are written in an alphabet of just these four bases, which are repeated many millions of times in different sequences along strands of DNA. The bases always occur in pairs that join the two strands of the DNA molecule. The strands take the shape of a twisted rope ladder called a *double helix.* Scientists have now broken the code and learned to read the messages written in the strands of DNA.

For any individual of any species, the sequence of base combinations on the ladder spell out a complex coded message that can transmit to an offspring all the instructions needed for every genetic trait. The coded sequence of bases contains a unique pattern setting forth the chemical specifications for some living creature. These sequences of DNA pairs are, by definition, genes, which govern the chemistry of life and determine all inborn characteristics, from blood type to eye color. Genes are packed into the 23 pairs of human CHROMOSOMES. See also GENOME PROJECT.

Genetic Engineering The use of scientific, biological techniques to manipulate or rearrange genetic material to alter hereditary traits. Also known as *gene splicing* or RECOMBINANT DNA technology. Once researchers understood the GENETIC CODE, the inevitable next step was to change the code or engineer the genetics. In re-

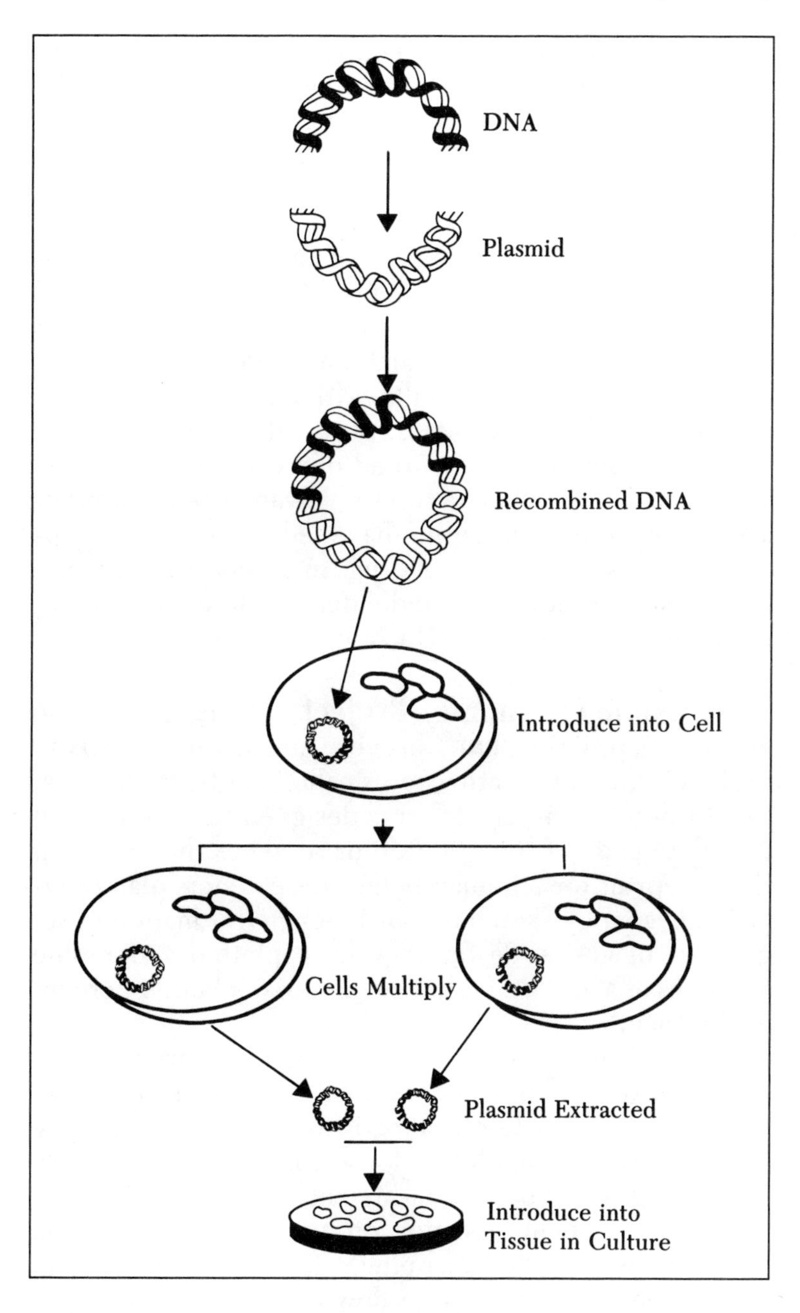

GENETIC ENGINEERING involves recombining a DNA strand with a plasmid DNA strand, inserting the recombined DNA into a bacterial cell, which then replicates. The replicated plasmids can then be removed from the bacterial cells and introduced into tissue cells in culture.

combinant DNA engineering, specialized ENZYMEs are used to snip a gene from one organism and splice it into another. If the transfer is successful, the recipient organism will carry out the instructions of the new gene.

The most serious of the issues created by genetic engineering is constituted by humankind's imminent power to interfere in the processes of heredity, to alter the genetic structure of our own SPECIES. The first step in this process will be gene therapy, which involves inserting GENES with correct information into CELLS that contain defective genes or inserting new genes that code for disease-fighting substances. As of this writing the ability of biotechnology to correct nature's tragic errors is still in its infancy. In July of 1990 the first official attempts at human gene therapy in the United States were endorsed by a panel of medical experts. Approved were two plans to use gene therapy against an immune deficiency in children and skin cancer in adults. Although these efforts involve relatively rare diseases, the work opened the door to development of a new breed of treatment for a wide variety of illnesses in the future. Advancements in other areas of genetic engineering have led to improved crops and livestock, production of drugs in microorganisms, and techniques for identifying individuals or screening for genetic diseases. See also **DNA**, and **GENOME PROJECT**.

Genome (jee-nome) Project A major scientific effort to map all the GENES on every human CHROMOSOME. Biology's first foray into what is called big-ticket science, the $3-billion, 15-year effort is designed to decipher the complete code of the 100,000 or so genes that make up the blueprint for a human being. Researchers plan to attain this goal by sketching out biochemical maps and sequences of genes. So far they have plotted geographic locations of about 4,000 human genes, or about 4 percent of the total.

The amount of DNA in each bundle of chromosomes in a human cell, if unfurled, would form an invisible thread reaching about six feet in length. Interspersed randomly along that six feet of information are about 100,000 genes. The biologists propose to locate each one of those genes, pinpoint its specific location on a specific chromosome (this process is called mapping), and ultimately decode the biochemical information down to the so-called "letters" of inheritance, the four basic constituents or base nucleotides of all genes. In the famous *double helix*, these letters are linked in pairs, whose sequence makes up the genetic code. These sequences involve three billion pairs in all. As of this writing, biologists have deciphered about 35 million. See **CELLS**, and **GENETIC CODE**.

GENOME PROJECT as symbolically depicted by a 23-foot chain DNA molecule made up of 500 telephone books. This exhibit illustrates some idea of the vast amount of information contained in a DNA molecule.
Source: Technology Center of Silicon Valley.

Geological Dating Scientists know the rate at which a radioactive ELEMENT, such as URANIUM or thorium, disintegrates to form the end product lead. By measuring the relative amounts of radioactive element and lead, scientists can calculate how long the process has been going on in a sample of rock. In this way, dates have been established for the eras, periods, and epochs of geologic time. See GEOLOGIC TIME SCALE, HALF-LIFE, and RADIOACTIVITY.

Geologic Time Scale Until the 19th century, nearly everyone believed that the EARTH was only a few

History of Earth—4.6 Billion Years

Precambrian Era—4 Billion Years	570 million	Present

Paleozoic, Mesozoic, Cenozoic

Era		
Paleozoic	570–230 million years ago	Development of Single Cells Into Multi-Cellular Life Age of Fishes
Period	*Million years ago*	
Cambrian	570–500	
Ordovician	500–435	
Silurian	435–400	
Devonian	400–345	
Carboniferous	345–280	
Permian	280–230	
Mesozoic	230–65 million years ago	Dinosaurs Age of Reptiles
Triassic	230–195	Pangaea begins to break up
Jurassic	195–140	
Cretaceous	140–65	The Great Dying
Cenozoic	65 million years ago to Present	Age of Mammals
Tertiary	65–2	
Quarternary	2 to Present	

thousand years old and that the appearance of Earth was fixed; that is, the continents, mountains, valleys, oceans, and rivers were the same as they had always been since the beginning of time. It was not until English geologist Charles LYELL, published his monumental *Principles of Geology* early in the 19th century that the concept of a much older and constantly changing and evolving Earth came to be accepted. Since that time geologists have learned to read the long history of Earth by studying the thousands of layers of sedimentary rock (called *stratas*), and the long history of changing life-forms whose remains are found in successive layers of rock. This geological STRATIFICATION along with radioactive dating have provided humankind with information on archaeological, paleontological, and geological events. For reasons of classification, scientists have divided the 4.6 billion-year history of Earth into four major time spans (each called an *era*) and further subdivided the first three of these into 11 subclassifications called *periods* as shown in the table above. The most ancient era is the PRECAMBRIAN, which shows minute evidence of life. Then comes the PALEOZOIC (ancient life), MESOZOIC (middle life), and CENOZOIC (modern life). The smaller units of time called periods are named for regions where fossils of that particular time segment were found. In addition, the Cenozoic era is further subdivided into smaller units of time called *epochs*. See also GEOLOGICAL DATING, and HALF-LIFE.

Geothermal Energy The term literally means "earth heat," and refers to those areas where underground heat is enough to provide a usable source of energy. Those geothermal areas that are apparent at the surface are those that leak steam or hot water, such as the hot springs systems of Yellowstone National Park or Wairakei, New Zealand. Geothermal energy has been harnessed in a number of places, most notably Iceland where natural steam is used to heat buildings and agricultural

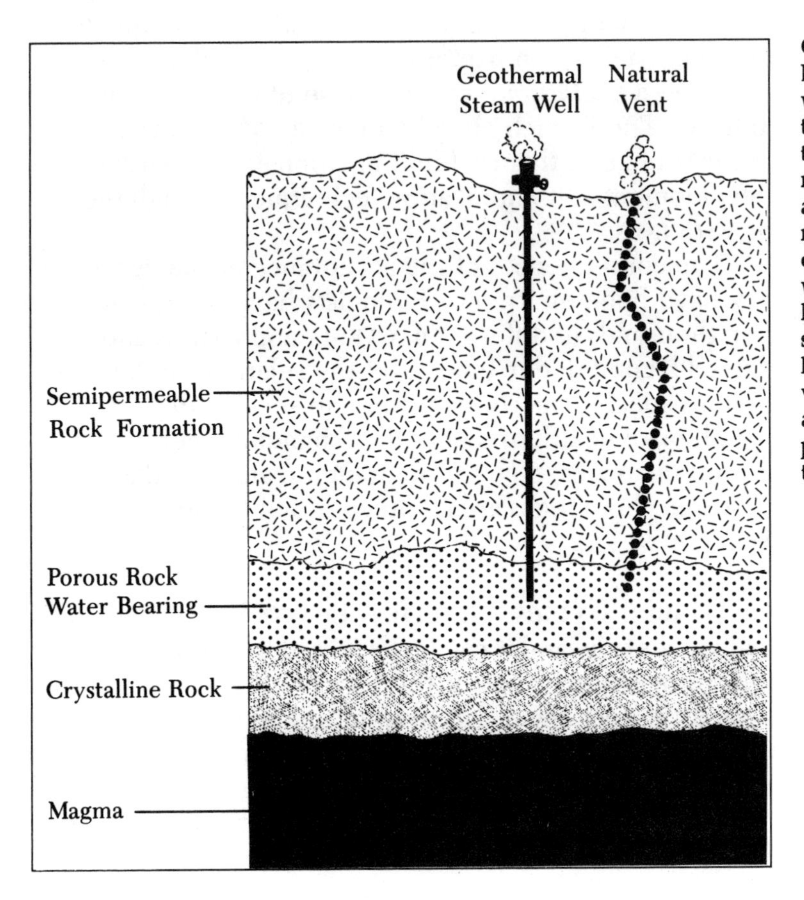

GEOTHERMAL ENERGY results when heat from the magma in the Earth's core radiates upward and reaches the porous rock layer, converting groundwater to steam or hot water. The resultant steam or hot water can be vented naturally, as in a geyser, or put to use as geothermal energy.

fields. The only geothermal power plant in the United States is located in northern California.

The future potential for greater use of geothermal energy is great. However, the expense of drilling steam wells and the problems associated with the acidity of natural steam which tends to corrode pipelines and equipment, have so far limited the use of this natural source of energy. The world's geothermal areas are potential sites of enormous amounts of energy and future decades will see expanded use of this resource. See **ENERGY, USES OF.**

Germ Theory The evolution of germ theory in the 19th century drastically changed the explanation of what causes diseases. Until that time, humankind puzzled over the cause of the many infectious illnesses affecting life. The development of the microscope in the 17th century led to the discovery of a new world of microscopically small life-forms. The connection between microorganisms and disease, however, was not immediately apparent and the germ theory was not accepted until 1876, when Robert Koch, a German bacteriologist, proved that a specific microbe, a *bacterium,* was the cause of a specific disease, anthrax. The French chemist Louis **PASTEUR** was another pioneer in germ theory, building upon earlier work in the development of immunization techniques through the use of **VACCINES.**

The consequences of the acceptance of the germ theory have been enormous. Biologists have by now identified thousands of different **BACTERIA** and viruses and have gained a good understanding of the relationship between various specific microorganisms and specific diseases. Today, human health practices, food handling methods, sanitation measures, quarantine, immunization, and antiseptic surgical practices are all outcomes of the germ theory. See **VIRUS.**

Glacial Cycles **EARTH'S CLIMATE** has changed radically over time and it can be expected to continue changing, primarily because of the effects of geological shifts, such as the advance or retreat of glaciers over centuries of time. These long-range changes in climate show up as **ICE AGES**, separated by intervals of long-lasting warmer times called *interglacial periods.* During the last 600 million years there have been 17 known periods of glaciation on the Earth. During these ice ages, enormous sheets of ice advance toward the equator, sometimes reaching as far south as the latitude of New York. The most recent cold period lasted from A.D. 1500 to 1900 and is called the Little Ice Age. Geophysicists who have studied glacial cycles have concluded that temperature changes of only a few degrees can have a profound effect on the Earth's climate.

Today, the glaciers of the world occupy about 10 percent of the Earth's surface. Most of this ice is contained in two great ice caps: Antarctica and Greenland, with the rest located mostly in mountainous regions throughout the world. It has been calculated that if all this ice were to melt, the sea level would rise about 330 feet. See ICE AGES.

Global Warming Whether as a consequence of the GREENHOUSE EFFECT or for some other reason, global warming may turn out to be the central environmental issue of the 1990s. The average TEMPERATURES on Earth in 1990 were the highest since record keeping began, continuing a warming trend first detected in the 1980s. Five of the six warmest years in more than a century have occurred since 1980. In descending order, the six warmest years on record are 1990, 1988, 1987, 1944, 1989, and 1981. Global average temperatures are now 0.6 degrees CELSIUS warmer than they were 100 years ago. There is no conclusive proof linking this heating to the greenhouse effect, but circumstantial evidence has convinced many scientists that this is the cause. In 1990, half of the U.S. Nobel laureates and members of the National Academy of Sciences wrote the president urging him to take action to prevent global warming.

Of concern is the faster rate of warming predicted by computer models of the climate for late in the next century. Increases of 2.5 to 5.5 degrees Celsius (4.5 to 9.9 degrees FAHRENHEIT) are projected. These increases may not seem very large until compared to historical changes in Earth's temperature. In North America, during the thousands of years that elapsed as humankind emerged from the last ICE AGE and the glaciers retreated northward, temperatures increased about this same (4 to 8 degrees Fahrenheit) amount. The predicted warming then is comparable to that since the last ice age. Moreover, this

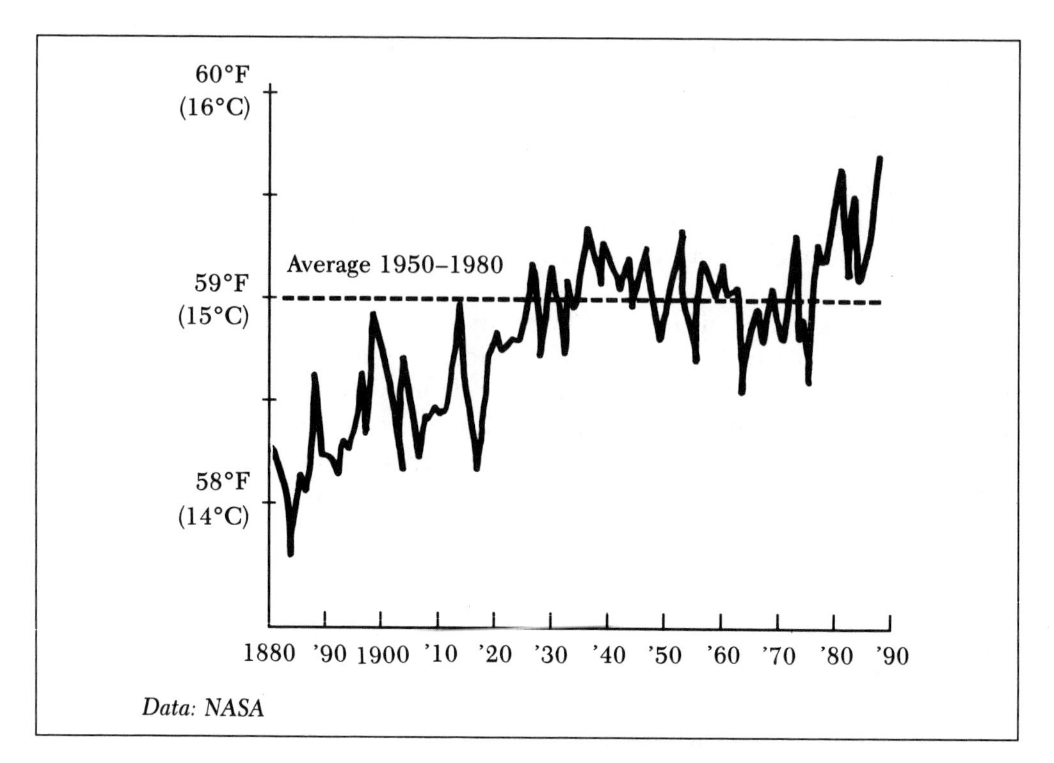

GLOBAL WARMING TREND as indicated by the global temperature fluctuations over the period 1880 to 1990 and as compared to the average global temperature over the period 1950 to 1980.

projected warming will take place over 50 to 100 years, rather than over many thousands of years. This pace of warming is likely to stress life on Earth far more than did the glacial pace of warming from the last ice age. See also CLIMATE, and GLACIAL CYCLES.

Gluons According to the quantum theory, one of the four forces of nature, specifically the strong nuclear FORCE, is carried or conveyed by fundamental units of ENERGY called gluons. Other carriers are PHOTONS, carrier of electromagnetism; BOSONS, the carrier of the weak force; and GRAVITONS, carrier of gravitational force. Some physicists lump together all the force-carrying fundamental units of energy under the term *gluons.* See QUANTUM PHYSICS.

Godel's Theorem In mathematics, a *theorem* is a proposition that is provable on the basis of explicit conditions. In 1931, the American mathematician Kurt Godel demonstrated that within any given branch of mathematics, there would always be some propositions that could not be proven either true or false using the rules and axioms of that particular branch of mathematics. Godel showed that one could prove every conceivable statement about numbers within a system but only by going outside the system to come up with new rules and axioms. In other words, according to Godel's theorem (sometimes called *Godel's proof*), all logical systems of any complexity are incomplete in that each of them contains more true statements than can be proved according to its own defining set of rules.

According to this theorem, COMPUTERS can never be as intelligent as humans because the extent of the computer's knowledge is limited by a fixed set of axioms, or truths, built into the machine by the designers, whereas humans can discover unexpected concepts and truths.

Grand Unified Theories (GUTs) Theories that attempt to unify the electromagnetic and weak nuclear forces with the strong nuclear FORCE into what is called a grand unified theory (or GUT). The difference between these forces in nature is attributed to the breaking of symmetrical relationships among forces immediately after the BIG BANG when the early universe expanded and cooled. As Stephen Hawking points out in his book *A Brief History Of Time*, the theories are neither grand nor fully unified, as they do not include GRAVITY. They are, however, considered a first step toward a complete, fully unified theory of physics. See PARTICLE PHYSICS.

Gravitons The fundamental unit (or *quanta*) of nature thought to convey gravitational FORCE, just as PHOTONS convey electromagnetism, GLUONS convey the strong nuclear force, and BOSONS convey weak nuclear forces. Gravitons are predicted by the quantum theory of gravity but as of this writing, no graviton has ever been detected. See GRAVITY, and QUANTUM PHYSICS.

Gravity One of the four fundamental FORCEs of nature, the others being *electromagnetism* and the two nuclear forces called the *weak* and the *strong* force. Although the effect of gravity is easily perceived, how gravity works is not completely understood. There are two different interpretations that physicists use: (1) Newtonian physics tells us how the force of gravity is proportional to the product of the two masses and inversely proportional to the square of the distance between them. In other words, the larger the mass the greater the force of gravity, and the force of gravity falls off quite rapidly as the objects are moved farther apart. A doubling of the separation distance, for instance, reduces the force to one-quarter of its initial figure. (2) Einstein's general theory of relativity tells us that gravity is the consequence of the curvature of space induced by the presence of a massive object.

Newtonian physics accounts for the observed orbits of PLANETS and MOONS, the motion of COMETS, the motion of falling objects at the Earth's surface, weight, ocean tides, and the Earth's slight equatorial bulge. Newton's concepts prevailed as a scientific view of the world for 200 years. Albert EINSTEIN's theories of relativity did not overthrow the world of Newton but, rather, modified some of its most fundamental concepts. See **GENERAL THEORY OF RELATIVITY**, **GRAVITONS**, and **NEWTON**.

Great Attractor Many astronomers believe that the MILKY WAY and its neighboring GALAXIES seem to be falling toward some enormous but unseen mass, which they have named the "great attractor." This is an example of speculative cosmology. See also **DARK MATTER**, and **GREAT WALL**.

Great Wall Recently discovered huge chain of GALAXIES some 400 million light-years from Earth and stretching for 500 billion light-years across the sky. The *great wall* is a puzzle to scientists because according to the BIG BANG theory, which most scientists accept, MATTER should be evenly distributed in the universe rather than collected in clumps as seems to be the case. See **INFLATIONARY EXPANSION**.

Greenhouse Effect The process of trapping HEAT within the EARTH's ATMOSPHERE because of the billions of tons of CARBON DIOXIDE and other gases that humankind emits into the atmosphere each year, and the resultant gradually increasing global TEMPERATURES, is called the greenhouse effect. The gases are an inevitable by-product of the burning of fossil fuels—coal, oil, and natural gas. In the upper atmosphere, these gases act like a blanket, trapping heat and radiating part of it back down to the surface of Earth. The level of carbon dioxide in the

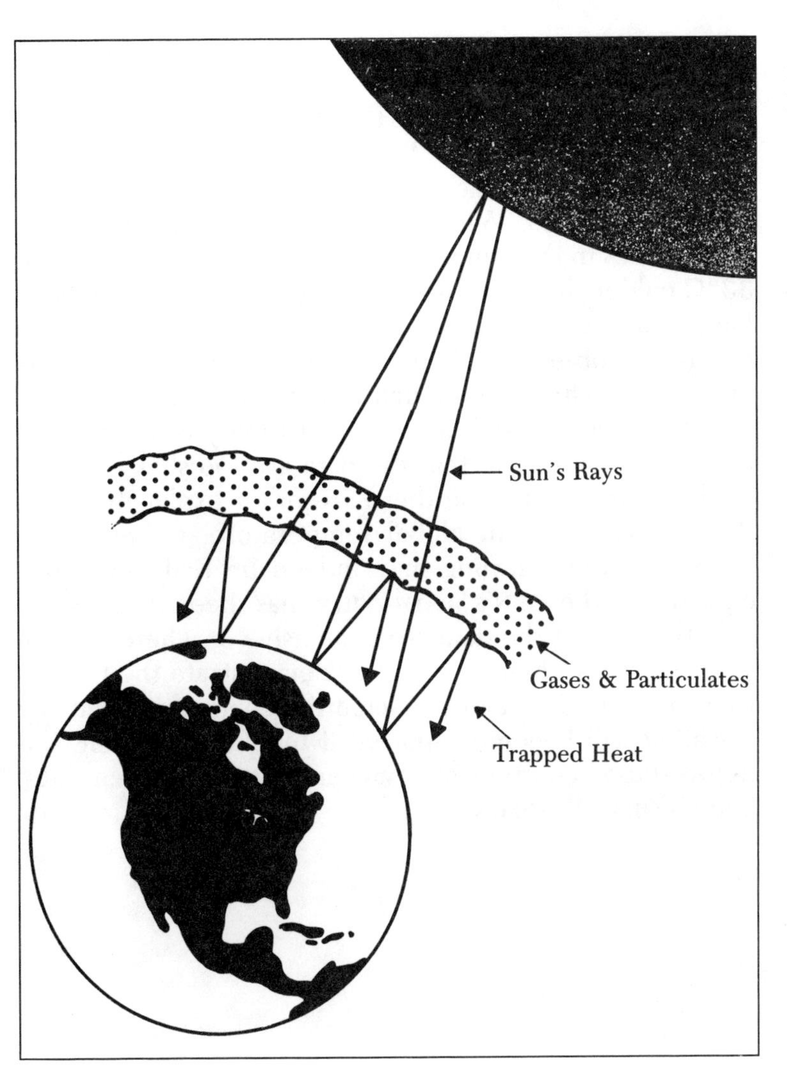

GREENHOUSE EFFECT. Gases and particulates form a "blanket" around the Earth that traps heat within the atmosphere, much as the glass in a garden greenhouse does, and leads to gradually increasing global temperatures.

atmosphere has increased by 80 percent since the Industrial Revolution. About two-thirds of this increase is due to fuel burning. The remainder of the carbon dioxide buildup is from the massive DEFORESTATION that has occurred during the past 200 years. Predictions of future CLIMATE warming are based on estimates of how fast greenhouse gases will continue to build up in the atmosphere. These predictions are inexact, but if the current rate of population growth and per capita fossil-fuel use is projected, in about the year 2031, humanity will have doubled the atmospheric concentration of carbon dioxide as compared to its value at the start of the Industrial Revolution. Climatologists forecast that at this rate, a warming trend of about 1 degree FAHRENHEIT per decade should be expected.

It should be understood that the greenhouse effect is not a new phenomenon. A variety of gases, most importantly water vapor, have warmed the Earth's surface for billions of years. Without these INFRARED-absorbing MOLECULES in the atmosphere, the planet would be about 33°C colder than it is today. The greenhouse effect is, in fact, what made life possible on Earth. The phenomenon that is the subject of scientific concern today is a runaway greenhouse effect—too much of a good thing.

The runaway greenhouse effect theory is not without its critics, who argue that the data are too shaky to conclude anything because there are other explanations for the apparent rise in global temperature. Other critics think that the warming may indeed be real but that a change in the SUN's RADIATION has been the cause. Whether or not the data are conclusive or whether more years of research are needed to demonstrate the theory, there is little doubt in the scientific community that the warming will become apparent if fossil-fuel burning and deforestation continue at anywhere near the current rate. See **GLOBAL WARMING**.

Half-Life The time it takes for half of a given quantity of radioactive material to decay (release ENERGY). An ATOM decays when it disintegrates, when it changes from instability to stability. All radioactive substances tend to form stable substances in time and in that process they emit RADIATION. Half-lives can vary from less than one-millionth of a second to millions of years. The half-life of any particular substance is constant and unaffected by physical conditions such as pressure or temperature. RADIOACTIVITY, therefore, can be used to estimate the passage of time by measuring the fraction of nuclei that have already decayed. For example, the fraction of carbon remaining in a fossil sample can be used to estimate how long ago the fossil was formed.

The radioactive substances that pose the greatest harm to humankind have neither very short nor very long half-lives. The short lose all activity so quickly that they are not dangerous. The very long half-life substances take so long to lose their radioactivity that their radiation is virtually harmless. A few examples: Iodine 131 has a half-life of 8 days; cobalt 60, 5.2 years; cesium 137, 30 years; and carbon 14, 5,570 years. Potassium has a half-life of 1.3 billion years. See CARBON DATING, and ISOTOPE.

Hanford Nuclear Weapons Plant, Radiation from Evidence now shows that for more than 40 years the U.S. Government knowingly leaked radioactive emissions into the atmosphere from their nuclear weapons plants without informing the citizens of the areas at risk of the dangers involved. In the case of the Hanford nuclear plant in Washington State, results of a 1990 Department of Energy study show the release of dangerous amounts of radioactive emissions—400,000 CURIES of radioactive iodine into the atmosphere in the 1944 to 1947 period alone. The bodily absorption of 50 microcuries is sufficient to raise the risk of thyroid cancer. The 1979 accident at THREE MILE ISLAND resulted in the release of only 14 curies of radioactive iodine. Of the 270,000 residents of Washington, Oregon, and Idaho at risk, 1 in 20 absorbed an estimated dose of 33 rads to their thyroid glands some time during the last 40 years. Between 500 and 1,500 rads will kill most people. (To relate rads and curies, see RADIATION.) Radiation dangers from the other nuclear weapons plants at Idaho Falls; Rocky Flats, Colorado; Fernald, Ohio; Savannah River, Georgia; and Oak Ridge, Tennessee, have yet to be quantified. See CHERNOBYL.

Hardware, Computer The physical machinery of a COMPUTER system, which might include the keyboard, the central processing unit (CPU), the monitor, tape or disk drives, and peripheral equipment such as printers and control devices (e.g., a joystick or a mouse). In contrast, see SOFTWARE, COMPUTER, which includes the programs

HARDWARE, COMPUTER. The physical machinery of a computer system that includes the keyboard, the central processing system, the monitor, disk drives, and peripheral control devices such as a mouse. By mid-1990 there were some 57 million personal computers in use in the United States.
Source: IBM photo.

or sets of instructions that tell a computer exactly how to proceed to solve a specific problem or process a specific task.

HDL (High-Density Lipoprotein) The so-called "good" cholesterol. HDL, in simple terms, picks up cholesterol in the arteries and brings it back to the liver for reprocessing or excretion. Average levels for adult Americans, measured in milligrams per deciliter (one-tenth of a liter) of blood, are 45 to 65. Levels below 35 mg/dl may signal coronary risk. Unfortunately, it is harder to increase HDL than to lower total or LDL cholesterol. Preliminary results of tests conducted by Stanford University do suggest that reducing body fat—either by dieting or by exercise—may be the key to increasing HDL. See **CHOLESTEROL**, and **LDL**.

HDTV See **HIGH-DEFINITION TELEVISION**.

Health, Human To stay in good operating condition, the human body requires a variety of foods and exercise. The amount of food energy (**CALORIES**) required varies with body size, gender, activity level, and metabolic rate. In addition to energy needs, the human body requires substances to replace the materials of which it is made. Regular exercise is needed to maintain a healthy heart/lung system, to maintain muscle tone, and to keep bones from becoming brittle.

Human health depends on at least two other variables: avoidance of excessive exposure to substances that are harmful to the body—tobacco, alcohol, and drugs being the most obvious—and societal efforts to maintain the safety of air, water, soil, and food—the environmental influences on human health.

Heat A form of energy that results from the random motion of **MOLECULES**. Heat can be defined as the total **ENERGY** contained in the molecular motions of a given quantity of **MATTER**. When matter is heated, its molecules vibrate faster, causing it to expand. When matter is cooled, the molecules move slower, causing it to contract. When a solid is heated to a point where the vibrations are strong enough to break the bond between neighboring molecules, the solid melts and becomes a liquid. If the liquid is heated further, the movement of the molecules become energetic enough to free them from the liquid altogether, and the liquid is said to boil. Heat is capable of being transmitted from regions of hotter temperatures to regions of cooler temperatures: through solids and fluids by means of conduction, through fluids by means of CONVECTION, and through space by means of RADIATION. The scientific study of the movement of heat, and the manner in which heat can be converted into mechanical work (and vice versa), is called THERMODYNAMICS. In the metric system, the CALORIE is a measure of heat. One calorie is the quantity of heat (energy) that enters (or leaves) 1 gram of water when the TEMPERATURE of this water is raised from 14.5°C to 15.5°C. See BRITISH THERMAL UNIT (BTU), CONSERVATION LAWS, and THERMODYNAMICS, FIRST AND SECOND LAWS OF.

Hectare A metric unit of measurement, usually of land, equal to 10,000 square meters, which is equivalent to 2.471 acres. An area of a flat surface, be it a floor, wall, or farmland, is always measured in square units, such as square inches, square feet, or square miles and so on—except for acres or hectares, which are already squared units of measure. To convert acres to hectares, multiply by 0.4. To convert hectares to acres, multiply by 2.5.

Heisenberg's Uncertainty Principle In 1927 the German atomic physicist Werner Heisenberg demonstrated the uncertainty inherent in QUANTUM PHYSICS. He showed that the closer one gets to observing the velocity of a PARTICLE, the further one inevitably gets from measuring its position, and vice versa. In other words, when studying subatomic particles the process of observation itself changes the object being studied. Imagine, for instance, a very high-powered microscope that could make an ELECTRON visible. If we were trying to measure the position of the electron using this microscope, we would have to illuminate it to see it. However, an electron is so small that a single PHOTON of light falling on it would change its position. The uncertainty principle holds in any measurement we attempt, whether it be of TEMPERATURE, pressure, or electric current. The act of measuring changes what we are trying to measure. In most ordinary measurements, the change in the subject of the measurement is too small to make any difference. Not so in the subatomic world, where the consequence of this principle is that there are certain kinds of information that cannot be learned in specific detail.

Uncertainty can be thought of as a barrier. Since humankind can only perceive reality through the medium of the senses, there is a limit to what humanity can know. This is not a happy concept to many scientists, but in the field of subatomic physics at least, the uncertainty principle is widely accepted.

Heliocentric Solar System The fact that the EARTH orbits the SUN rather than the ancient belief that an immobile Earth was the center of the universe. First postulated in the 16th century by Nicolaus COPERNICUS, later confirmed when GALILEO used one of the first telescopes to verify Copernicus's concept. Later Johannes KEPLER refined Galileo's observations and showed that the Earth's orbit of the Sun was not a perfect circle but an ellipse. These observations were made in the 16th and 17th centuries, but a 1989 National Science Foundation survey showed that only a little over half of the Americans questioned knew that the Earth revolves around the Sun and takes one year to do so. See ASTRONOMY, and GALILEI, GALILEO.

Heliopause The outermost limit of the SUN's magnetic field and thus perhaps the most fitting candidate among the many definitions of the true edge of our solar system. *Voyager 2*, which encountered and photographed NEPTUNE in mid-1990 and is continuing its space sojourn, is expected to cross this boundary in 10 to 20 years. See VOYAGER SPACECRAFT.

Helium A colorless, odorless, tasteless, inert gaseous ELEMENT often used to inflate and provide lift for balloons. In liquid form, it is often used in cryogenic (low TEMPERATURE) research. The BIG BANG theory of the origin of the universe postulates a universe filled with radiation and consisting of about one-quarter helium. Observations have confirmed both aspects of this model. The theory, and models based upon the theory, state that roughly one-quarter of the universe will be transformed from HYDROGEN to helium in the first 20 minutes of cosmic evolution. Observational confirmation of the fact that the universe is, as far as can be told, one-quarter helium thus supports the Big Bang model.

Hemoglobin (Hee-muh-gloh-bin) The OXYGEN-carrying particles of blood. Hemoglobin both picks up oxygen from the lungs and delivers it to the tissues. When carrying oxygen, hemoglobin is bright red. Without oxygen, hemoglobin is dark blue, almost black. So-called *blue babies* suffer from oxygen lack, and whenever pulmonary function is unsatisfactory, a person will appear blue. Blood is half PLASMA and half corpuscles (units or particles), of which some are white CELLS and far more are red cells. It is the red cells that contain hemoglobin.

Hertz (Hz) A unit of frequency equal to one cycle per second. RADIATION is described in terms of its wavelength (distance between successive wave crests) and its frequency (the number of crests that arrive per second). Hertz is the international unit for measuring frequency, having replaced cps (cycles per second). See ELECTROMAGNETIC SPECTRUM.

Heuristics An educational approach in which an investigator is encouraged to make educated guesses based on knowledge gained in his or her research. Sometimes called the art of good guessing, the use of heuristics is the core concept of ARTIFICIAL INTELLIGENCE. Heuristics allow COMPUTER's to deal with situations that cannot be reduced completely to a mathematical formula and may involve many exceptions. Heuristics pertains to a trial-and-error approach to problem solving and is used when an algorithmic or step-by-step approach is not possible. See ALGORITHM.

High-Definition Television (HDTV) The goal of high-definition television is to increase the number of lines in a television picture, thus sharpening it. Today's television pictures in America and Japan are composed of 525 scanning lines, whereas those in Europe have 625. More scanning lines provide more details in the image on the screen. HDTV proposals call for 1,000 or more lines. A television image is produced on your home screen when a beam of electrons scans the screen, illuminating tiny *pixels* (smallest elements of a video display system). Because high-definition television has more scanning lines, it will need more data than can be transmitted over today's television channels. In other words, a new broadcast system is required. The Federal Communications Commission has ruled that the American HDTV system cannot be allowed to make the nation's existing TV sets obsolete. The trick then is to develop better-definition TV without changing the current broadcast system. One approach to this challenge is signal compression—squeezing the 30 megahertz signal band width needed for HDTV into the 6 megahertz band that normal TV broadcasting uses. This process has been likened to squeezing a large elephant into a small bathtub.

There are two approaches to the development of HDTV. The Japanese and the Europeans have based their efforts on the development of ANALOG systems that use wavelike transmission systems. Several of the American systems now under development are based on DIGITAL transmission systems, which offer advantages in fidelity picture and sound. Some futurologists foresee a national network of FIBER OPTIC cables that can carry enormous amounts of digital information in the form of LASER-light pulses, transmitting HDTV signals (and other information signals) to American homes.

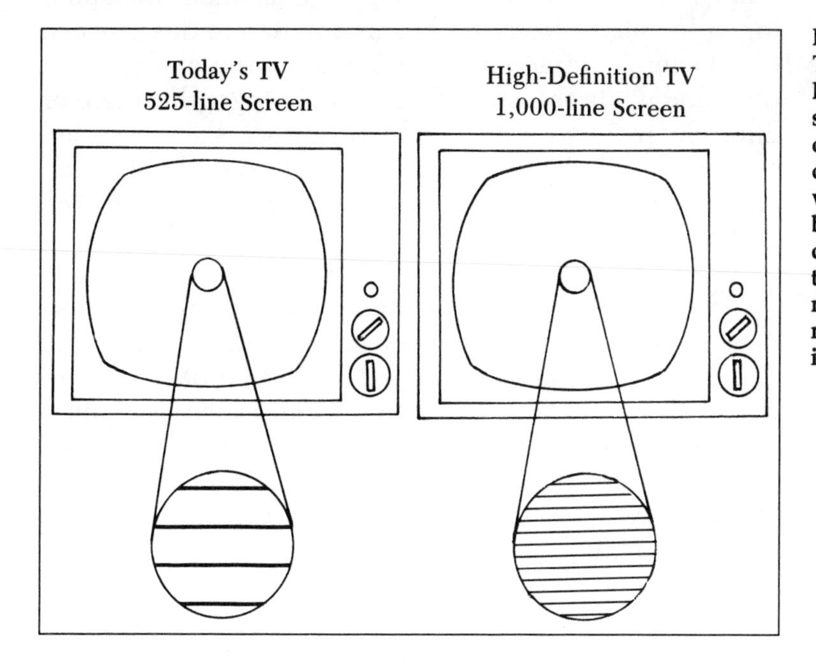

HIGH-DEFINITION TELEVISION, or HDTV, promises sharper pictures as compared to today's television by virtue of the number of lines used in creating the picture. More scanning lines provide more details in the image.

HIV (Human Immunodeficiency Virus) AIDS is the disease, and HIV is the AIDS-carrying virus. HIV destroys the very immune-system CELLS that would coordinate the body's defense. HIV is also notoriously variable, so to be effective against it, a VACCINE would have to be able to act against many strains. HIV merges with the genetic material of the cells it infects. Although the presence of antibodies to a disease normally means that the immune system is defending against the virus, HIV antibodies apparently do not fight off the disease. It is possible for someone to test positive for HIV and still not have the AIDS disease. See ANTIBODY; and VIRUS, BIOLOGICAL.

Holistic The term refers to the theory that whole entities have an existence other than the mere sum of their parts. Usually used in connection with medicine, in which case it means the care of the entire patient in all aspects—physical, mental, emotional—it is also used in reference to therapeutics outside the mainstream of scientific medicine, such as chiropractics. Holistic medicine often involves nutritional measures.

Holography The technique of producing three-dimensional images that appear to have depth just like a real object. LASERs are used to record an image on a photographic plate. Two laser beams are used in this process. One beam is diffracted by, or lights up, the object being recorded while the other falls on a photographic plate or film. The patterns produced by this process are recorded and when the plate or film is illuminated by light of the same frequency, a three-dimensional image is seen. See LIGHT.

Homeostasis Self-regulation, either by living organisms or by electrical-chemical-mechanical systems, to maintain a constant norm, or balanced ENVIRONMENT. Homeostasis can be defined as the maintenance of equilibrium within an internal environment. Controlling factors such as TEMPERATURE, OSMOSIS, and maintenance of chemical balance contribute to homeostasis. See FEEDBACK CONTROL.

Hominid The species of primates that evolved into modern human beings (HOMO SAPIENS). At a point in ev-

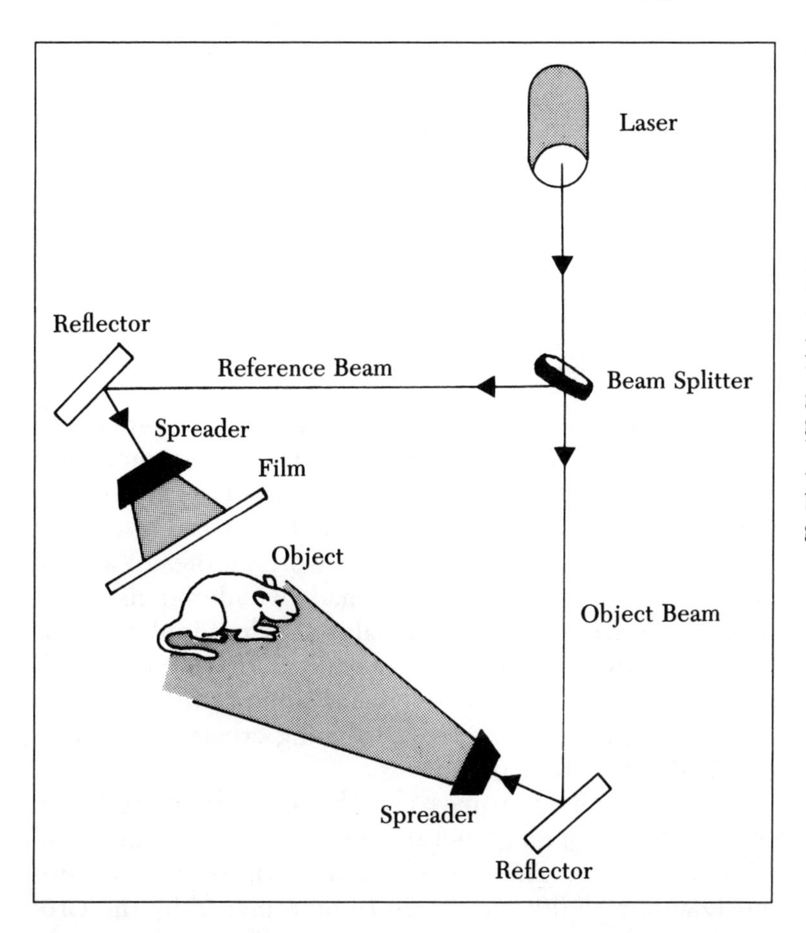

HOLOGRAPHY. The technique of producing three-dimensional images that appear to have depth like a real object. The light from a laser is split into two beams. One beam lights up the object being photographed and the other beam goes to a photographic plate. When developed, the plate or film becomes a hologram.

olution, a branch of primates split into two populations. One branch evolved into modern apes, and the other branch evolved into modern human beings. All of the individuals on the human side of the division are classified as hominids. Thus, we are not descended from the apes but rather have an ancestor in common with apes. The earliest undisputed hominid is called *Australopithecus Afarensis.* Fossils of this species, including the famous Lucy found in Ethiopia, date to about 3.5 million years ago. See AUSTRALOPITHECUS, and EVOLUTION.

Homo erectus A taxonomic designation, or classification, of early humanlike creatures who first appeared on the scene about 1.5 million years ago. *Homo erectus* were the first HOMINIDS to expand out of Africa and into Eurasia. They were also the first hominids to use fire and to develop many new stone tools. During the more than one million years that this species spread into widely separated parts of the world, their brain size was increasing dramatically. Early Homo erectus brain size was 800 cubic centimeters (cc), which is double that of chimpanzees and more than half that of modern humans. Later fossil data indicate a growth of brain size to 1,300 cc, which is the modern average human brain size. There is controversy among anthropologists concerning whether Homo erectus evolved into modern humans or were replaced by a species with greater survival skills. At any rate, the species became extinct around 400,000 years ago. See **EVOLUTION**, and **HOMO SAPIENS**.

Homo sapiens Modern humans who first appeared about 600,000 years ago. This term is the taxonomic designation for modern man, the only extant species of the genus *Homo*. Better hunters than *Homo erectus,* these early humans cooked their meat, wore clothes of animal hide, constructed shelters, and made wooden tools. From what is known about their social life-style it is almost certain that they communicated verbally. Most important was a brain capacity of 1,200 to 1,600 cubic centimeters and the classification of man as a thinking creature as distinguished from other organisms.

Humanlike creatures may have evolved first into **NEANDERTHAL** man (generally accepted as a subspecies of Homo sapiens). The Neanderthals disappeared mysteriously about 35,000 years ago to be replaced by the Cro-Magnons, famous for their cave paintings found in southern France. Cro-Magnons, an early form of modern humans, evolved about 40,000 years ago. The subspecies is called *Homo sapien sapien* (or wise man). These human beings were hunter-gatherers for about 25,000 years. They domesticated animals 18,000 years ago and invented agriculture about 12,000 years ago. The earliest permanent settlements appeared 9,000 years ago and the first real cities only 5,000 years ago. See **EVOLUTION**, **HOMINID**, and **HOMO ERECTUS**.

Horsepower (hp) A unit of power equal, in the United States, to 745.7 **WATTS** or 33,000 foot-pounds per minute. Other countries use a different conversion factor. Horsepower became a convenient unit to use in the early days of the evolution of the automobile because the horseless carriage was often compared to horse-drawn carriages. A typical car today requires about 20 hp to propel the vehicle at 50 miles per hour.

Hubble, Edwin P. The pioneer American astronomer who in 1929 revolutionized modern ASTRONOMY by establishing that the universe is expanding. Hubble first determined that the universe was organized into GALAXIES. He then observed that LIGHT from distant galaxies appeared to become redder with time. He determined that this REDSHIFT indicated that galaxies were receding from EARTH (and from each other). The amount of redshift increases as the speed of recession increases, which means that the farther away a galaxy is, the faster it is moving away from Earth. Hubble calculated that the speed of recession increased proportionally with the distance at about 15 kilometers per second per million LIGHT-YEARS. This is known as HUBBLE'S CONSTANT.

As a result of Hubble's observations using the 100-inch telescope atop Mount Wilson, and from more recent measurements, astronomers now estimate that the outer limits of the EXPANDING UNIVERSE may lie 15 billion light-years or more from Earth. If we convert this distance to miles (remembering that LIGHT travels at 186,000 miles per second), we find that the edge of the known universe is an almost inconceivable 90 billion trillion miles away. The discovery that the universe is expanding was one of the great intellectual achievements of the 20th century and led in turn to the BIG BANG concept of the beginning of the universe. See DOPPLER EFFECT.

Hubble's Constant Our universe has been expanding since the BIG BANG and this expansion is such that the farther an object, such as a galaxy, is from us, the faster it is moving away. Astronomers assume a direct relationship between an object's distance and its recessional velocity such that dividing the velocity by the distance always gives the same number. That number is called the Hubble constant. See HUBBLE, EDWIN P.

Hubble Space Telescope (HST) Placed in a 381-mile-high orbit by the space shuttle *Discovery* on April 25, 1990, the Hubble Space Telescope will be able to view 97 percent of the known universe, and it will be able to obtain pictures undistorted by the Earth's atmosphere. Compared with Earth-based observatories, the HST will be able to view celestial objects that are 50 times fainter, provide images that are 10 times sharper, and see objects that are 7 times further away. Unfortunately, the HST has severe problems with its mirrors and the images sent back by the telescope so far are no clearer than those obtained by ground telescopes. The discovery that the $1.5 billion telescope—the largest and most complex scientific instrument ever put into space—cannot perform a substantial portion of its scientific mission was a major setback to astronomical science. Some of the instruments on the spacecraft are unaffected by the main telescope problem. However, the wide-field and planetary camera that was to perform 40 percent of the scientific work will not be usable until, and if, repairs can be made. Scientists are working on means to sharpen the telescope's brighter camera images through ground-based COMPUTER en-

HUBBLE SPACE TELESCOPE carries a Wide-Field/Planetary Camera as one of its five astronomical instruments. In this drawing, a shuttle approaches the telescope so that the astronauts may perform periodic maintenance.
Source: NASA/JPL.

hancement, in hopes of minimizing the focusing problem that resulted from errors in manufacturing the primary mirror. The telescope's wide-field camera and other instruments will be fitted with corrective lenses early in 1993 according to present plans. NASA hopes that this will restore the Hubble to its original planned capabilities.

Critics of the cost of the Hubble Telescope, such as humorist Dave Barry of the *Miami Herald*, point out that "it would have been cheaper to take a regular telescope and put it on top of an 87-mile-high pile of $50 bills." See ASTRONOMY.

Hybrid The result of the crossbreeding of two different varieties of plants or animals—the product combining the traits of both varieties. Hybridization is often used in agriculture to produce superior types of plants (hybrid corn, for instance) or animals. Often a hybrid organism, plant or animal, grows more vigorously than its purebred counterpart. This phenomenon is known as *hybrid vigor.* Unlike genetically engineered new species, hybrid species cannot be counted upon to pass on desirable traits.

The term is also used to refer to something of mixed origin or composition. In automobile design, for instance, a vehicle that combines an internal-combustion engine with an electrical propulsion system is called a hybrid. Various hybrid automobile concepts have been proposed, primarily to overcome the deficiencies of the pure electric vehicles—limited range and performance. See GENETIC ENGINEERING.

Hydroelectric Electricity that has been generated by the conversion of the ENERGY inherent in falling water. Water stored behind a dam can be used to drive turbines whose shafts are connected to an electric generator. The electricity thus generated is conveyed over power transmission lines to wherever it is needed. About one-fifth of the electricity used in the United States today—about 60,000 megawatts—comes from hydroelectric sources. It is theoretically possible to increase this figure four- or fivefold but this would involve the construction of many more dams, each of which would upset the local ecology. The environmental impact of this choice might be a cause of concern to some people. See ENERGY, SOURCES OF; and ENERGY, USES OF.

Hydrogen The most abundant ELEMENT in the universe, hydrogen is a colorless, highly flammable gaseous element, the lightest of all gases. Hydrogen carries the atomic number 1 in the PERIODIC TABLE OF THE ELEMENTS and is the simplest atom, having one PROTON as a nucleus and one ELECTRON in orbit. A MOLECULE of water consists of two ATOMS of hydrogen and one atom of oxygen (H_2O). The commonest, most abundant substance on the face of the Earth, water is found in only minute traces anywhere else in our SOLAR SYSTEM. The hydrogen bomb (sometimes called the H-bomb) derives its destructive power from the fusion of the nuclei of various hydrogen isotopes in the formation of helium nuclei.

Hydrogen is currently the fuel of choice for space vehicles (it's being used in the Space Shuttle) and is known to have great potential for aircraft because it packs more chemical energy in a pound of fuel than any other known substance. Hydrogen's extremely light weight is not an advantage for ground vehicles, but it does have other advantages. Because it does not contain carbon, it does not produce hydrocarbon exhaust emissions when burned. Also, combustion engines using hydrogen can be run much leaner (less fuel and more oxygen) than gasoline ones; therefore, emissions can be reduced. But there are also problems with hydrogen. It is difficult to store and refuel safely, because to maintain it as a liquid, it must be kept under pressure and at very low temperatures. Furthermore, because of its low density, it requires a very large tank—three times larger than a present-day fuel tank for the same range. Last, and most important at current prices, it is far more expensive to produce than gasoline or most of the other suggested alternative fuels. In the very long term, however, if a premium is placed on air quality, hydrogen's drawbacks may be outweighed by its cleanliness. See FUELS, ALTERNATE.

Hydrosphere All of the water in, on, or above the Earth's surface: in the oceans, rivers, or lakes, under the ground or in the air. See also, ATMOSPHERE, EARTH'S; and LITHOSPHERE.

Ice Ages Long-range changes in EARTH'S CLIMATE show up as ice ages separated by long-lasting warm (or interglacial) periods. During the past 600 million years, there have been 17 known periods of glaciation on Earth. During these periods, enormous ice sheets advanced southward from the Arctic toward the equator, sometimes reaching as far south as the latitude of New York City. When the ice sheet was at its farthest extent, it covered over 17 million square miles of land in both polar regions or some 30 percent of the Earth's land surface. (See diagram on page 150.) This is three times the amount of land covered by ice sheets today. During the four most recent major ice ages, the Earth's TEMPERATURE was about 11°F colder than today's average. Three of these four ice ages lasted about 100,000 years. The fourth major ice age extended from about 40,000 to 10,000 years ago. The most recent cold period, called the LITTLE ICE AGE, lasted from A.D. 1500 to 1900.

Just what causes or ends an ice age is the subject of much scientific speculation. The prevailing theory is that changes in the Earth's position relative to the SUN seem to have launched ice ages by influencing the amount of solar radiation the Earth receives. There is consensus, however, in the conclusion that it takes only a small change in temperature to bring on or terminate an ice age. If a global temperature drop creates summers cool enough so that the previous winter's snow never melts, then a feedback system causes several years' snow to build up and glaciers to form. See ASTRONOMICAL CYCLE, GLACIAL CYCLES, and in contrast, GLOBAL WARMING, and GREENHOUSE EFFECT.

Ice Caps, Melting of Monitoring of the polar ice packs by means of SATELLITE photographs indicates that Antarctica has shed more than 11,000 of its 5 million square miles of ice since the 1970s. Whether this ice melting can be attributed to GLOBAL WARMING and the

ICE AGES. Maximum extent of glaciers in the last ice age, about 18,000 years ago. Some glaciologists believe that rising concentrations of the "greenhouse" gases may delay the onset of the next ice age and convert the current interglacial period into a "superinterglacial" with the highest global temperatures in a million years.

GREENHOUSE EFFECT is not known. If global warming is occurring, its effects will be first noted at the poles and for this reason close monitoring of the arctic regions is being undertaken. Scientists are trying to determine if the changes detected by satellite in 1990 are random, cyclic, or whether they are indeed an early warning of coming climate shifts.

The Antarctic accounts for 90 percent of the Earth's ice. Most of the rest covers Greenland. Enough of this ice melted between the last two ice ages, 125,000 years ago, to raise the global sea levels at least 20 feet above their present level. A 1990 study by the National Academy of Sciences forecasts a rise of roughly 1.5 to 3.5 feet in the next century. A sea level rise of only a few feet could be catastrophic for coastal areas worldwide.

Imaging Technologies In medicine, any of a number of machines or devices capable of representing or optically reproducing pictures of organs or systems within the human body. X-ray technology, for instance, provides physicians with photographs taken by radiography of the internal functions of the body. Recent advances in medical imaging technology, or machine vision, have greatly improved the ability of physicians to see inside the body without the trauma of exploratory surgery. As a result, more progress has been made in diagnostic medicine in the past 15 years than in the entire previous history of medicine. See **CAT SCAN**, **DIGITAL SUBTRACTION ANGIOGRAPHY (DSA)**, **MAGNETIC RESONANCE IMAGING (MRI)**, **POSITRON EMISSION TOMOGRAPHY (PET)**, and **SONOGRAPHY**.

Immune System The human body's internal defense system. White blood **CELLS** surround invaders—**BACTERIA** or viruses—and produce specific antibodies that will attack them. The antibodies either directly render the invader innocuous or immobilize it until white blood cells can literally eat it. Some of these antibodies remain in the human body after a viral or bacteria attack. They have the capability of quickly producing many more like antibodies, and for years afterward, or even a lifetime, the immune system will be ready for that type of organism and be able to limit or prevent that particular disease. On the other hand, some viral diseases, such as **AIDS** (Acquired Immune Deficiency Syndrome), destroy critical cells of the immune system, leaving the body helpless in dealing with multiple infectious agents and cancerous cells.

Immunization **VACCINES** are used to fool the body's immune system into thinking that an attack is occurring. The microbe in the vaccine cannot reproduce, but its presence causes the immune system to produce the appropriate antibodies. These will circulate in the blood for many years, ready to prevent a serious disease. See **ANTIBODY**, **INTERFERONS**, **PASTEUR**, and **VIRUS, BIOLOGICAL**.

In Situ (in-sy-too, or in-sit-too) In science, this term refers to experiments conducted in the actual location of the phenomena under investigation. Literally, the Latin term means "in its original place." In contrast, see **IN VITRO**, and **IN VIVO**.

In Vitro (in-veetro) In science, this Latin term refers to experiments carried out in an artificial environment outside of a living organism, such as in a laboratory test tube. Literally, the term means "in glass." See also **In Situ**, and **In Vivo**.

In Vivo (in-vee-vo) In science and medicine, the term refers to experiments carried out using living animals or humans. Literally, the Latin term means "in the living body." See also **In Situ**, and **In Vitro**.

Incandescent Refers to emitting visible **LIGHT** as a result of being heated. Incandescent lamps were the first type of electric light to be developed. Electrical **ENERGY** passes through a wire, heating it until it glows or *incandesces.* The wire is called the filament and today usually is made of tungsten. Incandescent lamps are low in initial cost but they burn out rapidly, are low in efficiency compared to other types of lamps (e.g., fluorescent), and have a high heat output. See **Fluorescence**.

Inertia **Newton**'s first law of motion states that every body (or **MASS**) persists in a state of rest or of uniform motion in a straight line unless compelled by external **FORCE** to change that state. This law is also called the *principle of inertia.* Another way of looking at the quality of inertia is to consider it as resistance to motion, action, or change. To move a body at rest, enough force must be used to overcome inertia; the larger the body at rest, the more force required. See **Acceleration**.

Infant Mortality One key indicator of a nation's health; overall infant mortality rates in the United States have declined over the years to 9.1 deaths per 1,000 live births in 1990—about half the rate of 1970 but twice that of Japan's. The United States consistently ranks 15th to 17th among the 30 or so nations reporting these data. Infant mortality, related to poverty, is higher in the United States than in many less affluent countries. The mortality rate for black infants in the United States remains twice as high as that for whites. Black infants die at a rate of 18.7 per 1,000 live births. Researchers have always thought that blacks had the nation's worst infant mortality rate because pregnant black women receive the poorest prenatal care. But a 1990 study by the National Center for Health Statistics suggests that other factors, yet to be determined, may play a larger role than expected. Mexican-American women, on average, receive prenatal care as poor as that for black women, but their infant mortality rate is half that of blacks. The study suggests that something beyond prenatal care and socioeconomic status may be responsible. See also **Life Expectancy**.

Infinity The quality or condition of being unbounded or limitless in space, **time**, or quantity. The term is often used to refer to something immeasurably large or small, or, as in mathematics, an indefinitely large number. In time, infinity refers to something that continues forever.

Inflationary Expansion According to one model of the beginning of **time**, the universe started off with a period of exponential or *inflationary* expansion in which it increased in size by a very large factor. This model, developed at MIT, speculates that the radius of the universe increased by a million million million million million (1 followed by 30 zeros) times in only a small fraction of the first second of time. As the universe expanded, it would cool, particle energies would diminish, and the rate of expansion would slow down. Astrophysicists think that the universe is continuing to expand today but at a decreasing, rather than inflationary, rate. See **Big Bang**, and **Expanding Universe**.

Information Processing, Storage, and Retrieval See **Computers**.

Infrared A region of the **electromagnetic spectrum** consisting of **radiation** of wavelengths just longer than visible **light** and shorter than those of **microwave** or radio waves. We usually experience infrared radiation in the form of **heat**. Night-vision weapons have been developed by the Department of Defense that use the principle of infrared radiation to provide military personnel with the ability to "see" in the dark. As was demonstrated in the Persian Gulf War, aircraft can be flown and weapons can be aimed in the dark using infrared night-vision equipment.

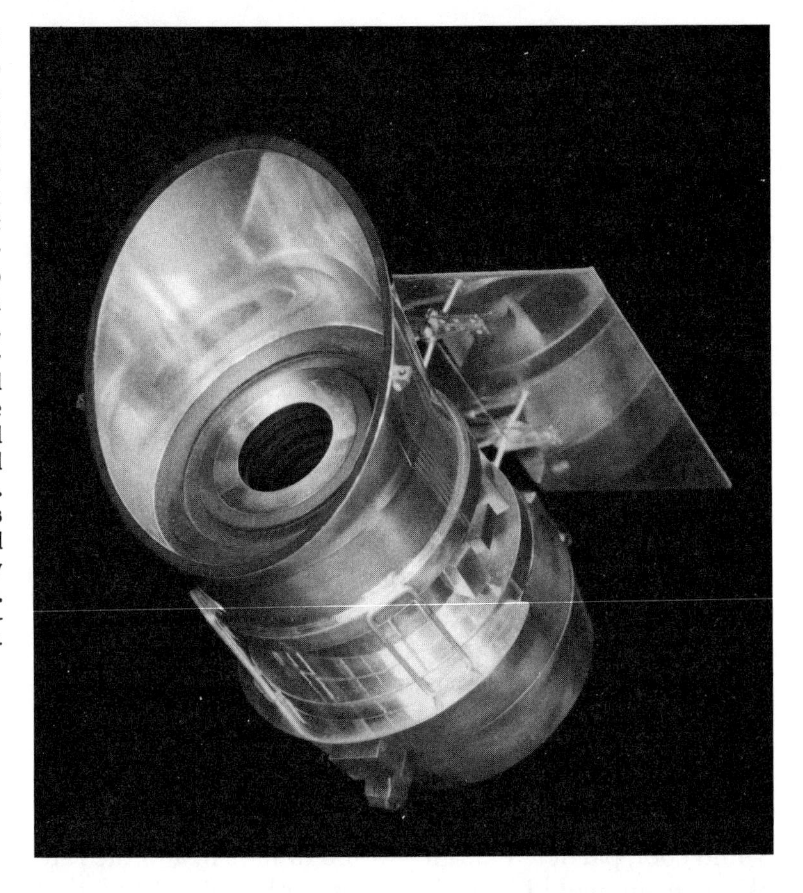

INFRARED ASTRONOMICAL SATELLITE (IRAS). This model of the completed IRAS satellite shows the aperture by which the infrared light entered. The telescope was cooled to eliminate its own heat radiation, which would interfere with infrared reception from the stars. The curved metal collar served as a sunshade. IRAS completed its important infrared survey of the sky in 1983.
Source: NASA/Ames Research Center.

Infrared Astronomical Satellite (IRAS) Because many astronomical objects—principally those made up of cool, solid MATTER—radiate most of their ENERGY at INFRARED wavelengths, this region of the ELECTROMAGNETIC SPECTRUM is of special interest to astronomers. Certain events, such as the birth of STARs and the condensation of planetary systems, can be seen to best advantage through infrared observations. However, at the surface of Earth the ATMOSPHERE blocks almost all infrared radiation. The solution to this problem is to make infrared observations from a satellite in space. This was accomplished in 1983 when a comprehensive sky survey covering a broad range of infrared wavelengths was completed by the Infrared Astronomical Satellite (IRAS). See ASTRONOMY.

Innumeracy A relatively recent coinage, innumeracy means mathematical illiteracy. It was the point of John Allen Paulos's popular book *Innumeracy: Mathe-*

matical Illiteracy and Its Consequences that we are largely a nation of innumerates, ignorant of both simple mathematics as well as very basic probability theory. It is one of Paulos's major points that the mathematically derived answer is frequently counterintuitive. For example, many people think that a slot machine that has just paid off a jackpot is less likely to pay off again as compared to a slot machine that has not paid off in some time. This belief, devoutly held by nearly all casino customers, is unfounded. The machine has no memory and the odds are the same on every pull of the handle. Applicable to many games of chance, this belief, called the *gambler's fallacy*, is persuasive to the innumerate because they believe that the greater the deviation from a MEAN, the greater the restoring force will be toward the mean. Another example of innumeracy, Paulos points out, is the inability to appreciate the difference between large numbers. For example, a million and a billion differ more than most people realize. A million seconds will go by in just under 12 days. But a billion seconds take nearly 32 years. Widespread mathematical ignorance in the United States is a cultural problem of some importance. A 1990 report issued by the National Research Council (NRC), a private group that advises Congress on scientific issues, states that although the fastest-growing occupations require a mathematical background, growing numbers of people entering the work force will not have the required training. By the year 2000, the need for mathematically trained teachers, scientists, engineers, and other workers for business, industry, and government is predicted to rise by 36 percent over today's figure. The NRC report said that there will be 21 million new jobs by the year 2000, with most requiring mathematical backgrounds and postsecondary education. See **PROBABILITY, LAWS OF**; and **SCIENTIFIC LITERACY**.

Insulation and R Values Material that prevents or limits the passage of HEAT or sound or electricity into or out of a body or region. In buildings, insulation limits the loss of heat or, conversely, limits the need for air conditioning. Insulation is measured in R values, which is a measure of the resistance to heat flow. The higher the number, the better the resistance. For instance, a standard attic ceiling is R-19, and a standard four-inch-thick insulated wall is R-11. A typical single-pane glass window (where most of the heat loss takes place) has a R value of only 1 whereas double-glazed windows have an R-2 rating. Newly available superwindows have an R-4 rating and are made by coating the inner surface of one pane with infrared radiation reflector material, such as tin oxide, and filling in the space between the panes with argon gas. Argon is a colorless, odorless, harmless, and chemically inactive gas currently used in fluorescent lamps.

Integrated Circuits (ICs) Sometimes called solid-state devices, integrated circuits are collections of electrically connected components or devices usually etched microscopically onto a small piece (or CHIP) of silicon, or some other semiconductive material. The function of an IC is to process and store electronic signals. Wiring a bunch of TRANSISTORS together in a complicated electronic circuit limited technological advances and complexity. One wiring error or one break in a tiny wire and the device did not work. Integrated circuits were developed to replace separate individual transistors with a unified, one-piece circuit. The miniature components in an IC are not individually made and assembled. Instead, the components are built up in layers of material in complex miniature patterns. See also MICROLITHOGRAPHY, and SEMICONDUCTOR.

INTEGRATED CIRCUITS (ICs) are collections of electrically connected components or devices usually etched microscopically onto a small piece (or chip) of silicon.

Interactive Computing COMPUTER programs that permit, or require, the user to respond to queries, or ask questions, while the program is being processed. The programs often produce an output and then wait for further input from the user before continuing.

Interferons Like antibodies, interferons are part of the body's defense against viruses. It has been found that human CELLS that are infected by a virus often produce special PROTEINS that diffuse into neighboring cells. The cells that receive these proteins become temporarily resistant to the virus because the protein *interferes* with the virus's attempt to reproduce inside the cell.

Early attempts to use artificially produced interferons to resist or help overcome viral infections, and even to fight cancer, were disappointing. Genetically engineered interferons were found to be ineffective against tumors and produced substantial side effects that made their use unlikely against minor ailments like colds. More promising has been the application of a class of interferon known as *alpha interferon,* which has proved to be a potent therapy for several serious diseases including hepatitis, a type of leukemia, and a rare form of blood cancer. Alpha interferon may also prove effective as an AIDS therapy when used in combination with AZT, the antiviral drug that is the only approved therapy for the treatment of AIDS, because the two drugs appear to fight the virus in different ways.

Recombinant alpha interferon activates GENES that produce proteins, which can prevent the replication of a virus. Alpha interferon also modulates the immune response, which can contribute to the antiviral effect, and inhibits some cancer cell growth. The growing number of applications for alpha interferon also holds out hope for new uses of other genetically engineered immune-system drugs. See GENETIC ENGINEERING, IMMUNE SYSTEM, and VIRUS, BIOLOGICAL.

Internal-Combustion Engine Today's automobiles, either standard Otto cycle (after Nicholas Otto [1832–1891], its inventor) or diesel, have reciprocating internal-combustion engines. They have pistons that go up and down as combustion occurs. A combustible mixture enters the cylinder chamber, is ignited, burns, and then leaves the chamber before the process is repeated.

Internal combustion refers to the fact that combustion occurs in the working fluid that supplies the output power. The air/fuel mixture burns and directly pushes on the pistons to supply power. External-combustion engines exist in experimental models. An example of external combustion is the Sterling engine, in which the combustion takes place outside the cylinder. The HEAT from the burning of the air/fuel mixture is transferred to a working fluid (usually hydrogen) inside the cylinder, which expands and contracts and does all the piston pushing. External combustion offers several advantages because it is thermodynamically more efficient. See **THERMODYNAMICS, FIRST AND SECOND LAWS OF.**

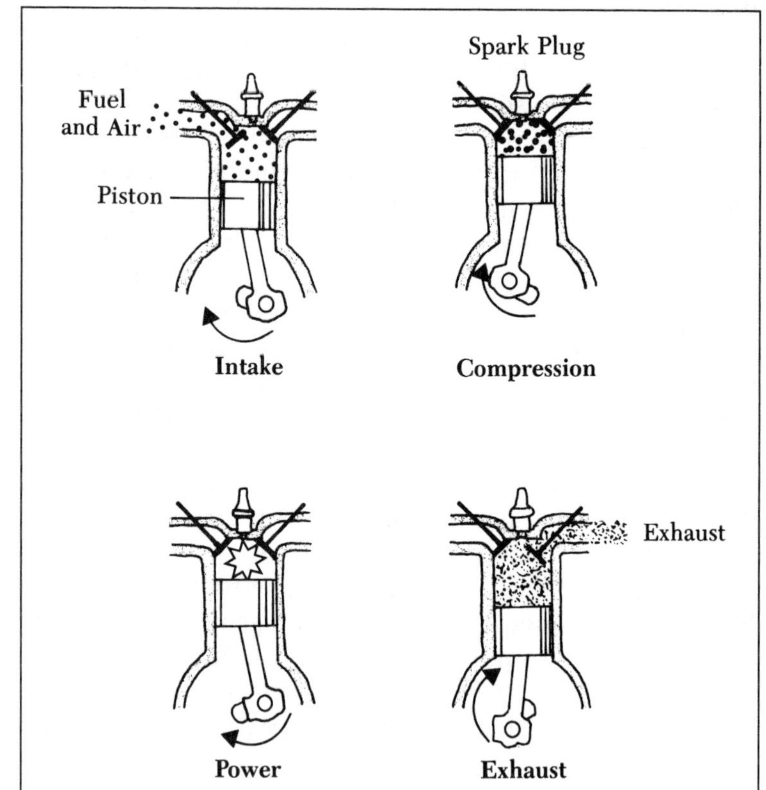

INTERNAL COMBUSTION ENGINE. The four-stroke engine uses a downward stroke of the piston to bring a fuel and air mixture into the cylinder through an intake port. An upward stroke compresses the mixture and a spark ignites the fuel, driving the piston down in the power stroke. An upward piston movement expels the exhaust gases.

Interstellar Travel/Communication The distances in deep space—beyond our SOLAR SYSTEM—are almost unimaginable. At the speed of our current space probes, *Pioneer* 10 or *Voyager* 2 for example, it would take 40,000 years to reach the SUN's nearest stellar neighbor, the triple-star system ALPHA CENTAURI. Manned interstellar space vehicles cannot be completely excluded, but the state of today's technologies are such that the energy requirements for interstellar travel are prohibitive. Considering the distances and energy requirements involved, it is not likely that humankind will be making interstellar trips anytime in the near future. For these same reasons, it is extremely unlikely that UFOs have ever visited our planet, despite the frequent reports in the checkout-stand tabloids.

Interstellar communication is a different issue. Here the hopes of astronomers like Carl Sagan and Frank Drake who believe in the possibility, however remote, of EXTRATERRESTRIAL INTELLIGENCE are viewed with respect (if not total agreement) by the scientific establishment. In terms of foreseeable technological developments, radio waves seem to be the best way to seek interstellar communication. So far attempts to detect radio signals beamed toward Earth from other intelligent civilizations have been unsuccessful. However, the number of stars that have been examined is less than .1 percent of the number that would have to be investigated if there were to be a reasonable statistical chance of discovering extraterrestrial civilizations. See also **DRAKE'S EQUATION**, **PIONEER SPACECRAFT**, **SETI**, and **VOYAGER SPACECRAFT**.

Inversion Layer Layers of atmospheric stability such that vertical motions of air are suppressed. Air pollution is reduced when the lower levels of the ATMOSPHERE are able to mix with cleaner air aloft. Under certain conditions, such as those times when warm air sits on top of cool air (an inversion of the norm), this vertical mixing cannot readily take place. A TEMPERATURE inversion describes an atmospheric condition in which vertical mixing is virtually impossible and concentrations of pollutants in the air can reach dangerous levels. See **STRATOSPHERE**, and **TROPOSPHERE**.

Invertebrates Animals that have no backbone or internal skeleton, such as one-celled protozoa or insects. A backbone is so important a feature of a skeleton that, in most classifications, all animals are divided into vertebrates and invertebrates. Vertebrates include fishes, amphibians, reptiles, birds, and mammals, all of which have a segmented bony spinal column.

Ion An atom that has an electric charge, either positive or negative. **ATOMS** are fundamentally neutral in that the number of negatively charged **ELECTRONS** orbiting the nucleus equals the number of positively charged **PROTONS** in the nucleus. Atoms, however, easily gain or lose electrons; that is, they become charged. An extra electron gives the atom a negative charge, whereas the loss of one electron leaves the atom with a positive charge.

Ion-drive engines have been proposed for interstellar space probes. An ion engine would use electricity to accelerate ions up to high speeds, as in the particle beam weapons proposed for SDI (Star Wars) programs. In this case, the particles would be used as fuel. They would be exhausted out the back of the spacecraft much like a rocket engine, using the reaction to propel the spacecraft. Although ion engines cannot produce enough power to launch a spacecraft, they could be used to produce power in space over the extended time (years) needed for interstellar travel. See **CHEMICAL BONDING**.

Ionization The process by which **IONS** are formed by the addition of **ELECTRONS** to or the removal of electrons from an electrically neutral atomic configuration. This can be accomplished by heat, electrical discharge, **RADIATION**, or by chemical reaction. See **ATOMS**.

I.Q. (Intelligence Quotient) Refers to the ratio of the mental age, as measured by a test of some kind, to the chronological age, this ratio being multiplied by 100 to remove the decimal point. In theory, a standardized test will measure an individual's ability to form concepts, solve problems, acquire information, reason, and perform other intellectual operations. In practice, however, no test has ever been devised that is not culturally biased. That is, I.Q. tests reflect the cultural background of the person or persons who set up the test. Because intelligence is always the interaction between genetic potentialities and environmental pressures, it is not one thing, but rather a name for a group of overlapping skills. What is measured by an I.Q. test is not necessarily the same as what is referred to in everyday language as intelligence, cleverness, aptitude, or wit. Still less does an I.Q. score give an estimate of the value or worth of a person. It has been said that a chimpanzee will start scoring well on I.Q. tests only when another chimpanzee has set up the test.

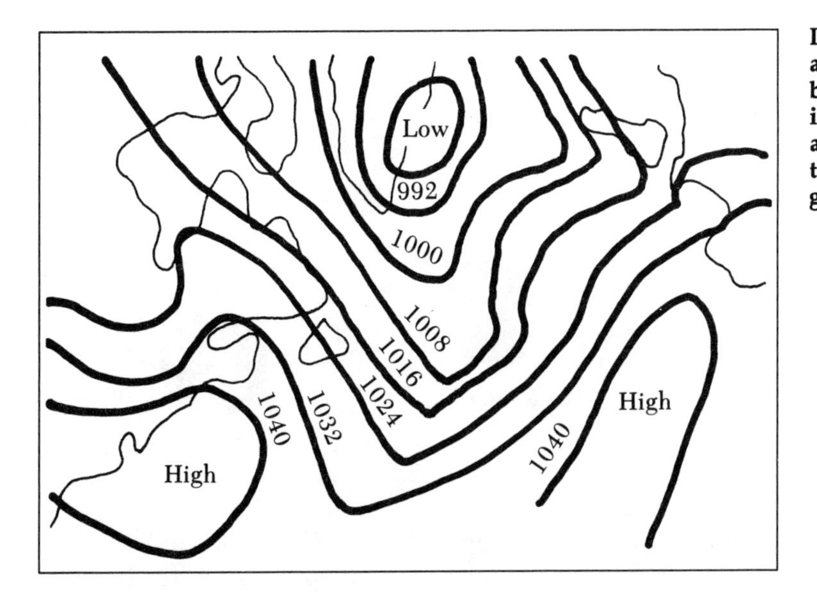

ISOBARS are lines along which the barometric pressure is the same. They are similar to contour lines on a geographical map.

Isobars Lines on a chart or map that connect points of equal barometric pressure; often used on weather maps to show high and low pressure areas and consequent weather movements. Isobars are similar to contour lines, which are lines of equal altitude on a geographical map. Just as contour lines are drawn at intervals of 50, 100, or a 1,000 feet according to the scale of the map, so isobars may be drawn at intervals of 1, 2, or 4 millibars (units of barometric pressure). Weather forecasters monitor barometric pressures—and record them on charts as isobars—to predict the movement of large masses of air and hence the coming weather patterns.

Isopoint A pointing device used on some laptop COMPUTERs in place of a conventional MOUSE or trackball. The isopoint is a pencil-like object located just below the space bar on the keyboard. Rolling it up and down moves the cursor (the sometimes blinking symbol that indicates position on a video display) on the screen up and down. Sliding it left or right moves the cursor left or right, and pressing it works like the button on a mouse. Like a mouse, an isopoint is a faster means of moving a cursor on a screen as compared to the use of the keyboard.

Isotope ATOMS of a given chemical ELEMENT contain the same number of PROTONs and ELECTRONS; however, they may contain different numbers of NEUTRONS.

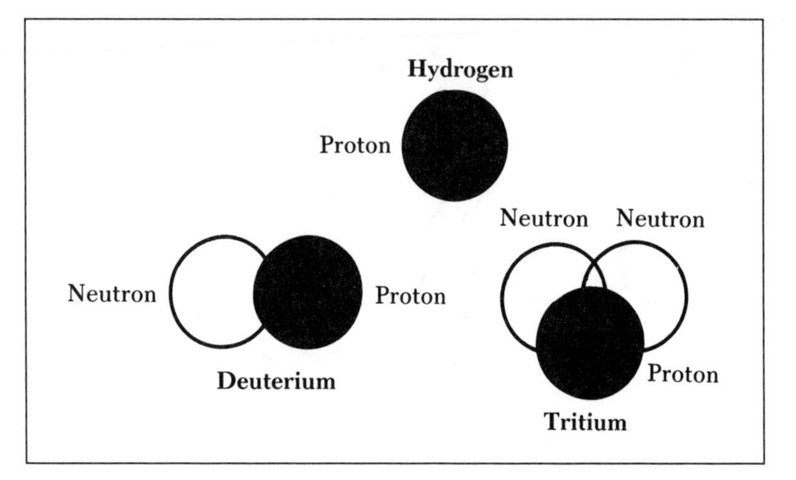

ISOTOPES refer to the members of a family of substances sharing the same position in the periodic scale but differing from each other in the number of neutrons they contain. Shown here is the nucleus of hydrogen and two of its isotopes. The hydrogen nucleus is made up of one proton, the deuterium nucleus is made up of one proton and one neutron, and the tritium nucleus is made up of one proton and two neutrons.

Atoms that have the same number of protons in their nuclei but a different number of neutrons will differ in MASS even though they have the same number of electrons. The term *isotope* (from the Greek words meaning "same position") refers to the members of a family of substances sharing the same position in the PERIODIC TABLE but differing from each other in the number of neutrons they contain. Isotopes are in effect varieties of a specific element; substances that are identical with one another in chemical properties but are different in RADIOACTIVITY. As an example, chlorine atoms have 17 electrons and 17 protons but can have either 18 or 20 neutrons. There are also several kinds (*isotopes*) of carbon atoms: carbon-12, carbon-13, and carbon-14. Most chemical elements, in fact, have more than one isotope. See SUBATOMIC STRUCTURE.

Java Man When originally found in Java in 1891, the famous Java Man bone fragments were thought to be the remains of what the popular press called the "missing link" between man and ape. It has now been reclassified as a FOSSIL of the extinct species HOMO ERECTUS. When first discovered, Java Man was thought to be a huge ape that lived in trees. Anthropologists now know that this early humanlike creature lived in Asia, Africa, and Europe for about a half a million years—during a period when the major ICE AGES were beginning. Java Man walked erect, made large stone axes and other tools, and used fire. Eventually Homo erectus gave way to the larger-brained HOMO SAPIENS. See EVOLUTION, and HOMINID.

Jet Engine As the power plant of almost all commercial aircraft today, jet engines suck in air at the front and eject it at high speeds out the back. The motion of the expanding air in one direction results in an equal motion, or thrust, in the opposite direction. The power for this forward thrust is generated by HEAT produced by burning jet fuel—kerosene or paraffin. The driving force in a jet engine is the same one that makes a toy balloon shoot forward when its opening is released and air escapes. The physical principle involved is called NEWTON's third law of motion: "For every action there is an equal and opposite reaction." See RAMJET, and TURBOJET.

Jet Lag Virtually all species of life have internal clocks that regulate their metabolisms in 24-hour, day-night, cycles called a CIRCADIAN rhythm. When we travel east or west for long distances quite rapidly we upset our internal biological clocks, and we have difficulty matching our activities with those of the people at our destination. Our internal clock tells our body that it is time to be sleeping, though everybody else is going on about their business. When we try to ignore this, our hormonal secretions do not match our activities, and we end up feeling tired and inefficient. When this occurs, we say we are suffering from *jet lag.* Researchers have discovered that the application of bright lights can sometimes be used to overcome jet lag by, in effect, resetting the internal biological clock. See **BIOLOGICAL CLOCK**, **CIRCADIAN**, and **LIGHT THERAPY**.

JET STREAM is a ribbon or core of fast-moving air, usually occurring at heights ranging from 7 to 10 miles (11–16 km), in which speeds of 50 to 100 miles (80–160 km) per hour are reached. Sharp differences in temperature occur across jet streams.

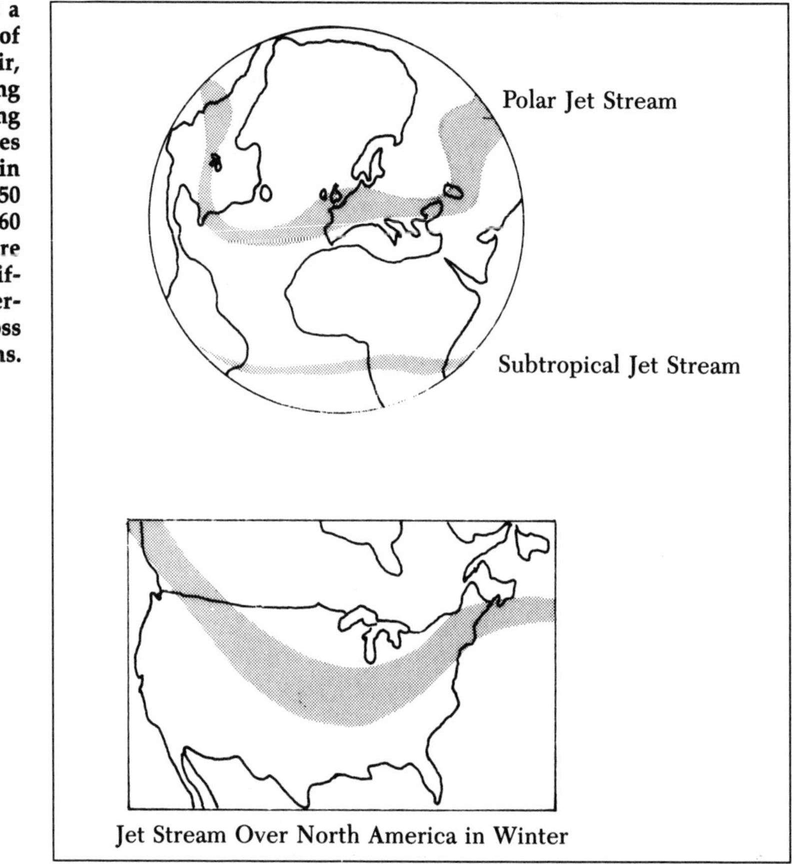

Jet Stream Over North America in Winter

Jet Stream A high-speed wind usually found just below the *tropopause* (the boundary between the **TROPOSPHERE** and the **STRATOSPHERE**), at 30,000 to 40,000 feet. Generally moving from a westerly direction at speeds often exceeding 350 miles per hour, jet streams affect both air travel and the weather. There are two main jet streams: one in the Northern Hemisphere at the general latitude of the United States; and one in the Southern

Hemisphere at the latitude of New Zealand and Argentina. Jet streams pursue a wavy course around the globe in each hemisphere and shift position from day to day. High-flying aircraft take advantage of the jet streams when they can and try to avoid them when they cannot. Jet streams also influence the weather by causing the movement of large air masses at the lower altitudes.

Josephson Junctions TRANSISTORS use a small voltage to act as the on/off switch necessary to the BINARY language of COMPUTERS. The amount of HEAT this voltage produces has been a crucial obstacle to miniaturizing computer components. Developments in SUPERCONDUCTIVITY have helped to minimize this heat obstacle. Josephson junctions, named for their developer, Nobel Prize-winning British physicist Brian Josephson, are speedy electronic switches that operate 1,000 times faster than a transistor and use 1,000 times less ENERGY. A Josephson junction consists of two superconducting films separated by an insulating layer that prevents current from flowing between the films when the junction is turned off and lets it flow when the junction is turned on. A Josephson junction can operate in less than 2 picoseconds (a picosecond is one trillionth of a second).

Joule (jewel) The International System unit of ENERGY, equal to the work done when a current of 1 AMPERE is passed through a resistance of 1 ohm for 1 second. (Ampere is the basic unit of electrical current whereas ohm is the standard unit of electrical resistance.) The Joule is a very small unit of energy or work used primarily in scientific investigations. A BRITISH THERMAL UNIT, or Btu, is equal to 1,054.8 Joules. The unit is named after James Joule, a British physicist who, in 1850, accomplished pioneer efforts in measuring work and quantifying discrete units of work and energy. Joule performed a series of experiments to study the conversion of mechanical energy to heat. He determined that heat and mechanical energy are always converted back and forth in precisely the same ratio. We know today that energy is never lost: When it disappears in one form, it always reappears in an equal amount in another form. This principle is known as the *law of conservation of energy* and it is one of the fundamental laws of classical physics. See **CONSERVATION LAWS**, **ERG**, and **THERMODYNAMICS, FIRST AND SECOND LAWS OF.**

Jovian The term usually refers to the planet **JUPITER** or planets like Jupiter in size and density. Jovian, then, means the giant planets that have a gaseous surface—**JUPITER**, **SATURN**, **URANUS**, and **NEPTUNE**. See **PLANETS**.

Juno One of the first **ASTEROIDS** to be discovered by early astronomers. By now more than 1,600 asteroids have been detected. Almost all of them are in the gap between **MARS** and **JUPITER**, an area referred to as the *asteroid belt.* Too small to be labeled planets, asteroids have diameters between one and several hundred miles. They are sometimes called *planetoids* and are thought to be the remains of moon-sized objects that never coalesced into full-fledged planets but instead smashed into each other, breaking up into pieces. Astronomers think that the strong gravitation effect of nearby giant Jupiter is the cause of this phenomena.

Jupiter The fifth planet out, Jupiter, is located 482 million miles from the **SUN** and is by far the largest planet in the **SOLAR SYSTEM**. Jupiter's mass is more than twice that of all the other eight **PLANETS** put together. It has an atmosphere hundreds of miles deep. Despite its monstrous size, Jupiter has a low density, only a quarter that of **EARTH**'s. Jupiter revolves so rapidly that its day is only ten Earth hours long. The planet's orbit around the Sun is a leisurely one, so that the Jovian year is 11.86 Earth years long. Like **SATURN**, the other gaseous giant, Jupiter emits more energy than it receives from the Sun and is composed of much the same material as the Sun—mostly hydrogen and helium. The temperature of Jupiter is estimated to be a very cold −130° Centigrade at the cloud tops.

American space probes *Pioneer 10* and *11* visited Jupiter in 1973 and 1974 followed by *Voyagers 1* and *2* in 1979. The **GALIEO** spacecraft is scheduled to orbit Jupiter in 1995. This spacecraft is expected to provide the first direct sampling of Jupiter's atmosphere and the first extended observation of this planet's largest **MOONS**. See **PIONEER SPACECRAFT**, and **VOYAGER SPACECRAFT**.

JUPITER as photographed by the *Voyager* spacecraft in 1979. The spacecraft was 17.5 million miles (28.4 kilometers) from the planet at the time. To the right of Jupiter is one of its four large moons, Europa.
Source: NASA photo.

Jupiter, Moons of Jupiter has 16 moons, the 4 largest of which are of special interest to astronomers. Each of these four will be a high priority target for the GALILEO SPACECRAFT, due to reach Jupiter in 1995. The outermost of the four large moons is Callisto. Completely ice covered and pitted with numerous craters, Callisto is about 3,000 miles in diameter. For comparison, EARTH's diameter is 7,909 miles. Next in from Callisto is a large moon called Ganymede, which is another ice ball; this one is laced with a strange network of grooves looking something like what used to be called the Martian canals. Ganymede is somewhat larger than Callisto with a diameter of 3,260 miles. Closer to Jupiter is the moon called Europa, which has been compared to a huge glass eye shattered by a hammer. Its surface is completely covered with canal-like cracks. Europa's surface, despite the cracks, appears to be as smooth as a crystal ball, with the cracks filled with ice. The smallest of the four large moons,

Europa is 1,940 miles in diameter. The fourth of the large moons, and the one closest to Jupiter, is named Io and photos sent back from the spacecraft *Voyager* revealed a fascinating moon world of active volcanoes. Io is now considered to be the most volcanic object in our solar system and as such it will be a special target for the *Galileo* mission. Io is 2,270 miles across, which makes it about the same size as our **MOON**. See **VOYAGER SPACECRAFT**.

Jurassic Period The second of the three periods constituting the **MESOZOIC** era, characterized by the existence of dinosaurs and the first appearance of primitive mammals and birds. The Jurassic period is named after the Jura Mountains in France, where the first **FOSSILS** of this period were found. Geologists date the Jurassic period as occurring between 140 and 195 million years ago. See **DINOSAURS, EXTINCTION OF**; and **GEOLOGIC TIME SCALE**.

K-mesons A subatomic particle with a MASS about halfway between that of an ELECTRON and a PROTON. Mesons are considered the FORCE inside the nucleus of an atom that holds the nucleus together. Because the protons in the nucleus all have a positive charge and particles of like charges repel each other, the nucleus of an atom should fly apart. The fact that it doesn't fly apart led scientists to search for the "glue" that holds the nucleus together. Two kinds of mesons were eventually found. One was called the *pi-meson,* later shortened to *pion.* The other is called a *kayon* or *K-meson.* If mesons are indeed the glue holding the nucleus of atoms together, then k-mesons are a type of glue.

Scientists once believed that ATOMS consisted of three PARTICLES: electrons, protons, and neutrons. Today, however, many more so-called subatomic particles have been found. Scientists now postulate that there are about 200 such particles. These new and strange particles have come to be called the *particle zoo.* In addition to meson, probably the best-known species in this subminiature zoo are NEUTRINOS, and PHOTONS. See also **QUARKS**, and **SUBATOMIC STRUCTURE**.

K-T Extinctions The time between the CRETACEOUS and the Tertiary periods—about 65,000,000 years ago on the GEOLOGIC TIME SCALE—has been labeled the *K-T boundary* (K is the geologic symbol for Cretaceous). At this point in time, a significant fraction of life on EARTH, including the last of the surviving dinosaurs, died out. The cause of this mass extinction has been the subject of much scientific debate. In 1980 a team of scientists proposed the hypothesis that the impact of a large ASTEROID or COMET crashing into Earth triggered a series of environmental disasters that led to the mass dying. Many researchers now accept that theory, but a vocal minority maintains that volcanic eruptions or slow CLIMATE change caused the widespread death. Current theory has identified the Caribbean Sea as the location of the impact or impacts. In this region, geologists found *shocked* quartz grains in a thick layer of clay deposited at the end of the Cretaceous period. Shocked mineral grains, which scientists often view as telltale impact evidence, appear in many K-T sites around the world, particularly in North America. The K-T layer on the Caribbean island of Haiti is unusually thick—25 times the size of any other such deposit, suggesting to many geologists that a major impact occurred within about 1,000 kilometers of where Haiti now lies. In addition, researchers have examined deep-sea sediments collected from sites in the area to the north of South America and have determined that at the level of the K-T boundary, the sediments contain evidence that a huge wave scoured the sea floor. See **NEMESIS**, and **DINOSAURS, EXTINCTION OF**.

Kaposi's Sarcoma A normally rare form of cancer that strikes many people with **AIDS**. Kaposi's sarcoma (malignant tumor) has mushroomed into a U.S. epidemic as the AIDS virus (**HIV**) became prevalent. It is 20,000 times more common in people with AIDS than in the general population. Researchers have yet to identify what triggers the sarcoma. It has been suggested that it is a sexually transmitted agent. This conclusion is based on epidemiological evidence that Kaposi's sarcoma is much more prevalent in patients who got AIDS sexually rather than in those who got AIDS from intravenous drug use or from contaminated blood products.

Keck Telescope Located atop Mauna Kea in Hawaii, the Keck Observatory will house the world's largest optical telescope when it begins operations in late 1992. The Telescope features a 10-meter primary mirror made of 36 hexagonal tiles. It is hoped that the Keck Telescope will be able to probe distant **STARS** and **GALAXIES** accurately, by collecting more **LIGHT** than any of its predecessors. Astronomers predict that when the **HUBBLE SPACE TELESCOPE** receives its badly needed pair of corrective lenses, it will be able to team up with the Keck. The Keck Telescope, with a primary mirror more than 17 times larger in area than the Hubble, will be able to better examine distant objects discovered by the Hubble.

Kelvin, Temperature Scale A Kelvin (K) is a unit of thermodynamic TEMPERATURE based on a scale where the zero point is equal to −273.16° Centigrade, or ABSOLUTE ZERO (often stated as zero Kelvin). Named for British physicist and mathematician Lord Kelvin (1824–1907), this thermometric scale is divided into tenths such that the freezing point of water is 273°K and the boiling point of water is 373°K. The Kelvin scale is often used in scientific work (e.g., in research in SUPERCONDUCTIVITY.) One of the advantages of this scale is elimination of the need for a plus or minus sign—all temperatures are above absolute zero.

Kepler, Johannes German astronomer and mathematician (1571–1630), the first man to discern the architecture of the SOLAR SYSTEM and formulate the phenomenological laws that govern the motions of the PLANETS. A contemporary of Galileo, Kepler was also an advocate of the Copernican theory of the universe—that is, a SUN-centered solar system, despite the fact that the orbits it predicted did not match the ones observed. It was Kepler who had the brilliant insight to make sense of COPERNICUS's heliocentric universe by determining that the planets orbited the Sun not in circles, as Copernicus had thought, but rather in perfect ellipses. He went on to formulate what are known today as Kepler's laws, setting forth the mathematical rules for the motion of the planets. See **GALILEI, GALILEO**; and **KEPLER, LAWS OF**.

Kepler, Laws of In order to summarize and understand the wealth of data on the motions of PLANETS that had been accumulated and codified by the great observational astronomer Tycho BRAHE and others in the 16th century, Johannes Kepler developed a set of laws: (1) Planets orbit the SUN in ellipses, with the Sun at one focus; (2) the line joining the Sun and a planet sweeps through equal areas in equal times; and (3) the cube of the MEAN distance of each planet from the Sun is proportional to the square of the time it takes to complete one orbit. Kepler's second law may be restated: When a planet is moving through the outer end of its ellipse, the line to the Sun will be longer but the planet will be moving more slowly. As the planet swings closer to the Sun, the line will get shorter but the planet speeds up and the area it sweeps in any period of time will stay the same. Kepler's third law may also be restated: If you cubed the average distance between the Sun and any planet, and if you squared the time it took that same planet to complete its orbit, the two resulting numbers would always have the same ratio, no matter which planet you were concerned with. In effect, Kepler's laws brought order and harmony to humankind's concept of the universe.

A complete understanding of the laws of planetary motion became possible only after Isaac NEWTON, building on Kepler's work of a half century earlier, derived his theories of universal gravitation in 1687. See also COPERNICUS, and GALILEI, GALILEO.

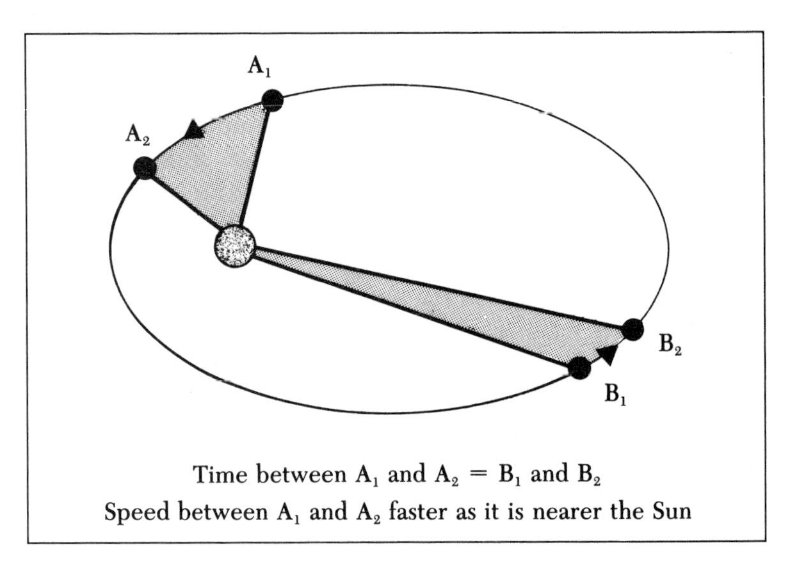

KEPLER, LAWS OF. Kepler's second law of planetary motion describes the speed of a planet traveling in an elliptical orbit around the Sun. It states that a line between the Sun and the planet sweeps equal areas in equal time. The speed of a planet increases as it nears the Sun and decreases as it recedes from the Sun.

Keratotomy, Radial An eye surgery technique in which the cornea is flattened by a series of cuts—like the spokes of a wheel—made by a tiny scalpel. It is used to correct problems in focusing, particularly ASTIGMATISM, and thus eliminates the need for eyeglasses. See also LASER, MEDICAL USE OF.

Kilo (K) Indicates 1,000 (10^3) of something. For example, a kilocycle is a unit equal to 1,000 cycles, a kilogram means 1,000 grams (about 2.2046 pounds), or a kilometer is equal to 1,000 meters (0.62137 mile). (Because COMPUTERS are built on a base-two or BINARY system, the term *kilobyte*, although loosely taken to mean 1,000 bytes, is actually 2^{10} or 1,024 bytes.) See NUMBERS: BIG AND SMALL.

Kilowatt-Hour (kwh) A unit of ENERGY equal to the expenditure of 1,000 watts of power for one hour. The toaster in your kitchen will use about this amount of

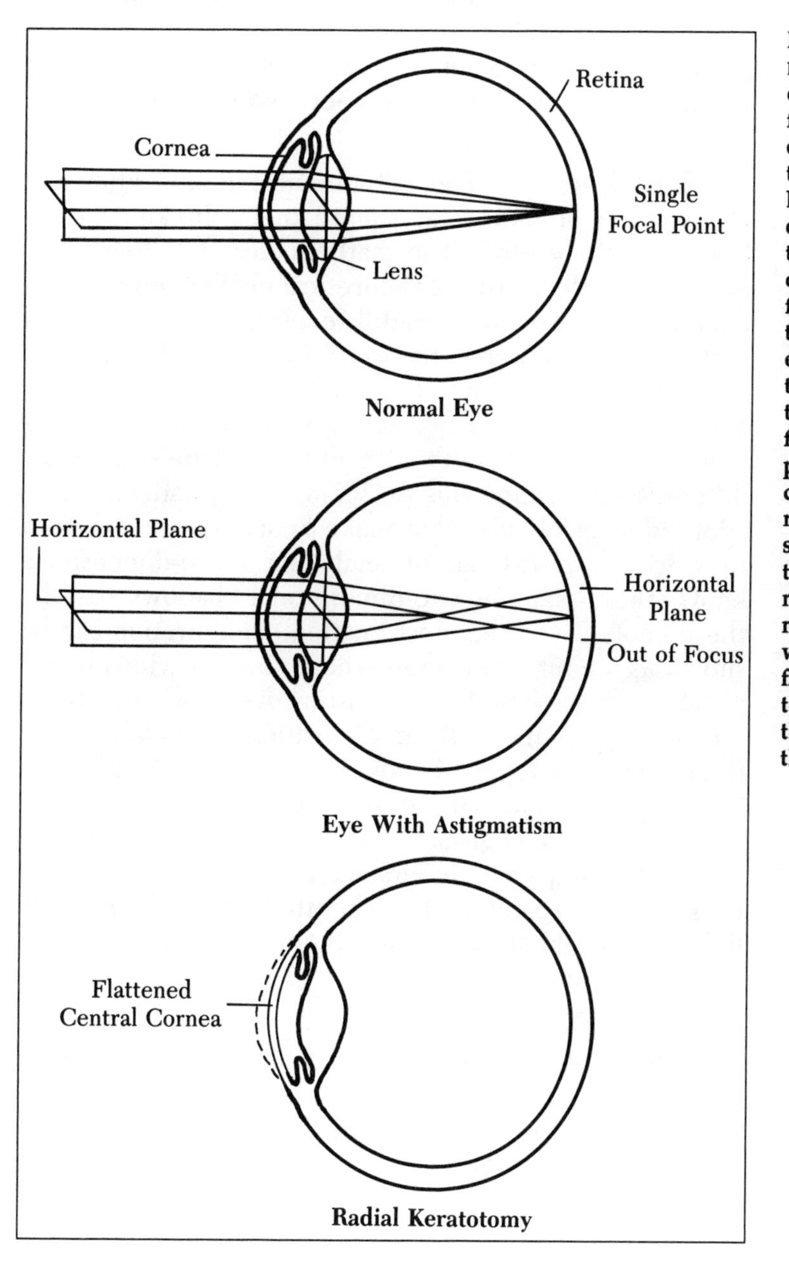

KERATOTOMY, RADIAL. The human eye is designed to focus light rays onto the retina at the back of the eye. Light rays are bent or refracted onto the retina by the cornea (the clear front window of the eye). Nearsightedness or astigmatism occurs when the cornea fails to focus the light rays properly. This condition can be corrected by a minor surgical process that causes the periphery of the cornea to bulge outward, thus flattening the center and refocusing the light rays onto the retina.

energy in one continuous hour of use. One kwh is equal to 1.34 horsepower-hours (hph) or 3,145 Btu (**British Thermal Unit**). See also **Watt**.

Kinetic Energy **Energy** appears in many forms, one of which is the energy inherent in moving bodies. Energy associated with motion (called *kinetic* from the Greek word meaning "to move") is equal to one-half the product of its **mass** and the square of its speed. See **Thermodynamics, First and Second Laws of**.

Knot Theory The mathematical identification and classification of knots (e.g., square knots, slip knots, bowline, half hitch, etc.). The mathematics of telling knots apart has evolved from a theoretical challenge for mathematicians to a possibly useful technique in the fields of molecular biology and theoretical physics. The mathematical challenge is to find a simple way to pin a label on a given knot so either two knots with the same label are equivalent or two knots with different labels are truly different. In tackling this problem, mathematicians have adopted a set of rules that make knots more convenient to work with. Instead of analyzing three-dimensional knots, they examine two-dimensional shadows cast by these knots. Even the most tangled configuration can be shown as a continuous loop whose shadow winds across a flat surface, sometimes crossing over and sometimes crossing under itself. By mathematically labeling loops, directions, and types of crossings, mathematicians have identified thousands of different knots.

Molecular biologists are using these esoteric mathematical techniques to understand how **DNA** strands can be broken, then recombined into knotted forms. The possible link between knot theory and theoretical physics is also being explored.

Knots, as a Unit of Speed In nautical usage, a knot is a unit of speed, not of distance, equal to one nautical mile per hour. A nautical mile is equal to about 1.15 statute miles. The term comes from the divisions on a log line marked off by knots and used to measure the speed of the ship through the water. The term has a built-in sense of "per hour" and it is therefore incorrect to say "knots per hour." It is also impossible to say that one place is so many knots from another, anymore than any two places on land could be said to be so many miles per hour from each other.

Kopff Comet Comets orbit the SUN just as the PLANETS do but in extremely elongated ellipses that rarely bring them close enough to EARTH to be seen. Comet Kopff orbits the Sun once every six years. Once a comet enters our SOLAR SYSTEM, the HEAT of the Sun vaporizes its icy material, and the resulting vapor and dust help to form the glowing tails visible in the night sky. The Kopff comet will be in close proximity to Earth in 2001 and NASA plans to launch a mission, called CRAF (Comet Rendezvous Asteroid Flyby), in 1996 to take advantage of this opportunity for close examination.

Krakatoa An island between Java and Sumatra in the south China Sea that was the scene of a great volcanic eruption in August 1883. The sound of this gigantic explosion is said to have been heard 3,000 miles away. Krakatoa is thought to have ejected 50 million metric tons of ash into the ATMOSPHERE, enough to darken the skies over hundreds of square miles. Fragmented rock and ash fell over an area of 300,000 square miles. *Tsunamis* (gigantic waves) 100 feet in height were caused by this eruption, and they killed 36,000 people on the shores of Java and Sumatra. Some climatologists think that this eruption may have changed the global CLIMATE for a period of several years.

Krypton One of several inert trace gases found in the ATMOSPHERE, each consisting of only a few parts per million. Krypton (literally "hidden") is used in high-power, tungsten-filament light bulbs. An ISOTOPE of this rare gas (krypton-86) has also been used as a standard measurement of length. Prior to 1960, the standard meter and standard kilogram were made of platinum-iridium alloy and kept under TEMPERATURE-controlled conditions to prevent expansion or contraction. In 1960, the scientific community abandoned material standards of length for a more precise method. The standard meter was redefined at that time as equal to 1,650,763.73 wavelengths of krypton-86.

LAN (Local Area Network) In the COMPUTER world, a network is a collection of computer/WORKSTATIONS, all having access to a common data base. Some workstations are terminals only and don't have their own microprocessors. When disconnected from the network they are just a piece of furniture that looks like a computer. Most workstations today, however, are stand-alone computers with their own microprocessors and their own memory systems. They are tied together into networks to facilitate communications as well as use common data. The network gives rapid access to all other users and the material in their systems.

Networks support anywhere from 10 to 30 or more users and the central computer to the network is called a *server*. The server is a computer with lots of memory and one or more high-capacity hard disks that operate under the control of a special network operating system to orchestrate the communication with the workstations, printers, and other devices on the network. Networks may be national in scope, such as the Defense Department's ARPANET or the National Science Foundation's ScienceNet; or they may be restricted to one business or one plant or one common field of interest, in which case they are called Local Area Networks or LANS.

Laparoscope A surgical device containing a LIGHT and a tiny television camera capable of magnifying an image 16 times. While one surgeon acts as cameraperson, the other surgeons watch the screen as they manipulate their instruments. This innovative technique eliminates the need for long abdominal incisions, which are needed in conventional abdominal surgery. In this new method, instruments and camera are both inserted through small punctures, and injury to abdominal muscles—the major cause of long recuperation times—is greatly reduced.

Laplace (luh-plahs), Pierre-Simon De Noted French mathematician and astronomer. Laplace (1749–1827) originated the first mathematical theory for the formation of the universe and the first mathematical theory for the phenomena of tides. He built on the work of NEWTON and, by redoing Newton's calculations, established that our solar system is basically stable. Laplace's other great contribution was the development of PROBABILITY theory, sometimes called the mathematics of games of chance. Through the mathematical definition of probablistic—or chance—events, he established the methods now used by modern mathematicians to deal with large quantities of data. Laplace was also a major contributor in establishing the metric system of measurement.

LASER (Light Amplification by Simulated Emission of Radiation) Basically, a laser is a way of making pure, or *coherent*, LIGHT. Ordinary light from a light bulb consists of many different colors (wavelengths) going in all directions and is called *incoherent* light, because it is unfocused and spreads and scatters over distance. A laser compresses light into a thin beam of one wavelength. The means of accomplishing this consists of a glass tube filled with a gas. When a laser is turned on, the injection of ENERGY, usually in the form of an electric current, excites the ELECTRONS in the gas into a high-energy state. These high-energy electrons release their extra energy as light. This light is amplified as it bounces back and forth between two mirrors. The laser light that is emitted from the device is a single color (depending on the gas used) and is a narrow, concentrated, powerful beam.

In theory, any kind of electromagnetic energy could be used to make a laser: INFRARED, ULTRAVIOLET, or even X rays, not just visible light. Visible light lasers are used in DIGITAL recording and fiber-optic communication systems. The intense HEAT of infrared laser beams can be used to cut metal. *Excimer lasers* generate pulsed ultraviolet beams that do not heat tissue, as other lasers do, making them particularly valuable for eye surgery. X-ray lasers were proposed as part of the SDI (Star Wars) defense system, but the energy requirements for a space-based laser weapon are now thought to be prohibitive. See LASERS, MEDICAL USE OF; and MASER.

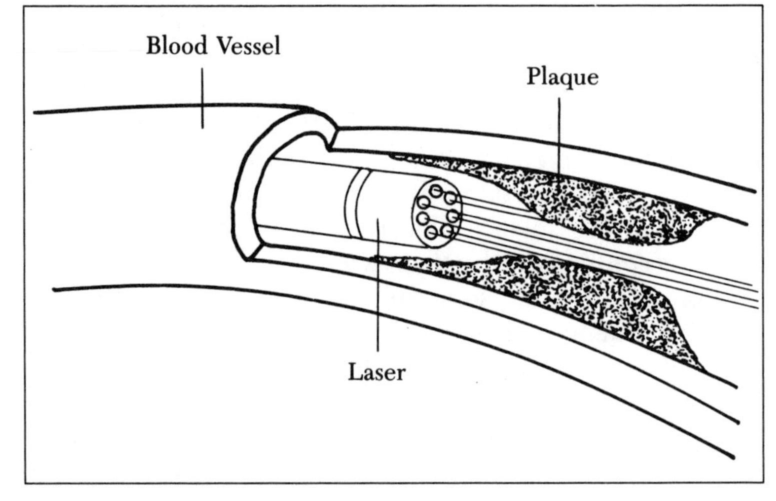

LASERS, MEDICAL USE OF. Laser angioplasty is a variation of balloon angioplasty in which a catheter with a balloon and a tiny laser is threaded through an artery and guided through the body until it reaches a blocked blood vessel. The laser beam vaporizes the plaque, and the balloon pushes the remaining plaque against the vessel wall.

Lasers, Medical Use of

Among the many uses for surgical lasers are the spot-welding of detached retinas in the eye, vaporization of abnormal growths and tumors, and halting of internal bleeding. New developments in laser technology are opening even wider areas of use, and laser therapy may become the standard procedure of the future.

Ophthalmologists are testing a new laser surgery that they hope will correct certain vision problems and eliminate the need for eyeglasses in many people. This procedure employs an *excimer laser* to sculpt the cornea—the transparent front outside coating of the eye—correcting nearsightedness, farsightedness, and astigmatism. Eximer lasers generate pulsed ultraviolet beams that do not HEAT tissue, as other lasers do. Instead, the beam breaks chemical bonds and allows cells to be washed away without damaging surrounding tissue. Doctors hope that this approach will let them reshape the curve of the cornea without scarring. The shape of the cornea determines how light is focused on the retina in the eye.

Laser angioplasty is a variation of balloon ANGIOPLASTY in which a catheter with a balloon and a tiny laser is threaded through an artery and guided through the body until it reaches the blocked blood vessel. In this procedure, the laser beam is used to vaporize the plaque, then the balloon is inflated to push the remaining plaque against the vessel walls. See LASER.

Latent Heat

The HEAT released when MATTER undergoes a change of phase; that is, from liquid to solid matter or from gas to a liquid. An example is the process of evaporation, which absorbs a large amount of heat. When water reaches the boiling point, its temperature does not rise further. Rather, the additional heat is all consumed in boiling away, or evaporating, the water. It takes nearly six times as much heat to evaporate a quantity of water as it does to raise its temperature from freezing to the boiling point. If the vapor subsequently condenses, the same amount of heat is given off. The heat involved in these processes, which is not accompanied by any change of TEMPERATURE, is called latent heat. Freezing and melting also involve latent heat.

Laws of Motion See NEWTON.

LDL (Low-Density Lipoprotein) The so-called "bad" cholesterol. Levels above 160 milligrams per deciliter (one-tenth of a liter) of blood are considered high and may signal coronary risk. See CHOLESTEROL.

Leakey Family of noted anthropologists; the late Kenyan-born Englishman Louis Leakey, and his wife Mary Leakey, whose work at the Olduvai Gorge in Tanzania and elsewhere revealed that humans probably evolved in Africa. The Leakeys discovered FOSSILS of human ancestors dating back more than 3.75 million years. Their son, Richard Leakey, has continued to make other discoveries in Kenya and Tanzania. In 1976, he discovered a nearly complete skull of HOMO ERECTUS in a geological strata one and a half million years old, making it the oldest known specimen of that ancestral creature. Richard Leakey is also a major figure in African wildlife management. A controversy in the anthropological world concerns the line of ancestry from early humanlike creatures to Homo sapiens (modern humans). Mary Leakey and her son, Richard, are at odds with Donald Johanson, the discoverer of Lucy in this ongoing debate. See EVOLUTION, HOMINID, HOMO SAPIENS, and LUCY.

Leap Second The U.S. Naval Observatory, whose atomic clock is accurate to a billionth of a second a day, has determined that the EARTH's rotation is slowing slightly. To keep the atomic TIME standard that we now live by synchronized with Earth time, a leap second has to be added when necessary. A leap second was added in 1990, making it the 16th since 1972 when this system was decided by international accord. If the Earth were to speed up, a leap second will be removed. See **LEAP YEAR.**

Leap Year Every time the Earth goes around the **SUN**, a year passes. This process takes about 365.25 days. Thus, after a year, the Earth has turned an extra quarter turn. After four years, the Earth has completed one whole extra rotation—an extra day. We therefore add an extra day to our calendars every fourth year, making it a leap year. Ordinarily, years that are divisible by four (e.g., 1992, 1996, 2000) are leap years.

There are exceptions to the divisible-by-four rule and these exceptions are necessary because a year is more precisely 365.2422 days long. Therefore to keep the calendar straight, we omit the leap year every 100th year. Thus, 1700, 1800, and 1900 were not leap years. One more adjustment is necessary to make the calendar match reality: Every 400th year we do have the leap year regardless, so do not fret, the year 2000 will be a leap year. See **LEAP SECOND.**

Left Brain/Right Brain The outer layer of the human brain (called the *cerebrum*) includes the cerebral cortex, which is the location of thought, memory, sight, speech, and other higher functions. The cerebrum contains two hemispheres, each of which controls different functions. In general, the right half controls the left side of the body and such functions as the perception of spatial relationships, whereas the left hemisphere controls the right side of the body and such functions as speech. Having two regions of the cerebral cortex capable of controlling bodily functions incurs the possibility of mixed-up commands. Messages from one half of the brain can reach the other half, but the human brain obviates the need for constant cross-reference by having one cerebral hemisphere dominant. Usually the left hemisphere is the dominant one. The issue of left-handedness is not directly related to dominant hemisphere; that is, left-handed people often have a dominant left hemisphere. Why the left side of the brain is dominant is not completely understood.

The left brain is popularly thought to be more analytical, rational, and logical than the right brain, which is thought to be more artistic and integrative. There is, however, only marginal scientific evidence to support these notions.

LEP (Large Electron-Positron Accelerator)

Located in Geneva and operated by the European Organization for Nuclear Research (CERN), this ACCELERATOR/COLLIDER has become the centerpiece of a major high-energy physics research center for the entire world. LEP is considered the largest scientific instrument ever built. Designed to expand the frontiers of physics, LEP is located in an underground tunnel 16.8 miles (27 km) in circumference; it is the biggest accelerator/collider in operation today.

Like the other large accelerators—Stanford's Linear Accelerator, Fermilab's Tevatron, and the proposed SUPERCONDUCTOR SUPERCOLLIDER—LEP was built to probe the nature of matter on a scale far smaller than that of an atom. The purpose is to provide physicists with answers to questions concerning the fundamental FORCEs of nature and what binds these subatomic parts together. LEP provides a means to bring two accelerated PARTICLES into collision with each other at very high speeds. This event releases enormous bursts of ENERGY that instantly condense into an array of particles, some of which may not have been seen since the BIG BANG. See also CERN, and PARTICLE PHYSICS.

Leptons

The fundamental, bedrock PARTICLES of all MATTER, as far as is known today, are QUARKS and leptons. Leptons (from the Greek *leptos*, meaning "thin") are the class of elementary particles that have no measurable size and are not influenced by the strong nuclear FORCE, as are quarks. ELECTRONS, MUONS, and NEUTRINOS are leptons. See SUBATOMIC STRUCTURE.

Levitating Trains

*Mag*netically *lev*itated (maglev) trains offer great promise for future transportation needs. Because of their speed, efficiency, passenger capacity, and safety features, maglev trains may well be competitive with air transport on the shorter runs. Capable of speeds greater than 300 miles per hour, these silent trains move without wheels down a guideway floating on an electronically generated magnetic cushion. The trains are both raised above the guideway and propelled magnetically. Propulsion takes place because each magnet on the train is attracted by a guideway coil of opposite polarity (the positive or negative state) immediately ahead of it and is repelled by a coil of the same polarity immediately behind it. Rapidly alternating the positive or negative state of the coils propels (pulls and pushes) the train forward. The frequency of the polarity reversals governs the speed. Reversing the poles of the magnetic field reverses direction, which provides braking.

Researchers are also investigating the use of superconducting magnets for use in maglev train systems. Prototype levitating trains are operating in Germany and Japan. U.S. routes under study include Chicago-Milwaukee, Southern California-Las Vegas, Dallas-Houston, Pittsburgh-Cleveland, and the San Francisco Bay Area. See SUPERCONDUCTIVITY.

LEVITATING TRAINS use magnets imbedded in the guideway and on the train to both lift the train cars off the guideway and to propel the train forward. Rapidly alternating the polarity of the guideway magnets pushes and pulls the cars along the guideway.

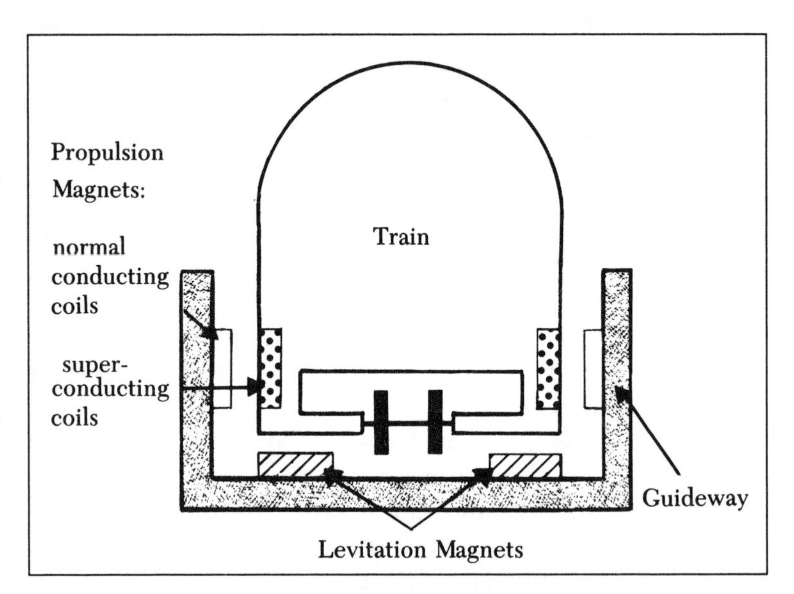

Life, Origin of

Present-day life-forms on EARTH evolved from common ancestors reaching back to the simplest one-cell organisms about three billion years ago. Details of the history of life on Earth are still being pieced together from the combined geological, anatomical, and molecular evidence. During the first two billion years of life, only microorganisms existed, some of them quite similar to BACTERIA and algae that exist today. With the development of CELLS with nuclei about a billion years ago, there was a great increase in the rate of EVOLUTION of increasingly complex, multicellular organisms. It is only during the last 600 million years—13 percent of Earth's 4.6-billion year history—that life blossomed into a vast variety of forms. Earth's present-day life-forms have evolved from these common ancestors. See DARWIN.

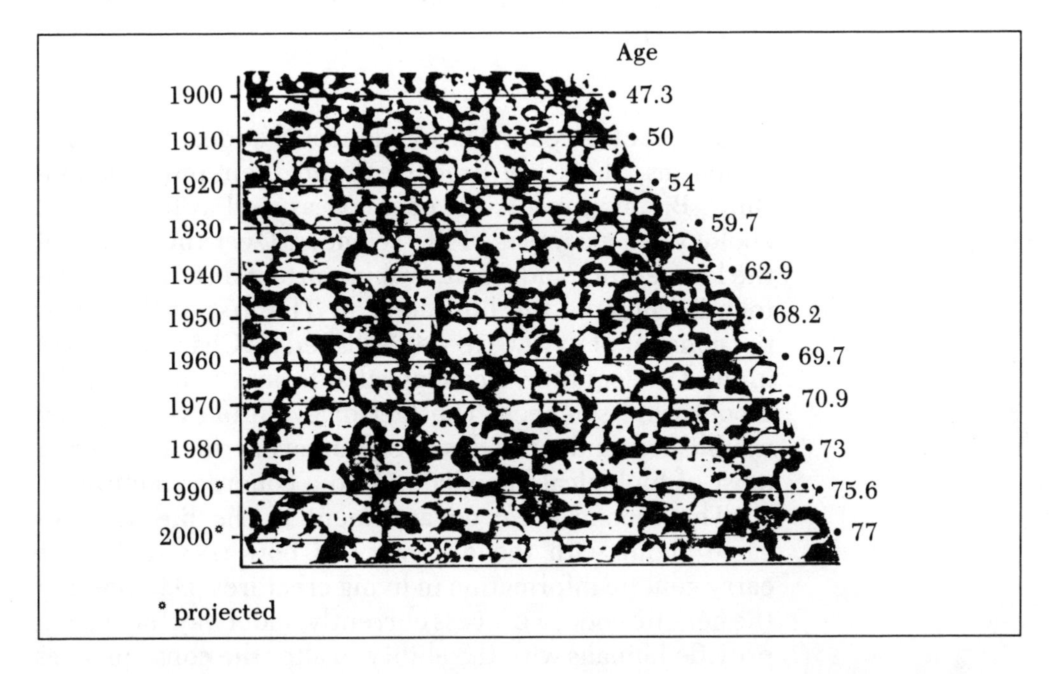

LIFE EXPECTANCY in the United States has changed dramatically over time.

Life Expectancy The statistically determined number of years that an individual is expected to live. This number has changed dramatically over time. Prehistoric humans, for instance, are thought to have had a life span of 21 years. By 1776 an American could expect to live 35 years. The median age in the country at that time was 16, meaning half the population was below age 16 and half was above (see MEAN/MEDIAN/MODE). By 1876 life expectancy had increased to 40 and the median age was 21. In 1990 life expectancy is 73 for men and 79 for women, with the median age standing at 38. In 1983, for the first time in U.S. history, there were more people over 65 than there were teenagers. By the year 2000, one in five Americans will be over 65. By 2030 that could mean as much as one-third of the entire population of the nation.

The above statistics are overall for the nation. When broken down into ethnic groups, a different picture emerges. In the United States, for instance, blacks have significantly shorter life spans than whites. According to U.S. Public Health Service statistics, the average black child born in 1986 will live an estimated 69.4 years, down from 1985. In contrast, the average white American's life span keeps getting longer. Blacks are increasingly dying of AIDS, pneumonia, and homicide. See also **INFANT MORTALITY**.

Life Sciences Those sciences that deal with living organisms, their life processes, and their interrelationships. Biology, medicine, ecology, as well as botany and zoology are examples of life sciences. Over the period of the last century and a half, two important discoveries profoundly affected the life sciences. The first of these was the concept of **EVOLUTION** established by Charles **DARWIN** in the 19th century, a concept that forms the basis for the modern life sciences. The fact of evolution is now well established and provides the framework for organizing most of biological knowledge into a coherent picture.

The second great advancement in the life sciences came about with the discovery of how **DNA** molecules carry genetic information in living creatures. Deciphering the genetic code, a process currently underway today, will provide humans with the ability to alter the consequences of heredity—in effect, giving humans a biological say in their own destiny. See also **GENETIC ENGINEERING**.

Light, Nature of That portion of the **ELECTROMAGNETIC SPECTRUM** that is visible to the human eye, we call light. Light varies in wavelength and thus appears as different colors to the human eye, from the longer wavelengths of red to the shorter wavelengths of violet. Red light has the lowest frequency, and the other colors increase in frequency through orange, yellow, green, blue, and violet. Red light has the longest wavelength at 0.7 micron (a *micron* is a millionth of a meter) and violet has the shortest wavelength at 0.4 microns. Ultraviolet energy is just outside the range of visibility and has a still higher frequency.

EINSTEIN's photoelectric theory explains that light exists in the form of tiny oscillating bundles of electromagnetic energy. These light bundles are called PHOTONS. Considering light as either a wave or as PARTICLES is called *wave-particle* duality. Whether light needs to be described as a particle or as a wave depends on the experiment being performed. No matter which description of light is used, the same answer will be derived. See LIGHT, SPEED OF; and QUANTUM PHYSICS.

Light, Speed of Light travels through a vacuum at about 186,000 miles, or 300,000 kilometers, per second. According to EINSTEIN'S SPECIAL THEORY OF RELATIVITY, the speed of light is finite; nothing can ever travel faster. Understanding that the speed of light is the speed limit of the universe is basic to understanding Einstein's concept of the universe. At the speed of light, Einstein tells us, TIME would stand still. At the speed of light, MASS would become infinitely large. This latter concept has been demonstrated in particle ACCELERATORS. As the particles move close to the speed of light, they pick up ENERGY (and mass) but no more speed. Einstein's theory is, in fact, verified every time an accelerator propels particles at very fast speeds.

If this concept is indeed true, why doesn't an automobile increase in size when the driver steps on the accelerator? The answer is that the effect is only significant at speeds close to the speed of light. At 60 miles per hour, an automobile is traveling so relatively slow (only 1/60th of a mile per second as compared to 186,000 miles per second) that the effect is not discernible. See also EQUIVALENCE OF MASS AND ENERGY ($E = mc^2$).

Light Engine A central lighting system for automobiles designed to replace the 80 or so light bulbs presently used as headlights, brake lights, turn lights, and instrument panel illumination. The General Electric Company is developing a way to "pipe" light from a central lamp using a FIBER OPTIC distribution system to whatever application the driver chooses. The network of fiber-optic cables takes the place of twisted wires, and there are no bulbs to burn out. A new type of lamp is proposed for the central source that can produce a brighter light while consuming less power. The new system, which is four times more efficient than incandescent bulbs, is expected to be ready for cars before 1995.

Light Therapy The human biological clock is sensitive to light and researchers have begun unraveling the complicated CIRCADIAN rhythm that governs sleepiness, hormone levels, and other daily cycles of the body. Scientists have found that they can reset the human BIOLOGICAL CLOCK backward or forward by administering certain amounts of light at various points in the circadian cycle. Airline passengers may one day use precisely timed doses of light to reset their biological clocks and thus overcome debilitating jet lag.

In addition, light therapy has shown some promise in the treatment of some AIDS patients. In this case, experimental treatment with a combination of ULTRAVIOLET RADIATION and a light-activated drug appears to have bolstered the IMMUNE SYSTEMS of a few people with AIDS-related complex. In the past, scientists have inactivated the AIDS virus (HIV) in a test tube by treating it with a light-activated drug and then exposing it to ultraviolet radiation. Researchers emphasize, however, that these are tentative findings based upon only a very small group of patients. See CHRONOBIOLOGY.

Light-Year A light-year is a measure of distance, not time. It is the unit of measurement used in ASTRONOMY and COSMOLOGY because the distances in space are so great that to use miles or kilometers would require pages of zeros. A light-year is the distance that light, moving at slightly more than 186,000 miles per second, travels in one year. A light-year then is roughly equal to six trillion miles. A trillion is a million million or a 1 followed by 12 zeros. The nearest STAR to EARTH, other than our own SUN, is ALPHA CENTAURI and it is so distant that it takes light 4.5 years to reach us. If we used miles to indicate this distance, it would come to 27 trillion, or just too many zeros to keep track of. The issue of appropriate measurement units may be illustrated by the example of using inches to measure the distance between Los Angeles and San Francisco. The answer is 24,520,320 inches but that is clearly an awkward and inappropriate unit to measure such distances. In space, earthbound units are too small to provide an appropriate indication of distances. For technical reasons, astronomers often use the PARSEC as a unit of distance in space. A parsec equals 3.258 light-years. See also ASTRONOMICAL UNIT, and LIGHT, SPEED OF.

Linear Momentum See MOMENTUM.

Lithosphere The outer or solid part of EARTH; the crust of the great land areas to a depth of about 100 km (60 miles) and the upper part of the mantle (that part of Earth between the crust and the core). See ATMOSPHERE, and HYDROSPHERE.

Little Ice Age Long-range changes in climate show up as ice ages separated at regular intervals by long-lasting warm periods. The most recent cold period marked by significant glaciation lasted from A.D. 1500 to 1900 and is called the Little Ice Age. During this period the average TEMPERATURE in Europe was nearly 3 degrees lower than today and mountain glaciers advanced all over the world. Crops failed and villages starved in parts of Europe. The ordeal of General Washington's army at Valley Forge took place during this period (1779–80) and New York harbor froze solid. Climatologists say that about every 2,500 years, a little ice age cools the globe. The cause of these little ice ages, which have about one-tenth the impact of a full ice age, is uncertain. Some scientists believe they may be related to variations in the Sun's energy output. See CLIMATE, and ICE AGES.

Local Group (of Galaxies) Galaxies of stars do not exist independently but are grouped into *clusters*. Clusters contain dozens of galaxies, or hundreds, or thousands, or even more. Our own galaxy, containing an estimated 200 billion stars, is part of a rather small cluster called the Local Group. The two best-known members of the Local Group are the MILKY WAY and the ANDROMEDA GALAXY. The Andromeda is larger than the Milky Way, containing up to as many as 1,000 billion STARS. Two other members of the Local Group are the Large Magellanic Cloud and the Small Magellanic Cloud, two *dwarf* galaxies containing about 20 billion stars apiece.

Logarithmic Scale Invented by John Napier (1550–1617) of Scotland, logarithms reduce problems in multiplication and division into ones of addition and subtraction, generally much simpler to perform. A powerful tool in the manipulation of numbers is the use of exponents—$100 = 10^2$, $1{,}000 = 10^3$, and $10{,}000 = 10^4$. If one multiplies $10^2 \times 10^3$, one has only to add the exponents to reach 10^5. Similarly, raising to a power or extraction of a root is a simple matter of multiplying or dividing exponents. Tables of logarithms developed by Napier made it possible to use this concept with any number. This seemingly small discovery greatly reduced the drudgery of astronomical and navigational calculations and marks the start of the era of computation. Scales, such as the RICHTER SCALE for measuring the intensity of EARTHQUAKES, are said to be logarithmic when a higher number is ten times greater than the one below. See **MATHEMATICS**, and **NUMBERS: BIG AND SMALL**.

Logic, Computer See **BOOLEAN LOGIC**.

Lookback Time Because of the finite velocity of LIGHT, the more distant is an object being observed, the older is the information received from it. A STAR five LIGHT-YEARS away from EARTH, for example, is seen as it looked five years ago. A galaxy one billion light-years away is seen as it looked one billion years ago. When we look through a telescope we are, in effect, looking backward in time. See **LIGHT, SPEED OF**.

Loop In COMPUTER programming, some instructions are designed to be executed repeatedly. When a program is flowcharted, or diagramed, the repetition forms a loop. When computer programmers specify the repeated execution of groups of statements, they often use the term *loop*, which is interchangeable with *repetition*.

Lucy Name given to the HOMINID fossil discovered by the American archaeologist Donald Johanson and his colleague, Tom Gray, in Ethiopia in 1977. Enough bones were found to make up about 40 percent of a complete individual, the most complete skeleton of an early ancestor of humans ever found. Lucy lived about three million years ago. She was a little creature about three and a half feet tall and, most important, she walked upright. This *bipedality* is an important anatomical characteristic distinguishing hominids from apes. See also AUSTRALOPITHECUS, EVOLUTION, and HOMO SAPIENS.

Lumen A measure of the total LIGHT produced by a lamp. For example, a 60-WATT light bulb delivers about 870 lumens. Many people think of wattage as a measure of light output whereas it is, in fact, a measure of electrical power going in to the bulb. The lumen-per-watt (LPW) value of a light bulb is similar to a miles-per-hour rating for a car, since it expresses what you can get out of a lamp in terms of what is put into it. A higher LPW indicates that a lamp is able to produce more light for the same amount of power and, therefore, is a more efficient light source.

Generally, larger bulbs are more efficient. For example, a typical 100-watt light bulb produces 1,750 lumens—slightly more than two 60-watt bulbs. See INCANDESCENT, and FLUORESCENCE.

Luminosity of Stars How bright a STAR appears to be is called its *apparent magnitude*—its brightness as seen from an observer here on EARTH. The more distant a star happens to be, the dimmer it appears to us. Because astronomers have other means of measuring the actual distance to a star, they can convert its apparent magnitude to its real brightness—the intensity of light that is actually given off. This true brightness is called a star's *luminosity*, or its absolute magnitude.

Lunar Eclipse When the light of the MOON is blocked by the intervention of the EARTH between it and the SUN. (See diagram on page 190.) This is not to be confused with a SOLAR ECLIPSE, which involves the blocking of the light of the Sun by the intervention of the Moon between it and a point on the Earth. In a lunar eclipse, the Moon seems to disappear from the night sky. In a solar eclipse, the Sun becomes invisible from some spots on the Earth.

Lunar Observer (Satellites) The proposed first step in a U.S. return to the MOON. The National Space Commission has proposed a permanently inhabited lunar

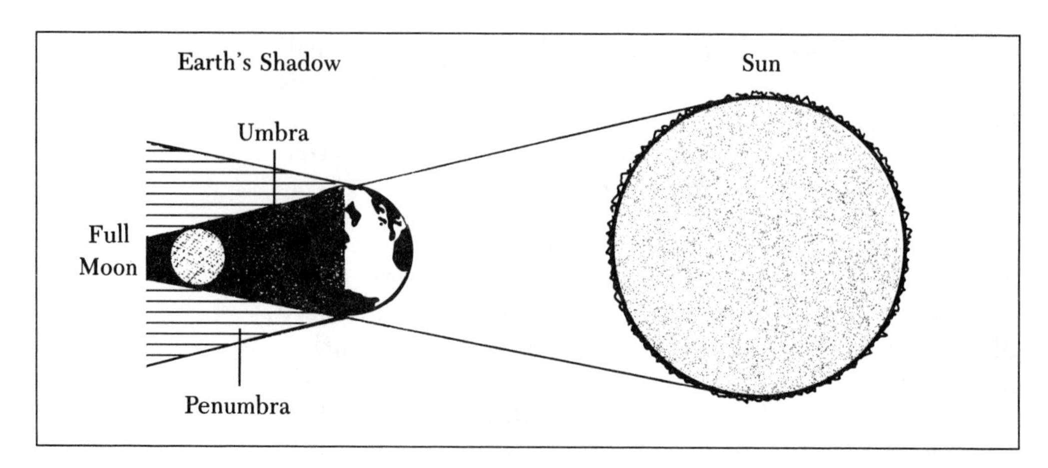

LUNAR ECLIPSE occurs when the Moon passes exactly into the Earth's shadow. The dark part of the Earth's shadow, called the *umbra*, takes about two hours to cover the moon. The total part of the lunar eclipse may last an hour or so.

outpost as a preliminary stage in the eventual human exploration of Mars. NASA's plans for the lunar outpost call for building habitats and an oxygen plant, as well as developing new construction technologies to use in the inhospitable lunar environment. The proposed *Lunar Observer* satellites would provide key facts about potential sites for the outpost. The *Lunar Observer* satellite would be the first spacecraft to follow a polar orbit around the Moon that would allow it to survey the entire lunar surface.

NASA is considering 2005 to 2010 for starting the lunar base. Providing engineering and environmental data in time to get it going by 2005 would require launching the *Lunar Observer* satellites by 1997 or 1998. The unmanned spacecrafts would map the Moon and study its surface for two years to help determine a proposed site for the lunar outpost.

Lunar Telescopes Astronomers have urged NASA to consider constructing a huge radio telescope on the MOON's far side. With the Moon blocking off human-made and natural emissions from Earth, this could allow scientists to observe faint stellar radio sources. Data would be sent back to Earth by means of a series of lunar satellites. Also on the astronomer's wish list is the Lunar Optical Telescope (LOT). Funding for these projects has yet to be approved, and neither may survive the budget process. See LUNAR OBSERVER.

Lyell, Charles British 19th-century geologist considered the "father of modern geology." In his book *Prin-*

ciples of Geology, Lyell proposed the idea, revolutionary in 1830, that **EARTH** is very old—much older than was supposed at the time—and that the processes that formed its geological features had been operating slowly over a very long time. He went on to postulate the idea that these same processes are still at work at their usual rate and that Earth is perpetually changing, albeit at a rate that is virtually imperceptible. This concept came to be known as **UNIFORMITARIANISM**. Prior to Lyell's book, most people thought that the Earth was young and that its most spectacular features, such as mountains and valleys, islands, and continents, were the products of sudden cataclysmic events including acts of God such as Noah's flood. This philosophy was called **CATASTROPHISM.**

It was Lyell who learned how to read rocks. He analyzed layers of rocks in mountains, including the cracks and folds in these layers, and determined that the only way they could have formed was through various gradual processes or frequently repeated events, such as **EARTHQUAKES** and volcanic eruptions. Lyell also recognized that different layers of rocks contained different types of **FOSSILS**. Lyell gave the distinct eras in geologic history the names that geologists use today—Miocene, Pliocene, and Pleistocene.

When Charles **DARWIN** set sail on the *Beagle* on its journey around the world, he carried with him a copy of Lyell's masterpiece. Darwin was clearly impressed with Lyell's observations and was influenced by his technique of theory backed with massive amounts of observational data. Darwin used this same approach in his own book, *On the Origin of Species*. See **EVOLUTION, GEOLOGIC TIME SCALE**, and **STRATIFICATION**.

Mach Number Ratio of an object's speed to the speed of SOUND (740 miles per hour at sea level). When a Mach number exceeds 1, the object is moving at supersonic speeds; Mach 2 means twice the speed of sound, and so on. The term is named after the Austrian physicist Ernst Mach (1838–1916), who first investigated the consequences of motion at such speeds. See **SOUND, SPEED OF**.

Magellan Spacecraft Launched in May 1989 from the shuttle *Atlantis*, the *Magellan* spacecraft was the first American interplanetary craft to be sent into space since 1978. Named after the 15th-century Portuguese explorer, the spacecraft's primary mission is to map 70 to 90 percent of the surface of the planet VENUS over a period of 243 EARTH days, the time it takes the planet to complete one rotation. After a 15-month voyage from Earth, looping the SUN one and a half times to gain gravity-assisted momentum, the *Magellan* spacecraft began mapping Venus in October of 1990—a month late due to unexplained communication problems.

Magellan carries an advanced RADAR system designed to penetrate the Venusian clouds and produce images of a world that has been glimpsed only briefly by previous spacecraft explorations. *Magellan's* radar is capable of unprecedented resolution, able to detect objects as small as 1,000 feet wide. Every three hours and nine minutes, the spacecraft completes an orbit of Venus, mapping a swath of the alien landscape with radar and transmitting the images to Earth. Each image covers an area 10 by 17 miles and shows details of the Venusian mountains, valleys, and plains never before seen. Images received so far have helped scientists to understand how geological forces have shaped Venus, a PLANET much like Earth in size and density but yet very different in climate. Venus is completely dry, lifeless and hot enough on the surface to melt

MAGELLAN SPACECRAFT. This artist's drawing shows the *Magellan* spacecraft with its solar panels pointed toward the Sun as it begins its orbit near the north pole of Venus. During its 1990 to 1995 mapping mission, Magellan will cover 70 to 90 percent of Venus.
Source: NASA/JPL.

lead. Is the hellish Venusian climate a result of a runaway GREENHOUSE EFFECT, and did Venus once have oceans? These are some of the questions scientists hope *Magellan* can help answer. See PIONEER SPACECRAFT.

Magellanic Cloud Two galactic clusters of stars that are the nearest independent STAR systems to the MILKY WAY galaxy. These galaxies are called the Large Magellanic Cloud and the Small Magellanic Cloud, because it was Ferdinand Magellan's crew who first reported seeing them when they sailed around the world in the 16th century. The term *cloud* refers to their resemblance to hazy white clouds in the night sky. See GALAXIES.

Maglev See LEVITATING TRAINS.

Magnetic Highways See ELECTROMAGNETIC INDUCTION.

Magnetic Resonance Imaging (MRI) An example of high-technology medicine, magnetic resonance imaging may prove to be the greatest advance in diagnostic medicine since the invention of X-ray technology in 1895. MRI works on the principle that HYDROGEN ATOMS subjected to a magnetic field line up like soldiers on parade. When a radio frequency is aimed at these atoms, the alignment of their nuclei changes. When the radio frequency is turned off, the nuclei realign themselves, transmitting a small electric signal when they do. Because the human body contains a lot of hydrogen atoms, an image can be generated from these small electric signals, showing tissue and bone marrow.

MRI has proven to be a sensitive diagnostic tool for cancer detection. It is able to detect some brain tumors so small that they do not show up on X rays. The MRI technique has also proven useful in measuring blood flow rates from specific locations in the brain in order to identify stroke-prone patients. In addition, MRI can be used to measure the effects of therapeutic drugs on the brain.

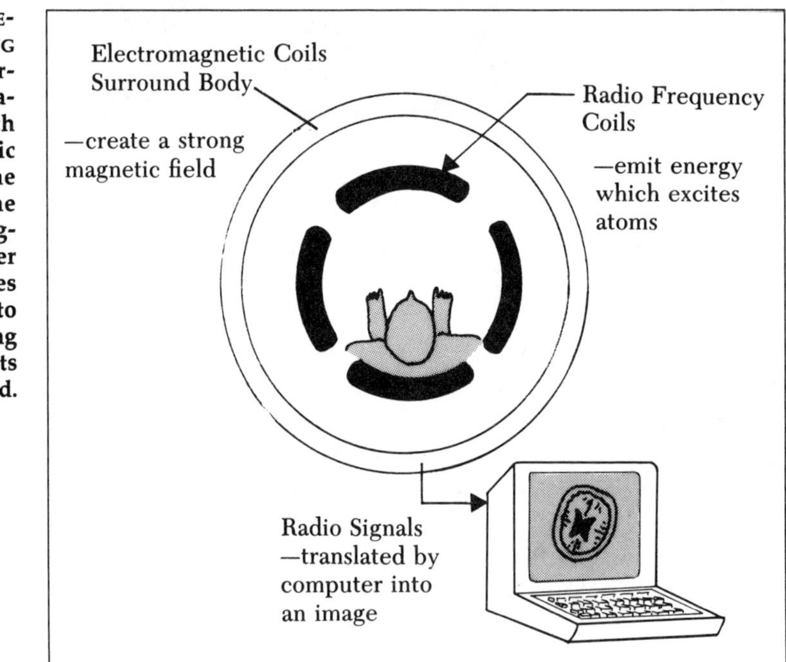

MAGNETIC RESONANCE IMAGING (MRI) involves surrounding a patient's body with electromagnetic coils that excite the atoms within the body to emit signals. A computer then translates these signals into images depicting the body parts being examined.

Of all the body-imaging diagnostic tools currently in use, MRI is considered to be the most promising because it combines the relative harmlessness of ULTRASONICS with the high-resolution imagery of CAT and PET SCANS.

Magnetoencephalography (MEG) A new technology designed to help physicians diagnose neurological disorders. The technology works by measuring the extremely faint magnetic fields produced when nerve cells in the brain fire electrical signals to communicate with one another. Magnetic fields are generated any time electricity flows, whether through a power line, a television set, or your brain. Biomagnetometers, which look like old-fashioned beauty-parlor hair dryers, measure brain-generated magnetic fields that are one ten-millionth the strength of Earth's magnetic field.

The heart of this new technology is called a *s*uperconducting *qu*antum *i*nterference *d*evice or SQUID, which measures weak magnetic fields by harnessing the peculiar properties of ultracold superconducting wire that carries electric currents without resistance. By measuring the magnetic fields with several different SQUID sensors at the same time, biomagnetometers are able to provide three-dimensional pictures of where in the brain any given NEURON electrical activity is occurring. When combined with images made with magnetic resonance technology, MEG promises to give neurologists a view of human brain function never possible in the past. See MAGNETIC RESONANCE IMAGING (MRI), and SUPERCONDUCTIVITY.

Magnetohydrodynamics (MHD) A theoretical revolutionary type of propulsion for ships and submarines that has no moving parts and is virtually silent. Magnetohydrodynamics involves magnetic fields (magneto) and fluids (hydro) that conduct electricity and interact (dynamics). In MHD, magnetic fields are used to move water through a thruster, thus providing forward motion without the use of motors, propellers, gears, or drive shafts. In theory, MHD works like this: A pair of electrodes on either side of the thruster pass an electric current through sea water, while at a right angle to the current is a magnetic field generated by superconducting magnets. The superconducting magnets, more powerful and more efficient than conventional magnets, exert a magnetic force on the sea water passing through the thruster's core, driving the water out the back and creating forward thrust.

This experimental technology is featured in the movie *The Hunt for Red October*, based on the thriller by Tom Clancy, in which an advanced Soviet submarine powered by this process is hunted by American and Soviet navies. It is not just fiction, however, as both Japan and the United States are conducting research on MHD propulsion. See SUPERCONDUCTIVITY.

Magnetosphere See VAN ALLEN BELT.

Mammography Special X-ray technique used to detect breast cancer at a very early stage, before a lump can be felt. X rays are passed through the breast to form an image on a special detector. The doses of RADIATION used in this technique are so low that there is very little risk involved. Breast cancer is the most common form of cancer to strike women, and mammograms are the best method of detecting it at its most curable stage. All women should have a base-line mammogram (for later comparison) between the ages of 35 and 39, screenings every two years between 40 and 49, and annual mammograms thereafter. Some studies even recommend that women in their forties be screened once a year.

Mars The fourth PLANET, counting outward from the SUN, in our SOLAR SYSTEM and our closest neighbor other than VENUS. Mars orbits the Sun at an average distance of 141 million miles, but because of the relatively large eccentricity of its orbit, this distance varies from 129 million miles at PERIHELION to 155 million miles at APHELION. A cold, dry planet swept by hurricane-force dust storms, with an atmosphere only one-hundredth as thick as Earth's, Mars is a small planet with a diameter almost exactly half that of EARTH and only a tenth as massive as Earth. A Martian day (one revolution of the planet) lasts 24.6 Earth hours, and its year (one orbit of the Sun) is equal to 687 Earth days. It is called the *red planet* because of the general appearance of the planet when viewed by optical telescope. What kind of climatic conditions prevail on Mars? Because it lies farther from the Sun than Earth, it is colder; but as it contains an atmos-

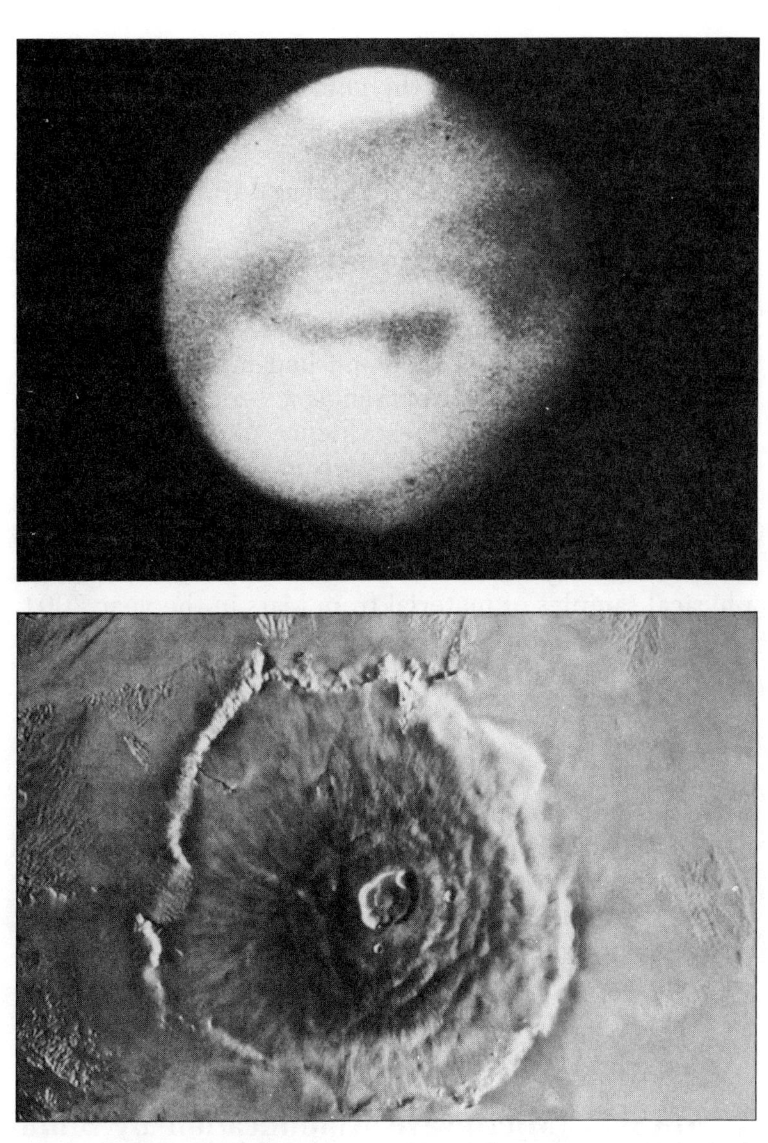

MARS. Top photo shows Mars as seen through a ground-based telescope. Lower photo shows the volcano Olympus Mons as seen from the *Viking* 1 spacecraft.
Source: NASA/USGS [U.S. Geological Survey].

phere, although a thin one, it is not as cold as our Moon. Average Martian temperature is −50°C. Although it is harsh and inhospitable by our standards, the Martian climate is in some ways similar to that of Earth's. Both planets have seasons (although Martian seasons are more extreme and twice as long as Earth's), and both planets have global wind systems marked by trade winds in the tropics and cyclonic storms in the midlatitudes. Major differences lie in the chemical makeup of the atmosphere. Whereas the dominant gases on Earth are NITROGEN and OXYGEN, the atmosphere of Mars is 95.3 percent CARBON DIOXIDE.

Scientists believe that early in the 4.6-billion-year history of the solar system, the climates of Mars and Earth were more similar than they are today. Once warm enough to support flowing water, Mars is now so cold that carbon dioxide freezes at the poles during the Martian winter.

Exploration of Mars by unmanned spacecraft has been extensive. America's *Mariner* 9 orbiter made the first accurate, full-planet map of Mars in 1971. The first landing on the red planet was accomplished in 1976 when the U.S. VIKING 1 and 2 orbiters sent spacecraft to the surface of Mars in search for some evidence of life—they found none. Planned future exploration includes the *Mars Observer,* which will orbit the planet collecting geological and climatological data sometime in 1992, and an unmanned landing craft designed to return to Earth the first physical samples of material from Mars in the year 2001. Landing people on Mars early in the 21st century is also a conceptual possibility and a joint U.S.-Soviet Union effort to accomplish this goal is in the preliminary planning stage.

Mars has two small MOONS, Phobus (from the Greek word for "fear") and Deimos (from the Greek word for "terror"), fitting names for the satellites of a planet named after the mythological god of war. Both moons are only about 120 miles in diameter. Phobus was the objective of two Soviet unmanned spacecraft launched in 1989. Unfortunately both spacecraft experienced technical problems and were lost. This double failure constituted a major setback to Soviet space exploration.

MASER (Microwave Amplification by Simulated Emission of Radiation) Masers are to MICROWAVES what LASERS are to LIGHT. Both devices concentrate diffuse waves of ENERGY into tightly focused beams of the same wavelength, intensifying their power. Focused beams of microwave RADIATION can be used for **RADAR**, because the maser's narrow beam produces sharper radar images than are now possible.

Mass Usually defined as a quantity of MATTER as determined by either its weight or by Newton's second law of motion. Ordinarily, we measure mass as weight and speak of it in terms of pounds or kilograms. But weight is only the result of gravitational force. If there were no gravity, an object would be weightless, but it would still contain the same amount of matter; that is, it would still have mass. Another way of measuring mass is to calculate the amount of FORCE necessary to move, or to accelerate the motion, of a given mass. If an object is placed on a perfectly frictionless horizontal surface, the amount of force necessary to overcome its inertia and cause the object to move can be measured. NEWTON found that the force (f) required to bring about a given acceleration (a) was directly proportional to the mass (m). If you double the mass, it would take double the force. Mathematically, this is expressed in the equation: $f = ma$. This equation is known as Newton's second law of motion. See NEWTONIAN LAWS OF MOTION.

Matter The substance or material of which any physical object is composed. In science, the term is used with regard to any physical substance (i.e., something that occupies space) whether solid, liquid, or gaseous. See MASS.

Mean/Median/Mode *Mean* is the same as *average* and is defined as the sum of all the numbers in a series divided by the number of different entries. The *median* is defined as the middle value in a distribution so that half of the items fall above it and half fall below it. Both mean and median are statistically useful numbers, but we must be careful to distinguish between them. *Mode* is the number or value in a series that occurs most often.

Taken by itself, the mean, or average, can give a distorted view. If, for instance, we take the annual salaries of all the players on a professional football team and compute the mean (add all the salaries and divide by the number of players), it will give an erroneous picture of the average earning. One two-million-dollar-a-year quarterback could be the extreme number that offsets the results. With salaries in particular, where there tend to be extreme scores in one direction or the other, the median is a better indicator of typical earning. If a company advertises that its average salesperson earns *x* amount per year, it might be better to take a look at the median income before applying for the job. As Benjamin Disraeli once pointed out, there are "lies, dammed lies, and statistics."

111222234455667		
Mean – the average of the numbers in the series above		3.4
– add all the numbers	= 51	
– divide by the number of numbers in the series	= 3.4	
51 ÷ 15 = 3.4		
Median – middle number in the series. There are seven numbers before 3 and seven after 3.		3
Mode – number that occurs most often		2

Melanoma The often lethal form of **SKIN CANCER**, which has increased in the United States by an alarming 1,250 percent since 1935. Once extremely rare, melanoma is now the country's ninth most common form of cancer, increasing faster than any other form of the disease. By the year 2000 an estimated 1 in 90 Americans will develop melanoma during their lifetimes, versus only 1 in 1,500 in the 1930s. One in four victims will be age 39 and under. One reason for this deadly epidemic is that people began spending more time outdoors in the 1960s and 1970s. Another reason is the depletion of the Earth's protective **OZONE** layer. **CHLOROFLUORCARBONS** (**CFCs**) and other synthetic chemicals are known to be the major cause of ozone depletion; a halt to all production and use of these harmful agents is the goal of world environmentalists. Most researchers now believe that there is a clear cause and effect relationship between ozone loss and increased melanoma mortality. They calculate that a 7-percent ozone loss worldwide could translate into nearly a 10-percent increase in skin cancer mortality.

Mendel, Gregor An Austrian monk who, in 1859, demonstrated that traits are passed from parent to offspring by means of discrete units that he called *factors*. These are what are now called **GENES**—the basic unit of heredity. Mendel used garden peas to demonstrate his thesis, and his work laid the cornerstone for modern genetics. See **GENETIC CODE**, and **GENETIC ENGINEERING**.

Mercury With a diameter of about 3,100 miles, Mercury is the smallest of the nine **PLANETS** in our **SOLAR SYSTEM**. It is also the closest to the **SUN** and because of this receives almost seven times as much **LIGHT** and **HEAT** as does the **EARTH**. This intense **RADIATION** raises the surface **TEMPERATURE** of Mercury's day side to as high as

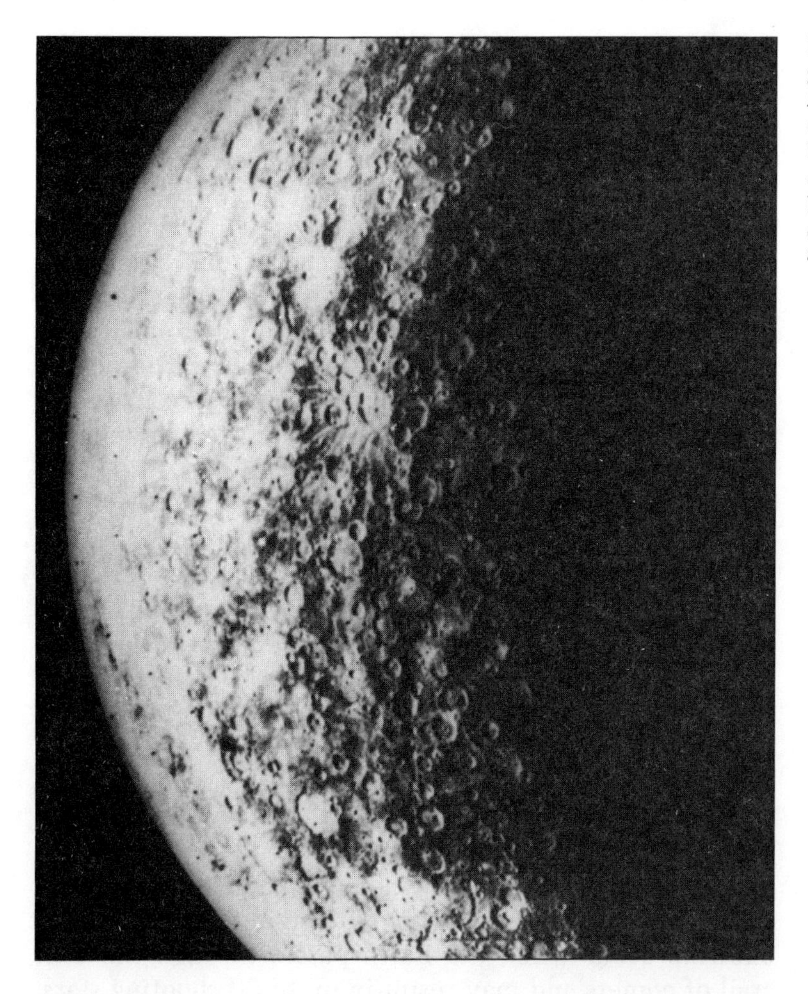

MERCURY as photographed by the *Mariner 10* spacecraft in 1974. Like Earth's moon, Mercury is an airless body covered by craters.
Source: NASA photo.

800°F (400°C). Mercury's orbit is highly eccentric, so that the planet approaches within 29 million miles of the Sun at PERIHELION and moves on out to 43 million miles at APHELION. Mercury rotates on its axis once every 59 Earth days and takes 88 Earth days for one revolution around the Sun.

Mariner 10 spacecraft photographs taken in 1974 have provided the best views of this very hot planet. Like Earth's MOON, Mercury is an airless body covered by mountains and craters. Mercury also shows some lines of cliffs, as though the planet shrank when it cooled soon after its formation.

Mesozoic Era Called the age of reptiles, the Mesozoic ("middle life") era occurred between 65 and 230 million years ago, when a long decline in the Earth's TEM-

PERATURE started. The Mesozoic era covered a span of perhaps 165 million years of Earth's 4.6 billion year history. The Mesozoic is divided into three ages: the TRIASSIC, the JURASSIC, and the CRETACEOUS. Dinosaurs arose in the Triassic age and reached their largest numbers in the Cretaceous. It was during the Jurassic that the earliest mammals and birds developed, each from a separate group of reptiles. The end of the Cretaceous age (65 million years ago) marks a period of time sometimes called the *Great Dying*, when as much as 75 percent of all life on Earth, including the dinosaurs, died out. The Mesozoic era was followed by the Cenozoic era, or age of mammals, which is the world we know today. See **GEOLOGIC TIME SCALE**, and **K-T EXTINCTIONS**.

Meteoroids, Meteors, and Meteorites A meteoroid is any of the small bodies, often remnants of COMETS, traveling through space. When such a body enters the EARTH's atmosphere it is heated to luminosity and becomes a meteor. If this body of stone or metal impacts the Earth, it is called a meteorite. Most of us have seen *shooting stars* in the night sky. They are caused by particles of space debris that enter Earth's atmosphere at very high speeds and burn up from the friction of the atmosphere at heights of about 50 miles above the ground. Most of these particles are no larger than grains of sand. Meteor showers occur when the Earth passes through the dust trail of comets and may result in up to 50 shooting stars. On an ordinary night when there are no "showers," an observer can expect to see between three and six meteors per hour. Though most meteoroids are destroyed well before they reach the ground, some chunks of matter, those the size of baseballs or larger, do survive their trip through the atmosphere.

In ancient China, meteors were held to be messengers from heaven, the importance of their message indicated by their brightness and speed. In this country, the Wintu Indians of early California believed that meteors were the spirits of departed shamans on their way to the afterlife. See also **ASTEROIDS**.

Methane (CH_4) A colorless, odorless, flammable gas and the major constituent of natural gas. Methane arises from rice paddies, forest fires, termites that abound in deforested areas, coal mines, landfills, and from the digestive tracts of animals, including the estimated 3.3 billion domestic cattle, sheep, goats, and camels that have proliferated around the world as human societies have expanded. Scientists now tell us that, because of human activity, methane in the atmosphere has doubled in the last 300 years, from 650 parts per billion by volume to 1,700. Methane is increasing at a high rate of 1 percent per year and is considered a major contributor to the GREENHOUSE EFFECT. Its present concentration could account for 20 to 25 percent of whatever global warming results from greenhouse gases (CARBON DIOXIDE would be responsible for 55 percent while CHLOROFLUORCARBONS are blamed for most of the remaining 20 to 25 percent).

Of the major greenhouse gases, methane may be the easiest to control. Its lifetime in the atmosphere is a relatively short ten years, and less reduction (only 10 to 20 percent) of it is necessary to stabilize or cut global levels. EPA scientists estimate that it would require a reduction of 50 to 80 percent to stabilize carbon dioxide levels, and an almost 100-percent cut to stabilize chlorofluorcarbons.

Some scientists now believe that there is a close link between methane levels and climate shifts over the past 160,000 years. They argue that fluctuations in carbon dioxide and methane concentrations, as determined from ice core samples, were probably responsible for 50 percent of the cases of dramatic warming and cooling of the Earth during the last two ICE AGES. See also CLIMATE.

Methanol A colorless, flammable liquid used as an antifreeze, general solvent, and fuel; also called methyl alcohol. Methanol is considered one of the possible alternatives to gasoline as a fuel for automobiles. Methanol is a high-octane fuel used at the Indianapolis 500 (where cost is not a factor). Air-quality regulators like it because it reduces smog-forming emissions by as much as 50 percent. Disadvantages include a shorter cruising range, limited production capabilities, and safety questions including corrosive qualities, as well as eye irritant emissions. See FUELS, ALTERNATE.

Microlithography A SEMICONDUCTOR manufacturing technology, microlithography involves the use of optical, electron beam, or laser-generated X rays to produce *masks* from which microchips are made. In this process, manufacturers use masks etched with the desired circuit design. Circular silicon wafers, which are eventually cut up into 100 or more chips, are coated with a light sensitive material, or *resist.* Technicians then shine visible LIGHT or X rays through the mask onto the wafer, exposing areas of resist. Washing the wafer with a solvent dissolves the unwanted resist and leaves a copy of the mask design on the wafer. A completed CHIP will be a sandwich of as many as 20 circuit layers. See INTEGRATED CIRCUITS.

Microwave An electromagnetic wave of extremely high frequency. A microwave oven uses these high-frequency waves to penetrate and heat food. Microwaves heat materials by vibrating the bonds between ATOMS, causing friction; the resulting HEAT causes food to cook in a relatively short time.

The phenomenon of microwave cooking was discovered by accident when a scientist doing some research involving microwaves discovered that a chocolate bar in his shirt pocket had melted and began to investigate.

Milky Way Our SUN and all the other STARS that we can see without a telescope are part of a much larger group, called the Milky Way Galaxy, which contains about a trillion (10^{12}) stars plus much gas and dust in between the stars. Most of the stars in the Milky Way are in a giant disk 100,000 light-years across; that is, light traveling at 6 trillion miles per hour would take 100,000 years to cross the Milky Way Galaxy. The fact that the Milky Way appears as a narrow band in the sky tells us that we live in a flat galaxy shaped something like a dinner plate. Our solar system resides about 30,000 light-years from the center of this giant dinner plate—two-thirds of the way out toward the rim. Despite the vastness of our galaxy, it is tiny compared with the dimensions of the universe at large. There are at least ten billion other GALAXIES in the universe like our Milky Way. Some are bigger, like the huge galaxy discovered in 1990, some smaller like the two MAGELLANIC CLOUD galaxies that are the Milky Way's neighbors. Galaxies are usually grouped into clusters and the Milky Way Galaxy is part of such a cluster, a relatively small one called the LOCAL GROUP. See UNIVERSE, EXTENT OF.

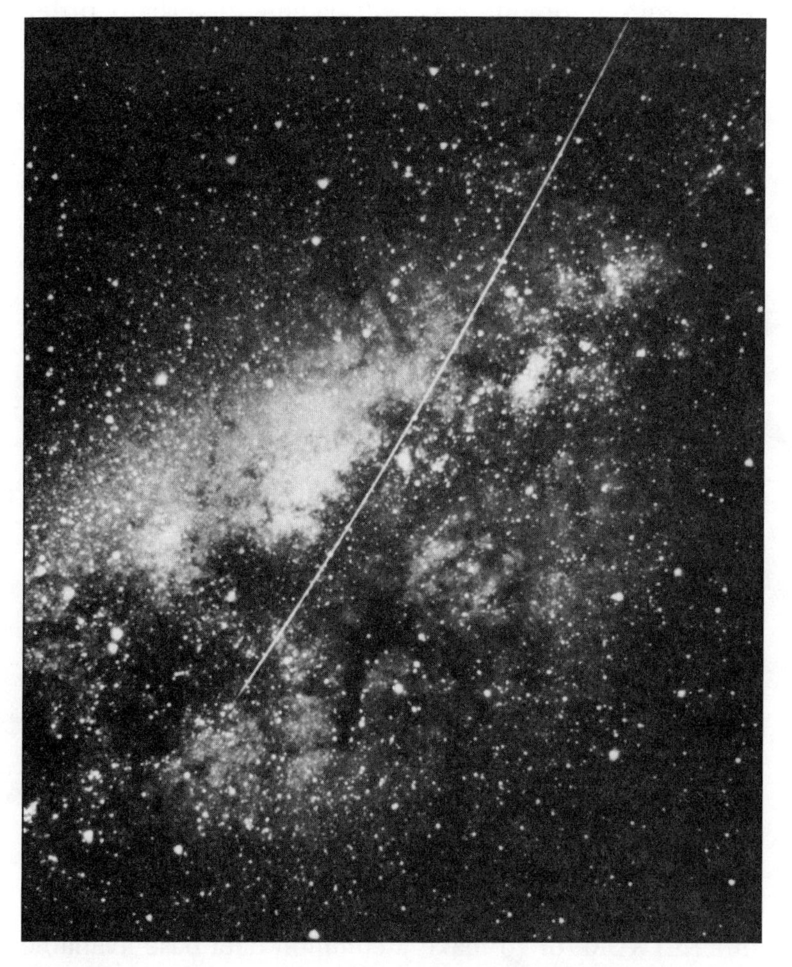

MILKY WAY Galaxy containing about a trillion (10^{12}) stars. The streak of light in this photo was caused by a meteor.
Source: NASA photo.

Mips *M*illions of *I*nstructions *P*er *S*econd is a standard measure of COMPUTER capability. The most powerful desktop computers today rarely exceed 20 MIPS, but if the current pace of development continues, future computers will be capable of much more. Affordable computer CHIPS capable of handling 100 MIPS should be possible by 1996. The consequences of this advancement will enable home computers to process instructions and represent data in ways not possible today. SPEECH RECOGNITION by computers, for example, is currently limited to special use or laboratory experiments. This is because it takes a very large amount of processing power to interpret speech or handwriting, both of which would be more natural means of communicating than typing or using a MOUSE. In the future, with 100 MIPS built into our television set/computer we will be able to verbally order the set to change channels or turn itself off.

Modeling, Mathematical The mathematical representation of a process, concept, or operation of a system, usually implemented by a COMPUTER program. Mathematical models are often used in simplified simulations of complex systems or phenomena such as in science or economics. CLIMATE forecasting is a good example of the use of mathematical models. The number and interaction of the variables—wind, CLOUDS, TEMPERATURE, and the like—are so complex that climatologists use computer models to imitate the actual global climate. Computer models also help scientists project consequences of hypothetical changes; "what if" scenarios can be considered without actually changing the process or system under examination. Sophisticated computer-based simulation models provide researchers important insights, but their outputs are only as valid as the assumptions that go into them. These assumptions, including the variables and the functional relationships of the variables, may not be a true reflection of reality.

MODEM (*Mo*dulator/*Dem*odulator) An electronic device used in the transmission of data to or from a COMPUTER via telephone or other communication lines. **MODEMs** are used to link a group of computers together in a NETWORK or to make a common data base available to a number of computers. **MODEM** capability is usually measured in units called *bauds*—equal to the number of pulses or bits of information per second. The higher the baud rate, the more capable the modem.

Molecule Made up of ATOMS, a molecule is the smallest physical unit of a particular ELEMENT or COMPOUND. For instance, a molecule of water (H_2O) is the ultimate, indivisible unit of water. It can, of course, be broken down into atoms of HYDROGEN and OXYGEN, but then it is no longer water. Branches of science that deal with MATTER or phenomena at this level are called *molecular* sciences. Examples are molecular genetics, molecular biology, or molecular astronomy.

Momentum The motion of a body or system equal to the product of the MASS of a body and its velocity. GALILEO observed that the velocity of a body, in the absence of any external force, is constant. A century later NEWTON made this his first law of motion (also called the principle of inertia): Every body persists in a state of rest or of uniform motion in a straight line unless compelled by an external force to change that state. See **GALILEI, GALILEO**; and **MOTION**.

Monera One of the five taxonomic kingdoms of living organisms. Monera comprise BACTERIA, blue-green algae, and related forms. The other four taxonomic kingdoms are ANIMALIA, FUNGI, PLANTAE, and PROTISTA. See **TAXONOMY**.

Moon A natural SATELLITE revolving around a planet. Of the nine PLANETS in this SOLAR SYSTEM, only

MOON as viewed by *Apollo 11* during its journey back to Earth after the historic lunar landing in 1969. When this picture was taken the spacecraft was already 10,000 nautical miles from the Moon. Aboard *Apollo 11* were astronauts Neil A. Armstrong, Michael Collins, and Edwin L. Aldrin.
Source: NASA photo.

MERCURY and VENUS do not have at least one moon. Our Moon orbits the EARTH (which at the same time is orbiting the SUN) at an average distance of 238,000 miles (384,000

MOON and its complicated path around the Sun. The Moon travels in an almost circular path around the Earth while at the same time the Earth moves in a nearly circular path around the Sun. Both motions combine to give the Moon a wavy path causing the Moon to appear to observers on Earth in phases—first a sliver of new moon, then a quarter moon, and finally a full moon.

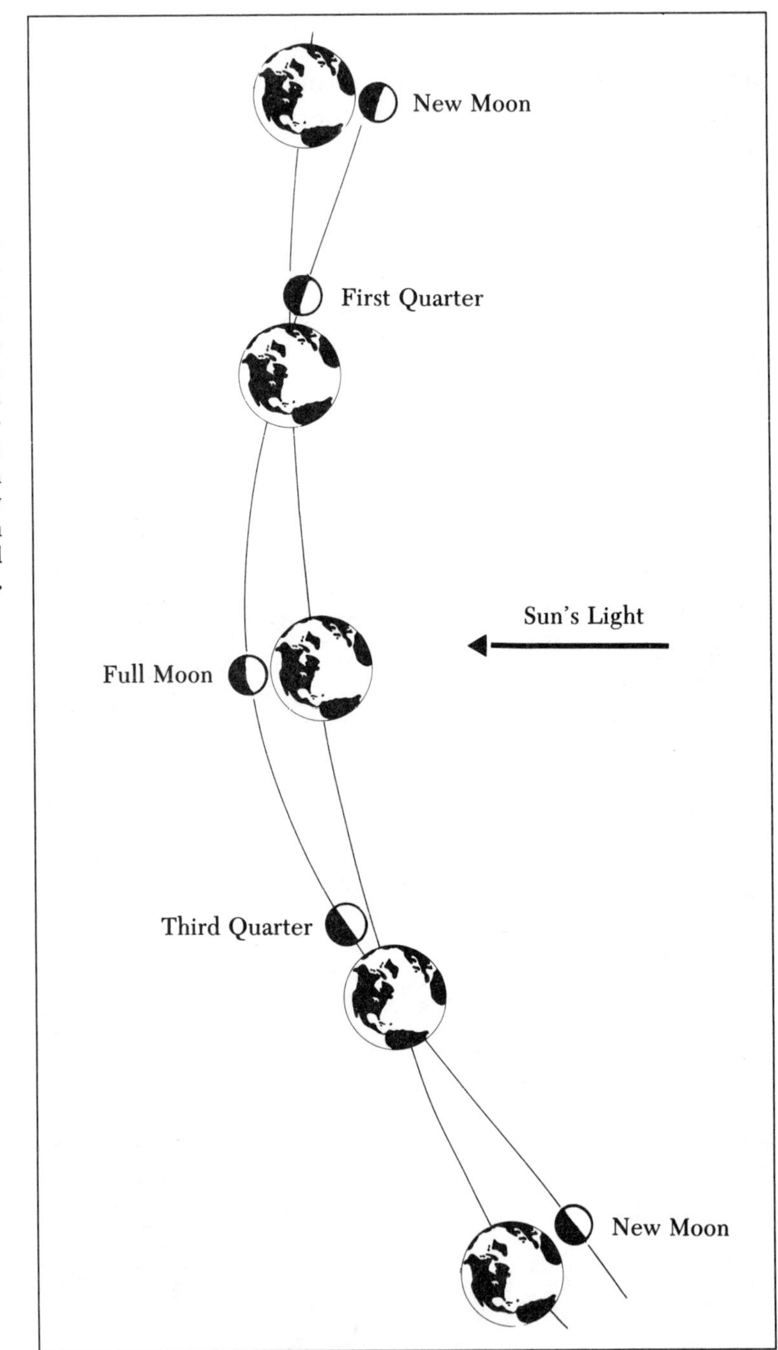

kilometers). Our Moon takes 27 days, 7 hours, and 43 minutes to complete an orbit (its SIDEREAL period). The Earth has moved in that time, so it takes 29⅓ days for the Moon to come back to the same place in our sky. During that 29⅓-day period, the Moon goes through a cycle of phases. The phase we see depends on the relative positions of the Earth, Sun, and Moon. When the Moon is on the far side of Earth from the Sun, we see all of the half that is sunlit and we call this phase a *full moon.* When the Moon is halfway around its orbit, in more or less the same direction from us as the Sun, we have what we call a *new moon.* The Moon's orbit around Earth is elliptical, so at different times it appears to be different sizes because its distance from the Earth varies. In fact, the diameter of the Moon is 2,160 miles, or just over a quarter that of Earth. The gravitational force on the Moon is only one-sixth of Earth's.

The main cause of TIDES on Earth is the gravitational pull of the Moon. The Moon's GRAVITY pulls water on the near side of the Earth toward the Moon and pulls the solid body of Earth away from water on the Earth's far side. Thus, in most places on Earth, we have two high tides every day. The Moon has been explored by a series of American unmanned and manned spacecraft, including the *Apollo* missions that brought 12 astronauts to the Moon during 1969 to 1972 (Soviet unmanned *Lunik* spacecraft have also visited the Moon). In addition to taking many measurements on the Moon, the U.S. astronauts brought back a total of 842 pounds (382 kilograms) of rocks to be analyzed on Earth. The many craters on the Moon were formed by meteorites—rocks from outer space hitting the Moon's surface, which is unprotected by any atmosphere. The lack of any atmosphere is judged to be the main cause of the extreme TEMPERATURE changes on the Moon, which range from 230°F during the lunar day down to around −230°F during the lunar night. No trace of water has been found on the Moon nor any evidence that water ever existed. There is no life on the Moon or any sign of chemicals that may be related to life. See **LUNAR ECLIPSE; METEOROIDS, METEORS, AND METEORITES;** and **SYNODIC PERIOD.**

Moons The natural SATELLITES orbiting around planets:

Planet	*Moons*
MERCURY	None
VENUS	None
EARTH	The Moon
MARS	PHOBOS and Deimos
JUPITER	Io, Europa, Ganymede, Callisto
SATURN	Mimas, Enceladus, Tethys, Dione, Rhea, TITAN, Iapetus, and more than a dozen smaller moons
URANUS	Miranda, Ariel, Umbriel, Titania, Oberon, plus at least ten smaller moons
NEPTUNE	TRITON, Nereid, plus smaller moons
PLUTO	CHARON

Motion Since everything in the universe is moving—ATOMS and MOLECULES, STARS and PLANETS, the EARTH and its surface—motion must be considered relative to whatever point or object we choose. A jet aircraft may be flying through the sky at a speed of 450 miles per hour with reference to the ground, but the surface of Earth is moving at about 1,000 miles per hour around the center of Earth. A passenger on the jet walking up the aisle of the aircraft will be moving at one speed relative to the airplane, another with respect to the ground below, and still another with respect to the Earth's center. See EINSTEIN, and RELATIVITY.

Mouse, Computer A COMPUTER control device consisting of a rolling ball and one or more buttons used to execute commands or select text or graphics. As the mouse is moved around a flat surface, the ball underneath moves and sends what are called *x* and *y* coordinate signals to the computer. Moving the mouse causes the cursor (the flashing indicator on the screen) to travel in a corresponding manner. The advantage of the mouse, as compared to the use of the keyboard, is the speed with which the cursor can be maneuvered.

Mouse, Genetically Engineered In 1988, the U.S. Patent Office issued a patent for a genetically altered mouse, the first time that a patent had been granted for an animal. The mouse had been genetically altered to aid in cancer research and as such was the first human-designed mutant. Critics of GENETIC ENGINEERING opposed the Patent Office action, claiming that the move would have profound economic, environmental, and ethical consequences. Since 1988, several other new species of mice have been custom-designed to aid in medical research. The most important of these new developments was the engineering in 1989 of a species of mice bearing human IMMUNE SYSTEMS. As a result of this important advancement, medical research into AIDS and other fatal diseases will be significantly accelerated. See GENETIC ENGINEERING.

Muon A short-lived elementary subatomic PARTICLE with a negative electrical charge. Muons are LEPTONS similar in most respects to the ELECTRON except that they are unstable and are 207 times more massive. A positively charge muon is the antiparticle of a negatively charged muon. See SUBATOMIC STRUCTURE.

Nanotechnology The technology of the very small (*nano* comes from the Greek word for "dwarf"). In nanotechnology, scientists envision using MOLECULE-sized machinery to control the structure of MATTER even at atomic levels. Proponents of this new science argue that designing machines on the nanometer (one-billionth of a meter) scale for directly assembling molecular and atomic components is possible at least in principle. Researchers at IBM's Almaden Research Center in San Jose, California, have used the tip of a SCANNING TUNNELING MICROSCOPE to crudely position individual molecules on a surface. Future nanotechnology could include nanomachines that extract pollutants from the atmosphere or that reverse biological mistakes such as cancer, perhaps by traveling into diseased cells and repairing them. Nanotechnology is in its infancy and there are skeptics who question the possibility for eventual useful capabilities.

Natural Gas A combustible mixture of gaseous hydrocarbons consisting of more than 80 percent METHANE together with minor amounts of ethane, propane, butane, NITROGEN, and sometimes helium. As an alternative to gasoline as a fuel for automobiles, natural gas offers a number of advantages. It has a higher octane rating (130) and costs 50 to 80 cents per gallon equivalent. It is also plentiful, clean, easily distributed, and safer than gasoline. Compared with gasoline, it produces lower levels of NITROGEN OXIDES and CARBON MONOXIDES and much lower levels of smog-causing hydrocarbons. Problems and limitations have so far prevented natural gas from supplanting gasoline. For example, natural gas contains less energy per gallon than gasoline, limiting the range for most cars to about two-thirds that of a full tank of gasoline. This means more frequent stops for fill-ups and the necessity to be near a refueling station. For fleet vehicles such as buses, taxis, and city delivery vans, this isn't as great a problem. They can refuel in their own garages. Vehicles that return to their home bases each day may be the first automobiles to be converted to natural gas. The main barrier to complete conversion of all U.S. automobiles is the large one-time cost—around $1,500. Few private cars are driven far enough to make conversion economically practical at today's cost of fuel. A private vehicle would have to consume 1,500 to 2,000 gallons a year, or more, to make the investment pay off—in comparison to the 500 gallons per year for the average motorist today. Moreover, the price of natural gas would probably rise, from either higher demand or tax hikes, if cars were converted in significant numbers. See **FUELS, ALTERNATE.**

Natural Selection The tendency of those individuals better suited to their environment to survive and perpetuate their species, leading to changes in the genetic makeup of the **SPECIES** and, eventually, to the origin of new species. An often-used example of this process is the finch birds of the Galápagos Islands. Those birds particularly adept at obtaining seeds or those able to get new kinds of food were the ones that survived and reproduced. A bird with a slightly longer and thinner bill, for example, could reach food that others could not. These birds, and their descendants, would gain in numbers at the expense of the other varieties of finch. **DARWIN** compared natural selection to *artificial selection* as used in animal breeding. Biologists have accepted the basic premise that species gradually change, but the mechanism for biological inheritance is imperfectly understood. Today the debate is no longer about whether **EVOLUTION** occurs but about the details of the mechanisms by which it takes place. Evolution is a fact amply demonstrated by the **FOSSIL** record and by contemporary molecular biology. Natural selection is a successful theory devised to explain the fact of evolution.

Neanderthal An early, humanlike creature, who flourished from about 130,000 years ago until about 35,000 years ago. Neanderthals were short, squat bipeds, the men averaging a little over five feet, the women a little shorter. Neanderthals walked erect, made tools, and lived in social groups. Neanderthals also introduced ceremonial internment of the dead, suggesting a highly developed religious system. Skeletal traces of Neanderthal man have been found all over Europe, Asia, and Africa. Most anthropologists speculate that we are not descended from Neanderthals. For a long time it was believed that Neanderthals were exterminated by our own species, HOMO SAPIENS, but there is no evidence to support this conjecture. One view is that Neanderthals mixed with and interbred with other humanlike creatures, including anatomically modern humans, and in the course of time were absorbed by such populations. See EVOLUTION.

Nebulae Term once used by astronomers for any celestial object that appeared *nebulous,* hazy, or fuzzy in a telescope's view. Nebulae is now used in reference to a cloud of interstellar gas and dust. *Bright* nebulae glow with light emitted by the gas of which they are composed. *Dark* nebulae consist of clouds of gas and dust that are not as illuminated.

Nematodes A type of worm often used in biological research projects. The term refers to any unsegmented worm having an elongated, cylindrical body, such as the common roundworm.

Nemesis (Death Star) The theoretical companion star to the SUN, thought by some astronomers to lie in orbit somewhere between PLUTO and ALPHA CENTAURI. The controversial Nemesis theory postulates that this so-far undetected companion star is in an orbit around the Sun that takes about 30 million years to complete. Proponents of this theory think that periodic visits close to our SOLAR SYSTEM from Nemesis could account for the mass extinctions of many SPECIES of life on Earth that have occurred at intervals in the history of this planet. The assumption is that, by passing through the OORT CLOUD of COMETS that lies beyond Pluto, Nemesis could cause a storm of millions of comets, some of which would impact EARTH with disastrous consequences. The companion star theory is, in fact, not illogical. The majority of stars in the MILKY WAY galaxy are *double stars* that orbit around each other constantly, so our Sun should be no exception. See DINOSAURS, EXTINCTION OF.

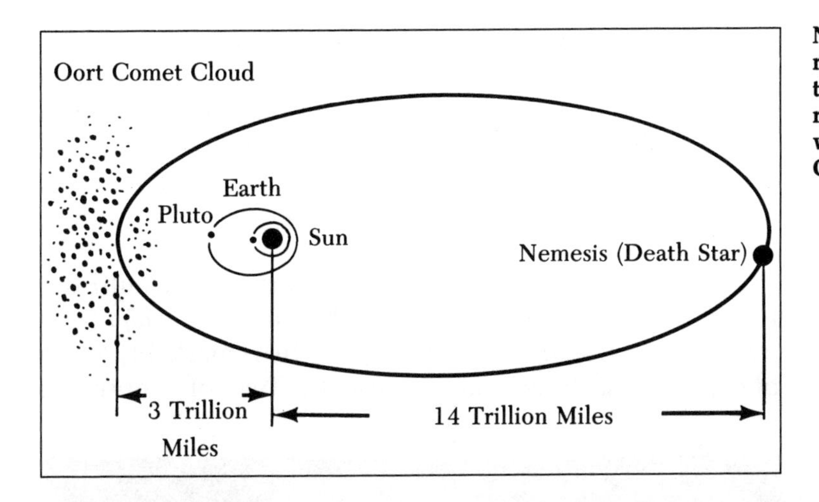

NEMESIS. The theorized orbit of a star that may cause a rain of comets when it passes the Oort Cloud.

Neocortex The outermost, evolutionarily most recent, layer of the brain. This critical part of the brain is the seat of consciousness and complex thought. Like the higher mammals and the other PRIMATES, humans have a relatively massive neocortex. It becomes progressively more developed in the more advanced mammals. The most elaborately developed neocortex is that of humans (along with those of whales and dolphins).

It is the neocortex that enables a person to be aware and respond to what surrounds him or her. The neocortex allows us to recognize one another, speak, and make plans. The neocortex is often discussed in terms of four major regions or lobes: the *frontal, parietal, temporal,* and *occipital* lobes. It is not clear that these subdivisions are separate functional units. Each seems to have many different functions and some functions are shared among or between lobes. Generally, the frontal lobe seems to be connected with deliberation and the regulation of action; the parietal lobes with spatial perception and the exchange of information between the brain and the rest of the body; the temporal lobes with a variety of complex tasks; and the occipital lobes with vision.

Neptune Although normally considered the eighth PLANET out from the SUN and, with a diameter of 30,700 miles, the fourth largest, Neptune is at this time the most distant planet in our SOLAR SYSTEM (PLUTO's eccentric orbit has at this time taken it inside Neptune's orbit). Neptune orbits at an average distance of 2.8 billion miles (4.5 billion kilometers) from the Sun and takes 165 EARTH years to complete its long journey. Neptune's day (one complete rotation) lasts 16 hours and 7 minutes. The planet is a huge ball of water and molten rock cloaked in an atmosphere of HYDROGEN and HELIUM mixed with METHANE. Neptune's great dark spots and some variable clouds near it are evidence of a powerful, turbulent weather system.

Little was known about this mystery planet until the U.S. spacecraft *Voyager* 2 encounter in 1989. *Voyager* 2 was able to obtain and relay to Earth remarkably clear photos of Neptune and its MOONS. Two of Neptune's

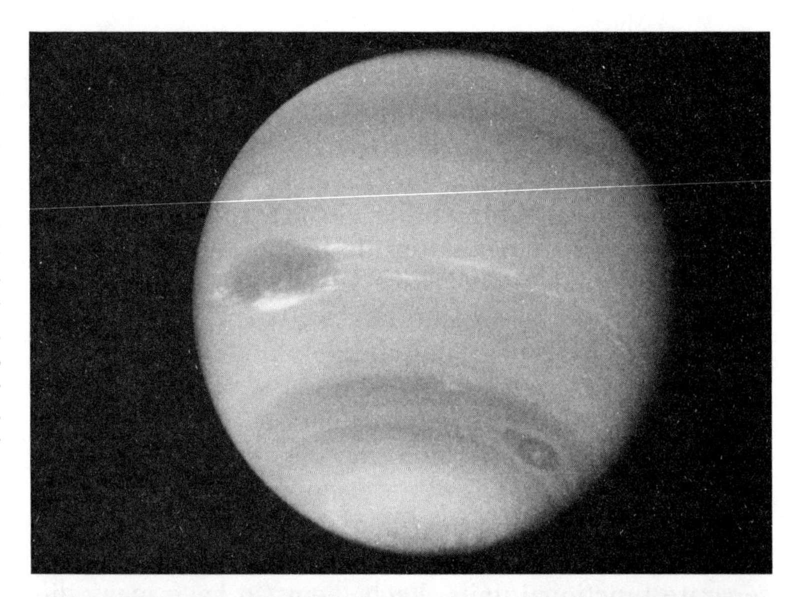

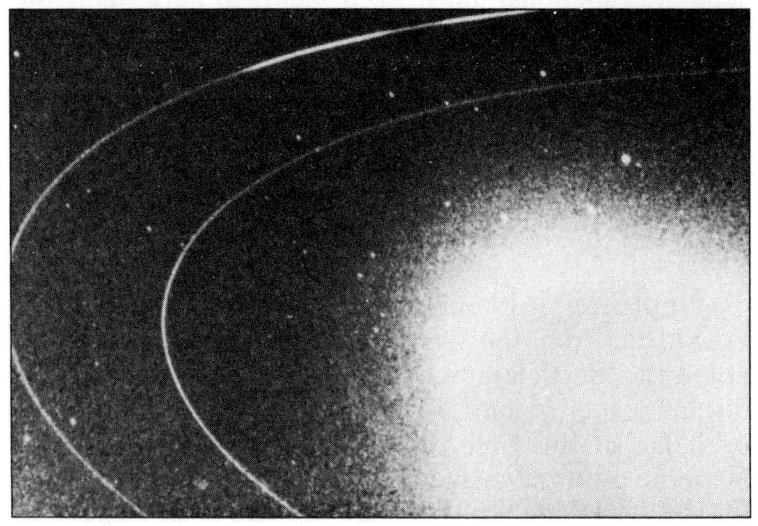

NEPTUNE and Neptune's rings as photographed by the *Voyager* 2 spacecraft in 1989. Neptune's great dark spots and some variable clouds near it are evidence of a powerful, turbulent weather system. Neptune's rings were not discovered until 1989. *Source: NASA photos.*

moons, TRITON and Nereid, were known from terrestrial observations. *Voyager* 2 found six new small satellites. See VOYAGER SPACECRAFT.

Neural Networks COMPUTER systems designed to mimic biological systems, neural networks are a growing branch of ARTIFICIAL INTELLIGENCE. An advanced form of computer programs that, like the human brain, can detect patterns in data, neural networks can learn new information and often make inferences from that information. Because neural networks can, in theory at least, use information in different ways in different situations, they are analogous to a biological organism that can behave differently in different environments.

Neural networks are good at solving problems involving multiple constraints such as image processing. Pattern recognition, for example, is easy for a neural network because such networks can process many things in parallel, like the human brain.

Neurons The impulse-conducting CELLS that are the functional units of the nervous system. The human brain contains about 100 billion of these switching elements. Most neurobiologists today believe that neurons are the active elements in brain function. For every neuron in the brain, there are roughly ten glial cells (from the Greek word for "glue"), which hold the neuronal structure together. An average neuron in the human brain has between 1,000 and 10,000 SYNAPSEs or links with adjoining neurons. It is now known that the brain contains microcircuits very small in size (about 1/10,000 of a centimeter) made up of neurons and synapses able to process data very rapidly. These microcircuits seem to increase in abundance with the complexity of a particular life form. They also develop late in human embryology. The existence of such microcircuits suggests that intelligence may be the result of the abundance of specialized switching elements in the brain. See NEUROTRANSMITTERS.

Neurotransmitters Any of several chemical substances that transmit nerve impulses across a SYNAPSEs, or gap, to other elements of the body such as another nerve, muscle, or gland. The most well known of the various neurotransmitter chemicals are acetylcholine, serotonin, and adrenaline. Some brain disorders are thought to be caused by the defective production of certain neurotransmitters, which results in faulty transmission of information from one nerve cell to another. An example is the relationship between the production of dopamine and PARKINSON'S DISEASE. See NEURONS.

Neutrino Massless, electrically neutral subatomic particle (actually a type of LEPTON) emitted by the SUN and other celestial objects. Neutrinos pass through most matter as if it were not there. An estimated ten million neutrinos will pass harmlessly through your body in the time it takes to read this sentence. Neutrinos travel at the speed of light, so by the time you finish this paragraph, the neutrinos that passed through your body will be farther away than the Moon. The name for this subatomic particle was coined by the great Italian atomic physicist, Enrico Fermi (1901–54), to describe a PARTICLE that was neutral in charge and had no MASS.

When a massive star in the large MAGELLANIC CLOUD exploded in 1987 (and thus became a SUPERNOVA), a wave of neutrinos was detected on EARTH, confirming the theory that supernova generate enormous quantities of neutrinos. The new science of observational neutrino astronomy thus was originated at that time. See SUBATOMIC STRUCTURE.

Neutron The ATOMS of all ELEMENTS except HYDROGEN have PROTONS and neutrons in their nuclei, with the total number of these PARTICLES roughly proportionate to the weight of the atom. (Hydrogen consists of a nucleus of one proton with one ELECTRON orbiting around it.) Like protons, neutrons are made up of QUARKS, the basic building blocks of nature. Neutrons have no electrical charge. See NUCLEUS, ATOMIC, and SUBATOMIC STRUCTURE.

Neutron Stars See PULSARS.

Newton, Isaac (1642–1727) One of the greatest scientific minds of all time, this English physicist provided thinking humankind with a new concept of the universe, a concept based on a mathematical view of the world that brought together knowledge of the motion of objects on

Earth and the distant motions of celestial bodies. In his *Mathematical Principles of Natural Philosophy,* published near the end of the 17th century and destined to become one of the most influential books ever written, Newton used a few key concepts (MASS, MOMENTUM, ACCELERATION, and FORCE), three basic laws of motion, and the mathematical law of how the force of gravity between all masses depends on distance to provide a rigorous explanation for both the movement of the PLANETS and STARS as well as the movement of objects on EARTH.

Newton's book, often referred to simply as the *Principia,* established the basic laws of physics. He was able to account for the observable orbits of planets and MOONS, the motion of COMETS, the motion of falling objects at the Earth's surface, weight, ocean tides, and Earth's slight equatorial bulge. In short, he made the universe understandable. It was a universe that ran automatically by itself according to the action of forces between its parts, and Newton's brilliant insight made it all clear for the first time. Newton demonstrated that the force of GRAVITY is proportional to the product of the two masses and inversely proportional to the square of the distance between them. In other words, the larger a body is the greater the gravitational pull, and as the bodies are moved farther apart the force of gravity diminishes rapidly. To carry out his calculations of bodies in motion, he invented what we now call *differential calculus.* Also, almost as a side issue, Newton formulated the mathematical laws governing the way colors combine to make up white LIGHT and showed how a prism can be used to demonstrate this phenomenon.

Newton's achievements were made possible by the pioneering work of three illustrious predecessors, COPERNICUS, KEPLER, and GALILEI, GALILEO. He acknowledged this debt in his often quoted remark, "If I have seen farther than other men, it is by standing on the shoulders of giants." Newton's concepts prevailed as the scientific and philosophical view of the world for more than 200 years. EINSTEIN's theories of relativity do not overthrow the world of Newton but rather modify some of its most fundamental concepts—particularly in regard to the motion of PARTICLES inside an atom.

Newtonian Laws of Motion Newton's three laws of motion express the relationship between FORCE and motion. They state that: (1) objects in motion or at rest tend to stay that way unless acted upon by an outside force (the law of inertia); (2) the ACCELERATION of a MASS by a force is inversely proportional to the mass and directly proportional to the force; and (3) for every action there is an equal and opposite reaction. See **NEWTON**.

Nitrogen A colorless, odorless, gaseous element that makes up more than 78 percent of EARTH'S ATMOSPHERE. Nitrogen is present in animal and vegetable tissue. Proteins, which compose the greatest part of our body's tissues, are compounds of nitrogen that comes from plants, or from animals that eat plants. Nitrogen is used to manufacture ammonia, nitric acid, fertilizer, dyes, and as a cooling agent. See **NITROGEN OXIDES**.

Nitrogen Oxides (No_x) A major air pollutant from the exhaust of INTERNAL-COMBUSTION ENGINES and other industrial sources. Nitrogen oxides have a direct and deleterious effect on our health. It has been shown that prolonged exposure to NO_x greater than 0.5 parts per million appears to be particularly hazardous for persons with asthma, chronic respiratory diseases, and cardiac disease. People living in the Los Angeles area are often exposed for long periods of time to NO_x concentrations above what is considered safe. Currently, the Clean Air Act limits the automobile tailpipe emission of NO_x to 1.0 gram per mile and calls for a 60-percent reduction in these emissions in many new cars by 1994 and in all new cars by 1996.

Nuclear Energy The United States now obtains some 20 percent of its electrical ENERGY from 111 nuclear reactors. Some other industrial countries rely much more heavily on nuclear power generation than we do: France, 70 percent; Belgium, 66 percent; South Korea, 53 percent. The U.S. Department of Energy (DOE) pointed out in mid-1990 that the United States is running short of electrical power and may have to learn to accept new nuclear power plants to avert a crisis. Energy Secretary James Watkins said that by conservative estimates the United States will need an additional 1,000 gigawatts (a *gigawatt* is equal to 1 billion watts) of electrical generating power by the year 2000, but only 40 percent of that is accounted for under current plans for new power plants.

The DOE has urged that the United States follow the French model to accelerate the protracted process by which licenses are granted to build new nuclear plants. France made its major shift toward nuclear power by choosing a single model plant with few variations in design. See NUCLEAR WASTE, and RADWASTE.

Nuclear Fission See FISSION, NUCLEAR.

Nuclear Fusion See FUSION, NUCLEAR.

Nuclear Waste The one intractable problem standing in the way of future nuclear power expansion is the issue of nuclear waste. More than 23,000 tons of intensely radioactive nuclear wastes (ironically called *spent fuel*) had accumulated around the nation's 111 operating nuclear power plants by mid-1990. The total will rise to 40,000 tons within a decade and nobody knows what to do with it. Permanent storage facilities were planned using a network of abandoned salt mines in Kansas and, more recently, a mountain repository in Nevada. Both sites proved less than ideal for technical or political reasons. By mid-1990 no permanent nuclear waste repository had been chosen, and construction is at least 20 years away. Newer nuclear plants have more than enough room in their on-site facilities to store all the spent fuel that the reactors can ever generate, but most of the nation's older nuclear facilities already face a storage crisis. If the waste storage issue cannot be resolved, America's demand for more and more electricity cannot be met by the nuclear option. See NUCLEAR ENERGY, and RADIOACTIVITY.

Nuclear Winter The doomsday prediction that a full-scale exchange of atomic weapons in the arsenals of the United States and the Soviet Union would cover the Northern Hemisphere's continents with such dense clouds of soot that even midsummer TEMPERATURES would drop below freezing and millions of people would starve as a result. Many more human casualties globally were predicted for this aftermath of nuclear war than were estimated for the direct effects of nuclear explosions. A distinguished group of scientists produced their first *nuclear winter* report in 1983, and it set off widespread controversy as other scientists accused the authors of baseless doomsaying and biased exaggeration. Critics downplayed the climatic effects of a nuclear exchange and coined the term *nuclear autumn* for their much more conservative estimates of casualties. The Department of Defense in particular was upset by the nuclear winter concept because, if indeed nuclear war was globally suicidal, then Pentagon plans for "fighting and prevailing" in a "protracted" nuclear war were nonsense.

As the Cold War era came to an end, the nuclear winter debate faded from the headlines, but in a new report published in January of 1990, the same group of scientists who developed the first picture of continents frozen and nations paralyzed in the aftermath of an all-out nuclear war strengthened their original warnings. The 1990 analysis, entitled "Climate and Smoke: An Appraisal of Nuclear Winter," detailed the results of five years of laboratory studies and field experiments that involved the deliberate torching of forest lands as well as complex computer calculations. The scientists concluded that hot smoke and oxides of NITROGEN rising into the upper atmosphere from burning cities, forests, oil refineries, and croplands would so disrupt global circulation patterns than more than half of the entire planet's protective OZONE shield would be wiped out in less than three weeks. The general conclusions that the nuclear winter scientists drew in 1983 were reinforced by the detailed calculations and controlled experiments reported in the 1990 analysis.

Nucleic Acid A chain of MOLECULES, bonded chemically, that control cellular activity. There are two kinds of nucleic acid: DNA and RNA. DNA is made up of nucleotides that contain deoxyribose, a sugar; phosphoric acid; and one of the following four bases: adenine, guanine, cytosine, and thymine. The bases bond with bases in other nucleotides to form a structure called a *double helix*. GENES are made up of DNA. RNA is made by DNA and plays a role in the synthesis of PROTEIN.

Nucleons The constituents of an atomic nucleus, consisting of PROTONS and NEUTRONS. See NUCLEUS, ATOMIC.

Nucleus, Atomic The positively charged MASS within an atom, composed of NEUTRONS and PROTONS, and possessing most of the mass but occupying only a small fraction of the volume of an atom. The nucleus is surrounded by a cloud of much lighter, negatively charged ELECTRONS. The number of electrons in an atom matches the number of charged PARTICLES, or protons, in the nucleus, and determines how the atom will link to other atoms to form molecules. The electrically neutral particles (neutrons) in the nucleus add to its mass but do not affect the number of electrons and therefore have no effect on the atom's link to other atoms (its chemical behavior). See SUBATOMIC STRUCTURE.

Nucleus, Cell Contains GENES in the form of DNA coiled into CHROMOSOMES. The nucleus of any living cell directs the growth, metabolism, and reproduction of that particular cell. The genetic information encoded in the DNA MOLECULES in the nucleus provide instructions for assembling PROTEIN molecules. The work of a cell is carried out for the most part by the many different types of protein molecules. See CELLS, and DNA.

Numerology A pseudoscience involving the study of the occult meaning of numbers, particularly the figures designating the year of one's birth, in the belief that these numbers have an influence on one's life or future. Like ASTROLOGY, this belief is unsupported by any creditable scientific evidence.

Numbers: Big And Small

Prefix	*Equivalent*	*Factor*	
tera	trillion	10^{12}	1,000,000,000,000
giga	billion	10^{9}	1,000,000,000
mega	million	10^{6}	1,000,000
kilo	thousand	10^{3}	1,000
hecto	hundred	10^{2}	100
deca	ten	10	10
deci	tenth	10^{-1}	0.1
centi	hundredth	10^{-2}	0.01
milli	thousandth	10^{-3}	0.001
micro	millionth	10^{-6}	0.000,001
nano	billionth	10^{-9}	0.000,000,001
pico	trillionth	10^{-12}	0.000,000,000,001

Ocean Thermal Energy Relatively small differences in water TEMPERATURE at different depths can be converted into useful electrical ENERGY. Pilot plants to do this have been built at various places around the world. The EARTH's oceans absorb vast amounts of sunlight, most of which is radiated back into the ATMOSPHERE. A small fraction of this HEAT can be used to generate electricity in those areas of the world where the temperature difference between the warmer water on the surface and the cold water below is at least 60 degrees Fahrenheit (15 degrees Celsius).

Ocean thermal energy conversion can be accomplished either with closed or open systems. In a closed cycle, warm seawater is pumped into the system from about 40 feet beneath the surface, where it is between 77 to 83 degrees Fahrenheit (25 to 28 degrees CELSIUS). This warmer water is pumped through a heat exchanger containing a fluid like ammonia, which vaporizes at low temperature. The vapor expands to drive a turbine that, in turn, drives an electrical generator. In an open cycle, warm seawater is pumped into an evaporator in which the pressure has been lowered to make the water boil. In this case, the water vaporizes to drive the turbine and generator. In both cases, electricity can be generated with little environmental damage resulting from the process. Costs, however, remain a problem: Electricity from the more efficient closed-cycle plants would cost about 15 cents a kilowatt hour, as compared to 6 cents for electricity generated by burning oil at $22 a barrel. See ENERGY, SOURCES OF; and ENERGY, USES OF.

Ohm's Law Named after German physicist George S. Ohm (1787–1854), who studied the resistance of various materials to the flow of electricity and who determined that resistance can be determined by taking the ratio of volts (the units of electrical force) to amperes (the units of electrical current passing through a circuit). Ohm's law states that 1 ohm (resistance) is equal to 1 VOLT divided by 1 AMPERE. Another way of stating Ohm's law: In any circuit the electric current is directly proportional to the voltage and is inversely proportional to the resistance. These ratios are often expressed by the following equations: $R = E \div I$, $E = I \times R$, and $I = E \div R$, where R represents resistance, E represents voltage, and I represents current.

Oncogene GENES that differ slightly from normal genes but that can cause tumors (the prefix *onco-* is from the Greek and is commonly used in medical terminology for "tumor"). The discovery that every CELL of almost every SPECIES contains genes that can become cancer-causing with only the slightest alteration is considered one of the most important developments in cancer research in recent years. These genes are called *proto-oncogenes.* After alteration, they become oncogenes. The alteration may involve as little as changing one "letter" in the GENETIC CODE. See **DNA**.

Oort Cloud Named for the modern Dutch astronomer, Jan Oort, this term refers to that area beyond PLUTO where a huge cloud of comets orbit the SUN like billions of miniature PLANETS. Made up of ices, and frozen gases of various materials such as water and METHANE, COMETS are sometimes nudged out of their distant orbits in the Oort cloud and fall into the inner SOLAR SYSTEM. When these objects get close enough to the Sun, the gas and dust in them are pushed out into a tail by the Sun's light and gases. We then see these objects as comets. See also ASTEROIDS, METEORS and NEMESIS.

Open Universe Theory or cosmological model in which the universe continues to expand forever. Early theory that the universe was static and unchanging gave way to the EXPANDING UNIVERSE concept in the late 1920s. It was Edwin HUBBLE who established the fact of an expanding universe in 1929 when he observed the REDSHIFT phenomena of receding galaxies. Whether or not the universe will continue to expand forever, or collapse back unto itself at some far off future time, or even oscillate between expansion and contraction is the subject of continuing cosmological speculation. See also **CLOSED UNIVERSE.**

Operating Systems Computer software that directs the flow of information and manages the flow of jobs through a computer system. An operating system is the base layer of software that controls a computer's internal functions. On top of it a computer user runs applications programs, such as spreadsheets and word processors. Examples of operating systems include DOS, OS/2, and System 7. See **SOFTWARE, COMPUTER.**

Optical Character Recognition (OCR) Simulating a human's amazing ability to read, a **COMPUTER**-based OCR system translates the jumble of black marks on a printed page into text characters in a computer file. At the present time most companies have to hire a typist to transcribe documents for electronic storage. OCR systems can help bridge the gap between paper and the computer system. A complete OCR system includes two main components in addition to the computer: a scanner to reproduce the printed document in electronic form, and OCR software, which analyzes the electronic document image and converts it into a text file.

Applications for OCR abound. Anyone who needs quick access to large amounts of textual information—lawyers, consultants, scientists, journalists, and teachers—could benefit from electronic storage of textbooks, professional journals, and reports. With the information on disk instead of in filing cabinets or bookshelves, one can use a text-retrieval program to find every reference to a particular topic in seconds.

Optical Fibers See **FIBER OPTICS.**

Organ Transplants Kidneys, hearts, lungs, livers, and other human organs and tissues are now being transplanted with increasing frequency and much improved survival rates. According to the United Network for Organ Sharing, these are the figures for 1989:

Organ	*Number of Transplants*	*Average Costs*
Heart	1,673	$ 162,000
Liver	2,160	$ 216,000
Kidney	8,886	$ 52,000

The success rate for heart transplants now stands at better than 75 percent. Liver transplantation has become less hazardous than in the past because of newly developed bypass techniques, and the current survival rate is more than 70 percent. The living related-donor success rate for kidney transplants is 85 to 95 percent. Cadaver (transplants from a dead body) donor success rate is 65 to 70 percent. Other organs being transplanted with increasing success include cornea, pancreas, skin, bone marrow, and bone. The reasons for the improvement of organ-transplant survival rates include improved surgical techniques, and new approaches to organ-rejection problems. The need for donor organs has increased as the success rate of transplantation improves. Some organs such as hearts and lungs are more difficult to obtain than others. Ethical questions surrounding the selection or prioritizing of organ transplant recipients is a subject of concern in medicine today. See BIOETHICS.

Origin of Life See LIFE, ORIGIN OF.

Origin of Species See DARWIN.

Origin of Universe According to the current scientific consensus, the entire contents of the known universe expanded explosively into existence from a single, hot, dense, chaotic mass between 10 and 20 billion years ago. Stars coalesced out of clouds of HYDROGEN and HELIUM, heated up by the energy of coming together, and began releasing energy from the FUSION of light elements into heavier ones in their extremely hot, dense cores. With time, many of the STARs exploded, producing new clouds from which other stars condensed. This process of star formation continues today.

Our SUN is a medium-sized star orbiting near the edge of collection of stars we call the MILKY WAY galaxy. This galaxy contains many billions of stars, and the universe contains many billions of such galaxies. The EARTH has existed for only about a third of the history of the universe and is in comparison a mere speck in a space so large that it is difficult for the human mind to comprehend. As the brilliant physicist Stephen Hawking has pointed out, belief in an expanding universe and the BIG BANG theory does not preclude belief in God as a creator, but it does place time limits on when he might have carried out this task.

Osmosis The process by which a fluid or gas passes through a semipermeable membrane or porous partition from one side to the other. **MOLECULES** of a liquid that are small enough to pass through openings in the membrane will move in both directions randomly, but if there are more of these molecules on one side than on the other, chance alone will dictate that more molecules from the higher concentration will pass through to the side with the lower concentration. Eventually the concentration of molecules will reach an equilibrium in the fluid.

Osmosis is the physical phenomenon that plant roots use to obtain water from the soil or that blood uses to get oxygen from the lungs. The popular phrase "learning by osmosis" refers to the alleged ability to absorb information without apparent effort. See **OSMOSIS, REVERSE.**

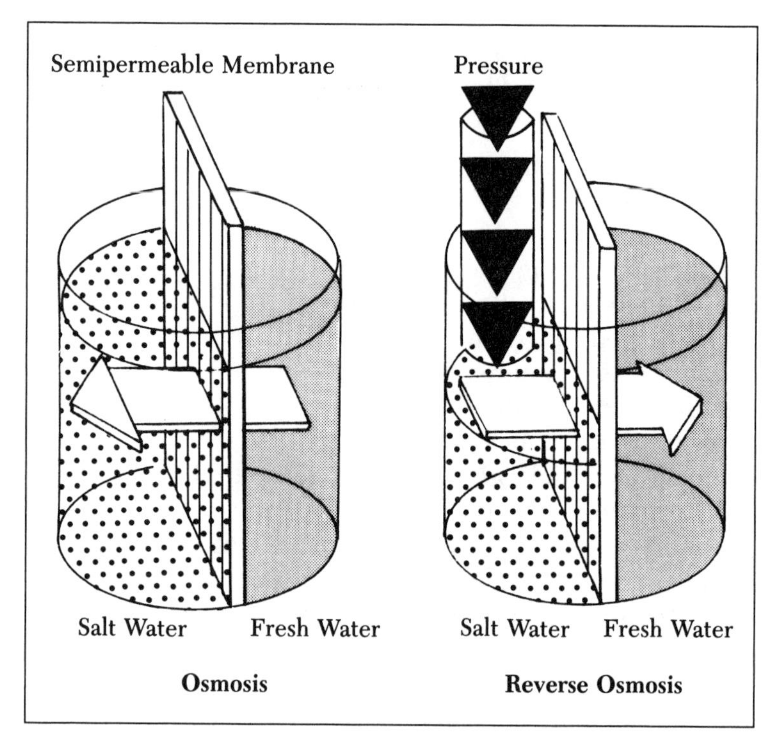

OSMOSIS and REVERSE OSMOSIS are shown here. Osmosis is the process by which a fluid (or a gas) passes through a semipermeable membrane or porous partition. In reverse osmosis, pressure is used to overcome the natural process of osmosis to force the liquid (or gas) through the membrane.

Osmosis, Reverse The process used in some water desalination plants whereby artificial pressure on salt water is used to overcome the natural osmosis phenomenon, forcing water MOLECULES from the salt water side to the fresh water side of a semipermeable membrane. Since only water molecules can pass through the membrane, the salt is left behind. Reverse osmosis systems use high-pressure pumps to squeeze seawater against a thin plastic membrane that allows pure water but not dissolved salts to pass through. These pumps use a lot of electricity, making the process energy intensive and expensive. See DESALINATION, and OSMOSIS.

Otto Cycle Engines The "four-stroke" INTERNAL-COMBUSTION ENGINE powering most of our cars today was invented by Nicholas Otto in 1880. The Otto cycle has four distinct strokes: intake, compression, power, and exhaust. Each stroke represents a sweep of the piston from top to bottom or vice versa, and each stroke thus represents one-fourth of the cycle. The intake downstroke sucks in a new charge of air and fuel. Then the upstroke compresses this mixture. Near the top, the spark plug ignites the mixture and, because the fuel/air expands as it burns, the piston is forced down, producing mechanical power. Last, the exhaust stroke purges the cylinder of the spent gases.

Oxygen A colorless, odorless, gaseous element that constitutes about one-fifth of the volume of the ATMOSPHERE on EARTH. Other PLANETS in this SOLAR SYSTEM have atmospheres, but Earth alone has oxygen as a major constituent. Life on Earth depends on the OXYGEN CYCLE.

Oxygen Cycle In this process, OXYGEN is released into the ATMOSPHERE by photosynthetic organisms (plants) and is taken up by aerobic organisms (animals) while the CARBON DIOXIDE released as a by-product of breathing or respiration is taken up by PHOTOSYNTHESIS.

Life on early EARTH—about 600 million years ago—was confined to the sea and fresh water while the land was drenched in deadly ULTRAVIOLET rays from the Sun. But plants were steadily increasing the amount of oxygen in the air, which in turn was forming OZONE high in the atmosphere and beginning to filter out the dangerous ultraviolet. Because of this process, it eventually became possible for animal life-forms to follow plant life ashore and live on land.

Ozone A form of OXYGEN with three ATOMS (as opposed to normal oxygen MOLECULES, which contain two atoms) that primarily exists as a thin gaseous layer between altitudes of 6 to 30 miles (10 to 50 kilometers) above the EARTH. Ozone is created when ordinary oxygen in the STRATOSPHERE is bombarded by ULTRAVIOLET RADIATION from the SUN. The ozone layer is threatened by the release of synthetic chemicals into the atmosphere. See **OZONE DEPLETION**, **OZONE HOLE**, and **OZONE SMOG**.

Ozone Depletion The degradation of the EARTH's stratospheric ozone layer, which was forecasted in the late 1970s and became evident in the mid-1980s, may eventually let in RADIATION severe enough to cause an epidemic of SKIN CANCER and cataracts and to reduce agricultural productivity seriously. The use of synthetic CHLOROFLUOROCARBONS (CFCs) is the principal cause of ozone depletion. These chemicals, used in refrigeration, air-conditioning, and as a foaming agent in the manufacture of plastics, are so stable that once released in the atmosphere MOLECULES can survive for more than a century. With time, they drift intact into the STRATOSPHERE. At high altitude they are exposed to intense ultraviolet radiation that breaks the molecules apart, producing free chlorine atoms. These chlorine atoms destroy enormous amounts of ozone.

Life on Earth exists in a delicate balance between various harmful influences on the one hand and defensive or restorative measures utilized by living creatures on the

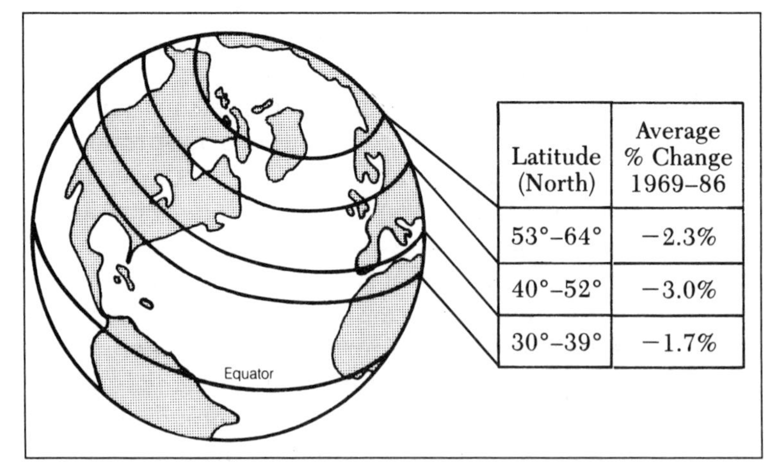

Latitude (North)	Average % Change 1969–86
53°–64°	−2.3%
40°–52°	−3.0%
30°–39°	−1.7%

OZONE DEPLETION, now confirmed by satellite photos, has prompted widespread concern.

other. Biologists believe that the existing balance between damage and protection has been achieved over a long period of time; readjustment to a relatively sudden increase in ultraviolet radiation may involve **NATURAL SELECTION** of the better protective devices in a population of organisms and would require many generations of the organism to accomplish.

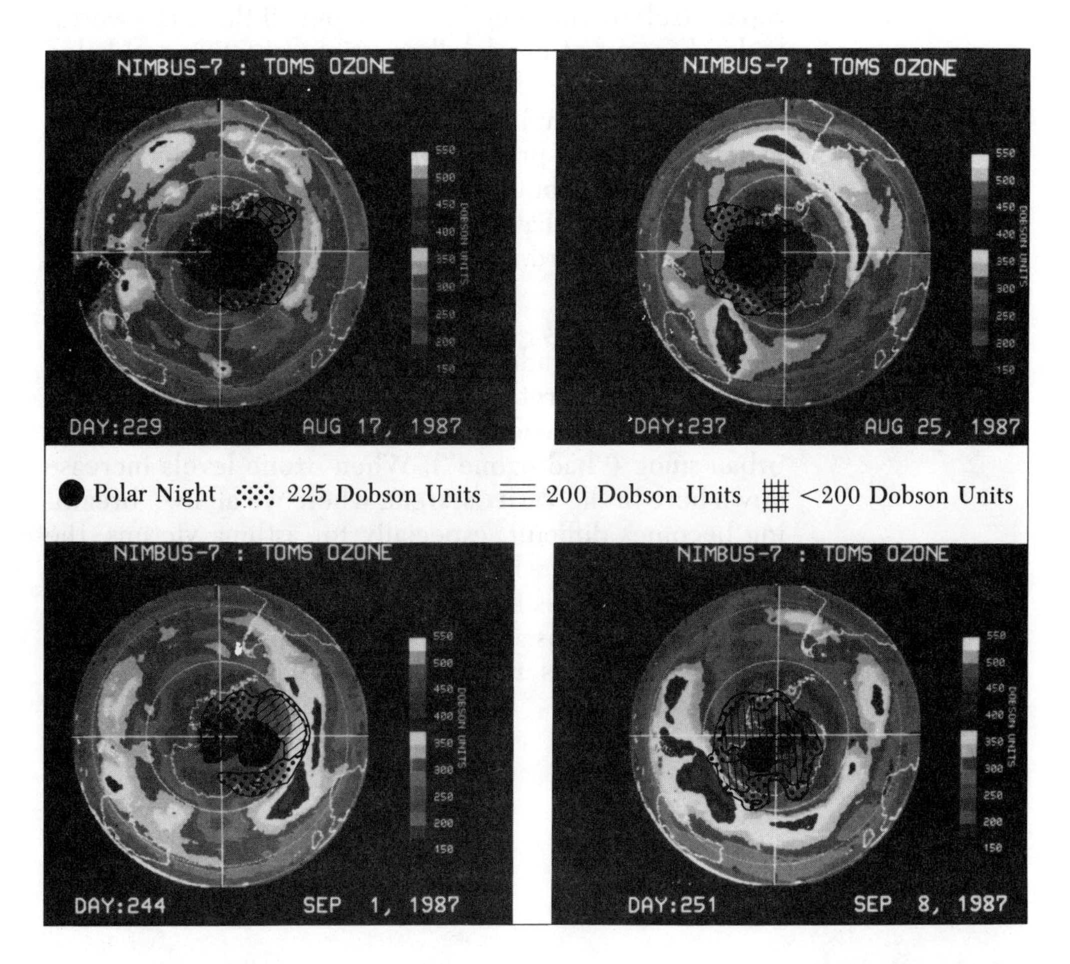

OZONE HOLE in the Antarctic area as detected by the Total Ozone Mapping Spectrometer (TOMS) instrument aboard NASA's *Nimbus* 7 satellite. *Source: NASA/Ames Research Center.*

Ozone Hole Prior to the discovery in 1985 of a winter hole in the ozone shield over the Antarctic region, ozone depletion theories were based primarily on computer modeling because little empirical data were available. Global ozone depletion could be explainable as natural variations. The discovery of the ozone hole changed the debate. The hole was definitely there, it could not be explained away as natural variations, and it was worsening. The loss was measured at 50 percent in 1985 and 60 percent in 1986, when the hole was judged to be as big as the continental United States and spreading toward South America and Australia. Scientists have now confirmed that the continuing decline in the global ozone shield is primarily due to the continued use of synthetic CHLOROFLUOROCARBONS (CFCs). NASA scientists have further concluded that the hole in the Antarctic layer can repair itself by the year 2100, but only if there is a worldwide phaseout of nearly all emissions of CFCs and halon gases early in the next century. The United States and the European Community countries have reached an agreement to phase out production and use of CFCs by the end of the century, but several large countries, including China and India, have not ratified this agreement called the London Accord.

Ozone Smog Confusing the issue of OZONE DEPLETION is the paradox between the stratospheric ozone layer needed to protect life on Earth ("good ozone") and the ground-level ozone that is the principal element of urban smog ("bad ozone"). When ozone levels increase down here in the TROPOSPHERE where we all live, breathing becomes difficult, especially for asthma victims, the elderly, and those who exercise. It is the same ozone, but what is good for us as a protective layer in the STRATOSPHERE is bad for us at the lower atmospheric level where ozone MOLECULES react adversely with our lungs and eyes.

Paleozoic Era The age of ancient life-forms—from a period beginning about 230 million years ago and covering a time span of perhaps 340 million years—prior to mammals. The Paleozoic era is subdivided into the following ages or periods:

Period	*Forms of Life*
Permian	Conifer forest (extinction of many marine invertebrates—creatures without a backbone)
Carboniferous	First reptiles; age of amphibians begins
Devonian	Age of fishes, insects, land animals
Silurian	Shellfish, first land plants
Ordovician	Primitive fish, seaweed, fungi
Cambrian	Age of marine invertebrates begins, shellfish, and so on.

See **GEOLOGIC TIME SCALE.**

Pangaea The name given to the single land mass that existed on **EARTH** 200 millions years ago, before it broke up into several continents, which then drifted to their present positions. Pangaea means "all lands" and refers to the single land mass that scientists now believe to have existed on Earth prior to a slow breaking apart with the various pieces moving only inches a year—a process known as **CONTINENTAL DRIFT**. See also **PLATE TECTONICS.**

Paradigm A model of reality—or a system of facts, theories, and philosophies—that is widely accepted and becomes the framework for thinking about a scientific problem. When new discoveries call into question an existing paradigm the result is called a *paradigm shift*. When in 1929, for instance, **EDWIN P. HUBBLE** and others discovered that the universe was expanding, the old model or paradigm of the universe was replaced with a new one. The rise of QUANTUM PHYSICS early in this century created a new paradigm in physics, replacing the strictly Newtonian concepts.

Parallel Processing The system used by very large supercomputers to handle extraordinarily complex problems. Ordinary home COMPUTERs have one microprocessor, usually on a single microchip, or INTEGRATED CIRCUIT. Processing is carried out serially, one step at a time. Parallel-processing computers, on the other hand, divide a problem into smaller pieces and assign each piece to one of many processors working simultaneously. The number of parallel processors used in a supercomputer define its capability. The real challenge, however, is not just to add more and more processors but to write software programs that can break a complex problem into parts that can be worked on simultaneously. See NEURAL NETWORKS.

Paranormal Pertains to the belief in claimed occurrences of events or perceptions without scientific explanation evidence or support. Belief in vague, anecdotal and often demonstrably erroneous doctrines such as ASTROLOGY, flying saucer kidnaping, channeling (or communicating with long-dead ancestors), mystical levitations, extrasensory perceptions (ESP), or any other purportedly supernatural phenomena. Widespread popular belief in various paranormal claims is evidence of widespread scientific illiteracy.

Parkinson's Disease The second most common (next to ALZHEIMER'S DISEASE) neurodegenerative, or brain damaging, disease, affecting one in 40 people over 70, or around a half a million people annually in the United States. Parkinson's disease develops when an as-yet-unknown process destroys certain dopamine-producing neurons in a region of the brain called the *substantia nigra.* Some researchers have recently hypothesized that this process somehow affects another brain region known as the *subthalamic nucleus,* causing its neurons to fire too rapidly and thus producing the muscular tremors and rigidity that plague Parkinson's patients. If confirmed, this new finding may lead to more effective treatment for persons with the disorder.

Parsec An astronomical unit of distance equal to 3.26 LIGHT-YEARS or a little more than 19 trillion miles. The term comes from the apparent shift in the position of STARS when viewed from opposite sides of the EARTH's annual orbit of the SUN—a phenomenon called parallax which is used to determine the distance to a star or planet. A parsec is defined as the distance at which an object (star or planet) must be to produce a parallax shift of one second of arc. See ASTRONOMICAL UNIT.

Particle Physics The branch of science that deals with the smallest known structures of MATTER and ENERGY. In atomic physics, the term PARTICLES is applied to the subunits that make up an atom, such as PROTONS, NEUTRONS, or ELECTRONS (protons and neutrons are made up of still smaller particles called QUARKS). The term particle is also applied to units of energy in accordance with EINSTEIN's relativity theory, which asserts that matter and energy are different forms of the same thing. High-energy physics deals with the application of large amounts of energy in the experimental investigation into the behavior of subatomic particles. See SUBATOMIC STRUCTURE.

Particles The fundamental or basic units of MATTER and ENERGY. All are classified as FERMIONS (particles of matter) or BOSONS (particles of energy or force). The term *particles* is somewhat misleading because it implies a portion or piece of something whereas in fact subatomic particles also evidence wavelike behavior. See PARTICLE PHYSICS, and SUBATOMIC STRUCTURE.

Pasteur, Louis (1822–1895) French chemist and bacteriologist whose name is most closely associated with the GERM THEORY of disease. Pasteur came to his discovery of the role of microorganisms through his studies of what causes milk, cheese, yogurt, or beer and wine to spoil. He demonstrated that spoilage and fermentation occur when microorganisms enter them from the air, multiply rapidly, and produce waste products. Pasteur showed that food would not spoil if microorganisms were kept out of it or if they were destroyed by HEAT (a process now called *pasteurization*). Pasteur went on to discover that a disease caused by germs could be prevented by the use of VACCINES—a very mild dose of the actual germ itself—that would cause the body to build immunity to that disease. Pasteur's identification of infectious disease with tiny organisms that could be seen only with a microscope was a milestone in medical technology.

Perihelion The point on an ELLIPTICAL orbit at which a PLANET or COMET is closest to the SUN. The prefix *peri* means "about" or "around" and *heli*(on) refers to the Sun. See APHELION.

Periodic Table of the Elements A table illustrating an arrangement of elements (substances that cannot be broken down chemically) into horizontal rows (called periods) in accordance with their atomic number, or number of PROTONS, and in vertical columns according to related groups. Devised by Dimitry Mendeleyev (1834–1907), a Russian chemist, the periodic table displays all the elements in such a way as to show the similarities in certain *families* or groups of elements. In addition to being a convenient way to display the elements, the periodic table also revealed gaps in the list of elements that correctly predicted the existence of elements that were discovered at a later date. As shown below, each element's atomic number appears at the top center of its box. This is followed by the elements name or symbol while the number at the bottom of the box indicates the atomic MASS of the element.

1	Atomic number
H	Symbol
HYDROGEN	Name
1.00797	Atomic mass

See ELEMENT.

PET Scan See POSITRON EMISSION TOMOGRAPHY (PET).

Phobos One of MARS's two MOONS (the other is Deimos). Phobus is a tiny mountain of rock that was probably captured from the asteroid belt by Mars' gravitational pull. It is about 15 miles in diameter and was the objective of two Soviet unmanned spacecraft missions in 1989. The two spacecraft, named *Phobos* 1 and 2, were scheduled to send landing craft to the surface of the Martian moon but both spacecraft were lost prior to their rendezvous with Phobos. This double failure constituted a major setback in Soviet space exploration efforts. The Soviets considered the *Phobos* missions as preliminary steps to the exploration of Mars.

Photodynamic Therapy An experimental treatment for lung, bladder, and esophageal tumors involving the therapeutic use of LIGHT. In this treatment, patients are given intravenous injections of a chemical that accumulates in cancer cells and becomes lethal to the cells when activated by light. The physicians then bathe the tumor in color-specific light rays, which activates the poison. When successful, this treatment destroys the cancer, leaving most normal tissue unharmed. This treatment is being used experimentally in an effort to compare its effectiveness against conventional treatments involving chemotherapy and surgery. The three tumors at issue can often be readily reached by a light source that is inserted through the mouth or urethra (the tiny tube through which the bladder empties). Photodynamic therapy (commonly called PDT) may find a niche in treating certain localized tumors that are known not to respond to chemotherapy and are difficult to remove surgically. The major limitation of photodynamic therapy is that it cannot yet be used to treat most tumors, which are deep and bulky and cannot be reached by light. See LIGHT THERAPY.

Photons Discrete units of electromagnetic ENERGY. Electromagnetic RADIATION (whether LIGHT, radio waves, or X rays) may be thought of as a flow of PARTICLES called photons. A photon is the smallest indivisible unit of electromagnetic radiation. The SUN emits huge quantities of photons, some of which are in the form of visible light, some of which are in some other form of energy. Radio or TV broadcasting towers emit photons that are received by our home antennas to produce sound and pictures. Photons move at the speed of light. See QUANTA, and QUANTUM PHYSICS.

Photosynthesis The process by which green plants utilize sunlight as a source of ENERGY to synthesize com-

PHOTOSYNTHESIS is the process by which green plants utilize sunlight as a source of energy. Through respiration, plants release water, and oxygen. Animal life would not be possible without the oxygen.

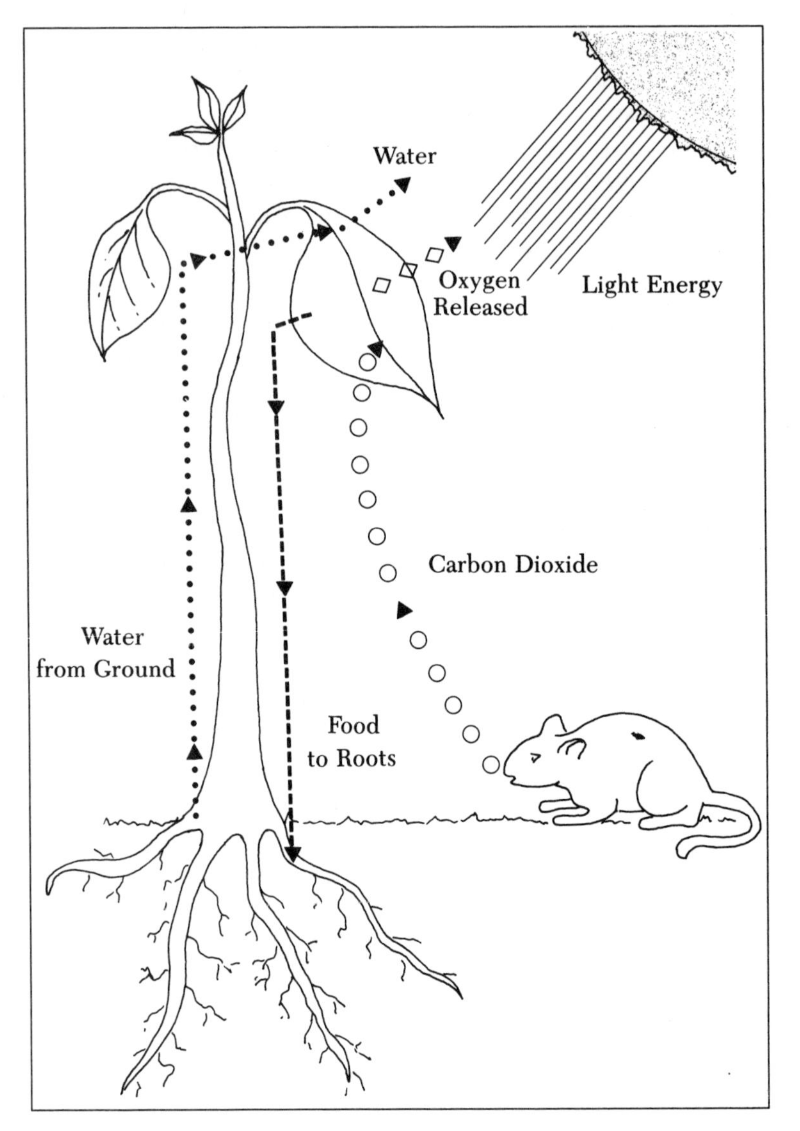

plex organic material, especially carbohydrates, from CARBON DIOXIDE, water, and inorganic salts. Plants make up 90 percent of all visible living organisms on Earth and they get their food energy from photosynthesis. When plants are eaten, they pass some of this energy on. The energy from plants thus sustains the remaining 10 percent of visible living things. Since fuel for photosynthesis is light from the Sun, the energy that is released when a plant such as a tree is burned is really stored SOLAR ENERGY. The energy in coal or oil is also stored solar energy resulting from dead plants or animals compressed under mountains of sediment for long periods of time. See OXYGEN CYCLE.

Photovoltaic Solar Cells Devices that convert the ENERGY of the SUN into electrical energy. The energy of the Sun is used to dislodge ELECTRONS from ATOMS of silicon or other material. The electrons, having a negative charge, and the atom that gave it up, having a positive charge, migrate to separate terminals, creating an electric current. The technology, which was started in 1954 and got its first boost from the space program, is now being used in a growing number of applications: from hand-held calculators and wristwatches to remote microwave communication relays and navigation buoys. Photovoltaic solar power is, however, far too expensive to compete with coal or oil for making electricity on a large scale. The average cost of electricity for residential customers in the United States is about 10 cents per kilowatt hour. Electricity from photovoltaic sources costs more than 25 cents per kilowatt hour.

Key to the future improvement of photovoltaic solar power is improvement in the efficiency of the cells; that is, the ratio of energy deposited as sunlight to electricity produced. Currently the efficiency ratio of silicon crystal cells is too low to make them competitive with other forms of energy. A more recent development is thin-film cells made by applying to glass small amounts of materials that convert LIGHT to electricity. Thin-film cells have a lower efficiency ratio but they can compete with silicon crystal cells because they are much easier to manufacture. See SOLAR ENERGY.

Physics The scientific study of the interaction of MATTER and ENERGY. Classical, or Newtonian, physics refers to the scientific studies made prior to the introduction of the quantum principle. Modern physics sees both matter and energy as being made up of discrete units, or QUANTA. See QUANTUM PHYSICS.

Phytoplankton Microscopic algae living on or near the surface of the sea which play the same role as grasses and other plants on land. Plant-eating creatures in the sea graze on phytoplankton just as cows and sheep crop grasses on land. The plankton or algae utilize the process of PHOTOSYNTHESIS to derive ENERGY from the SUN and grow just as land plants do. Phytoplankton is limited to the upper 120 feet of water because LIGHT is mostly absent below that depth and photosynthesis cannot take place. Phytoplankton is considered a first link in the FOOD CHAIN. Also, most of EARTH's OXYGEN supply is produced by phytoplankton. Scientists are concerned that the deteriorating OZONE layer in the STRATOSPHERE will cause an increase in ULTRAVIOLET rays reaching the surface of Earth which could limit the growth of this vital link in the food chain. Some researchers say that signs of this are already happening in the Antarctic regions. See **OZONE DEPLETION**.

Pioneer Spacecraft The *Pioneer 10* and *11* spacecraft are the first vehicles of humankind to venture beyond the limits of our SOLAR SYSTEM. Each carries a 6-by-9-inch engraved plaque, designed by Carl Sagan and colleagues, intended to convey information on the locale and nature of the builders of the spacecraft. The plaque shows a picture of two people, a man and a woman, and gives a map of the solar system that shows, in mathematical code, EARTH's position in the MILKY WAY galaxy. Launched in 1971 and 1972, respectively, *Pioneer 10* and *11* made successful observations of JUPITER and SATURN in 1973 and 1974. The two *Pioneer* spacecraft lived up to their name, returning the first close-up pictures of JUPITER as well as images of its four large MOONS. Both spacecraft are working well as they venture into interstellar space, and they are expected to send data back for another 10 to 20 years.

Placebo Any inactive substance (or procedure) given to a control group as if it were an effective treatment. It is used as a comparison for the substance (or procedure) being tested. Medical testing is complicated by the fact that placebos often seem to have the same effect as the drug being tested for about one-third of patients (this psychological phenomena is called the *placebo effect*). Patients seem to convince themselves that a substance has a beneficial medical effect, when in fact it has no effect whatsoever. Over time an effective drug will outperform a placebo.

Planck, Max German physicist (1858–1947) who originated quantum theory, generally considered one of the greatest achievements of 20th-century science. The basic principle of QUANTUM PHYSICS is that ENERGY is emitted not as a continuum, or continuously variable entity, but rather in discrete units, or tiny bundles, called QUANTA (the plural of quantum). Planck's concept of the quantum nature of energy strongly influenced Niels Bohr, EINSTEIN, and other modern physicists. See PLANCK'S CONSTANT.

Planck's Constant The formula for relating the ENERGY content of a QUANTUM (or unit of energy) to the frequency of the corresponding electromagnetic wave. In 1900, the German physicist Max Planck demonstrated that the energy in a quantum could be related directly to the frequency of a given RADIATION. The higher the frequency (or shorter the wavelength) of a particular radiation, the more energetic the quantum. Violet LIGHT (light being a form of electromagnetic radiation) radiates at a frequency twice that of red light and therefore a quantum of violet light is twice as energetic as a quantum of red light. Planck's concept cleared up the connection between temperature and wavelengths of emitted radiation. Planck expressed this direct relationship in an equation utilizing a very small number (6.624×10^{-27}), which gives the exact proportional relation between quantum energy and frequency. It is this number that is now known as Planck's constant.

Planck's constant has played a central role in modern physics and the theory of quantum mechanics. This number appears in numerous fundamental equations, including those defining the smallest amount, or quantum, of energy that a physical system can gain or lose. See PLANCK, MAX; QUANTUM PHYSICS; and WAVE THEORY.

Planets Celestial bodies more massive than ASTEROIDS but less massive than stars that revolve around a STAR and shine by reflected LIGHT. Our SOLAR SYSTEM contains nine planets and their MOONS. You can remember the

PLANETS are the nine major objects orbiting the Sun. They are less massive than stars and they shine by reflected light. The farther out the orbit, the longer a planet takes to complete it.

planet's names, in order of distance from the SUN, by the first letters of "My Very Educated Mother Just Sent Us Nine Pizzas": MERCURY, VENUS, EARTH, MARS, JUPITER, SATURN, URANUS, NEPTUNE, PLUTO. The first four of these celestial bodies are called the rocky planets; all have atmospheres of some sort, but Earth alone has oceans of liquid water. The next four planets—Jupiter, Saturn, Uranus, and Neptune—are known as the giant planets. These gas giants are composed of much the same material as the

Relative Size		Average Distance from Sun (million miles)	Revolution around Sun	Axial Rotation	Diameter (miles)	Maximum Surface Temp. (C°)
	Pluto	3,670	248 years	6.4 days	1,400	−230°
	Neptune	2,790	165 years	17 hours	30,690	−230°
	Uranus	1,778	84 years	11 hours	32,420	−210°
	Saturn	886	29.5 years	10 hours	74,800	−160°
	Jupiter	482	12 years	10 hours	88,500	−140°
	Mars	141	687 days	24.5 hours	4,208	25°
	Earth	93	365 days	24 hours	7,909	60°
	Venus	67	225 days	243 days	7,504	500°
	Mercury	36	88 days	59 days	3,024	400°

Sun. Tiny Pluto, the ninth planet, is so far from the heat of the Sun that its TEMPERATUREs approach the limits of coldness and this planet may consist mostly of ices.

Earth and the other eight planets of this solar system orbit the Sun more or less in the same flat plane; that is, as though they were all lying on a dish. Because of this, the Sun and planets all appear to travel along the same path across the sky. This path is called the *ecliptic*.

Plantae One of the five taxonomic kingdoms of living organisms. Plantae comprises all multicelled plant life. The other four taxonomic kingdoms are ANIMALIA, FUNGI, MONERA, and PROTISTA. See TAXONOMY.

Plasma, Blood A clear, almost colorless fluid consisting of blood from which the red and white blood cells have been removed. Blood plasma is 55 percent of whole blood and consists of water, proteins, salts, and most of the blood's cargo such as nutrients, hormones, waste products, and antibodies. Blood plasma is a substance used in transfusions. Often a patient needs the plasma portion of blood more than all the corpuscles. Unlike whole blood, which has to be of the right group, plasma can be mixed from all donors and is then useful for all in need. Unlike whole blood, plasma can be dried and stored. See ANTIBODY.

Plasma, in Physics A gas that at very high temperature has had all the ELECTRONS stripped off its ATOMS is called a plasma. In other words, plasma is a highly ionized (or electrically charged) gas containing an approximately equal number of positive IONs and electrons. In electrical storms, for instance, tiny regions of plasma are formed inside lightning. As soon as the ENERGY of the lightning flash is spent, the nuclei recapture electrons to become proper atoms again. In other words, plasma refers to a state in which matter consists of electrons and other subatomic PARTICLES without any structure of an order higher than that of the atomic nuclei. See IONIZATION.

Plate Tectonics Accepted geological theory, or model, that envisions the EARTH's surface covered by a number of relatively thin plates that move over the material below. Many geological facts that earlier appeared

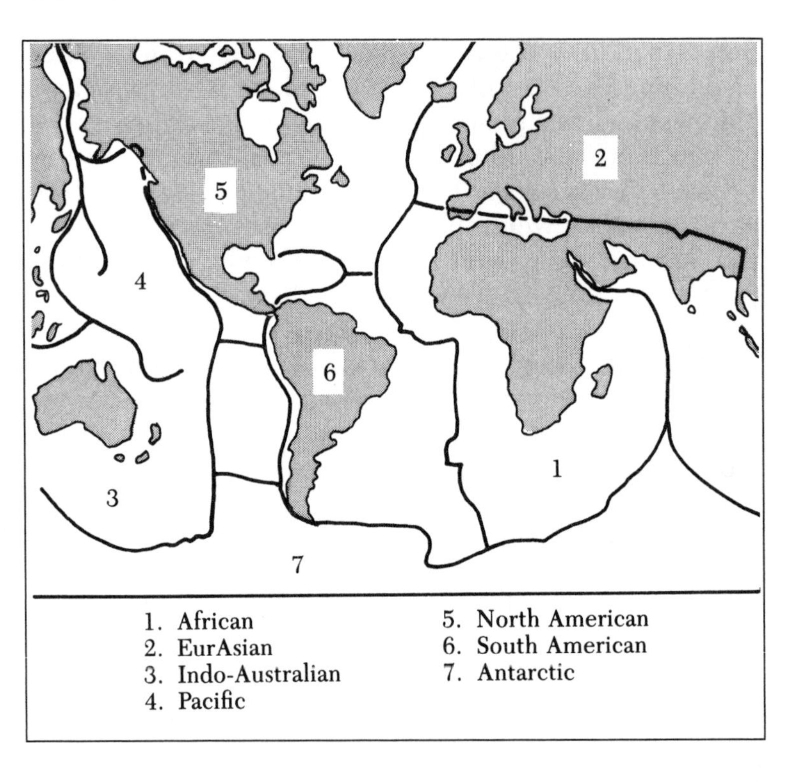

PLATE TECTONICS. According to this theory, the Earth has a rigid outer shell that is made up of a number of segments or plates that move slowly—fractions of inches per year—with respect to one another. The motions of these plates are a major cause of earthquakes.

to be unrelated are explained by this model. According to this theory, the Earth has a rigid outer shell, usually called the crust, about 60 miles thick, which is made up of a number of segments or plates—much like a set of irregularly shaped tiles on a floor. These plates move slowly, fractions of inches per year, with respect to one another. At some boundaries, the plates converge while at other boundaries they are moving apart. The motions of plates, especially where they slide laterally with respect to one another, are a major cause of EARTHQUAKES. Plate boundaries include the world's earthquake sites, volcanic regions, deep-sea trenches, and mountain ranges.

Geologists have identified seven larger plates—parts of the Earth's surface that behave as a single rigid unit—and several smaller ones. The large plates are: the African, Eurasian, Indo-Australian, Pacific, North American, South American, and Antarctic. Associated with plate tectonics is the theory of CONTINENTAL DRIFT, which says that the present continents were formed by the breaking up of one large supercontinent (PANGAEA) about 200 million years ago and have since moved to their present positions.

Pluto Although Pluto is usually the outermost known PLANET of our SOLAR SYSTEM, its peculiar orbit can carry it 50 million miles closer to the SUN than NEPTUNE—a condition that began in 1981 and will continue until 1999. Almost six million miles from the Sun, Pluto's orbit around the Sun takes 248 years. A small planet with a diameter less than 1,400 miles (only two-thirds the size of our Moon), Pluto has a MASS only about 1/500 the mass of Earth. Pluto is very cold, −370 degrees FAHRENHEIT (−223 degrees CELSIUS), and has one moon, named Charon, about half the size of the planet. Once thought to be a loosely compacted snowball, Pluto is now thought to contain a lot of rock in addition to the METHANE gas and ice previously identified. Astronomers have concluded that Pluto's thin atmosphere, consisting mostly of methane gas, sometimes reaches so far out that it envelops Charon. Some astronomers believe that Pluto is a former moon of Neptune that drifted into a separate orbit of its own.

Pluto does not define the outer edge of our solar system. Extending far beyond is a vast belt of cometlike material (the OORT CLOUD) that was part of the original primordial cloud from which our planetary system formed. See **BIG BANG**, **HELIOPAUSE**, and **MOONS**.

Positron In physics, a positron is an elemental PARTICLE having the same MASS and spin as an ELECTRON but having a positive charge equal in magnitude to the electron's negative charge. The term *positron* is a combination of *positive* and *electron*. A positron is the antiparticle of the electron.

Positron Emission Tomography (PET) A type of medical imaging, or scanning, device used by physicians to observe the biochemical changes taking place in the human body. Whereas a CAT scan shows an organ's shape and structure, but not how it is functioning, a PET scan provides metabolic portraits that reveal the rate at which healthy and abnormal tissues consume biochemicals.

PET imaging involves tracing the action of radioactive substances inserted into the human body. After inhaling or being injected with a small amount of radioactive ma-

POSITRON EMISSION TOMOGRAPHY (PET) scans provide metabolic portraits that reveal the rate at which abnormal and healthy tissue consume biochemicals. PET scans involve tracing the action of radioactive substances inserted into the human body.

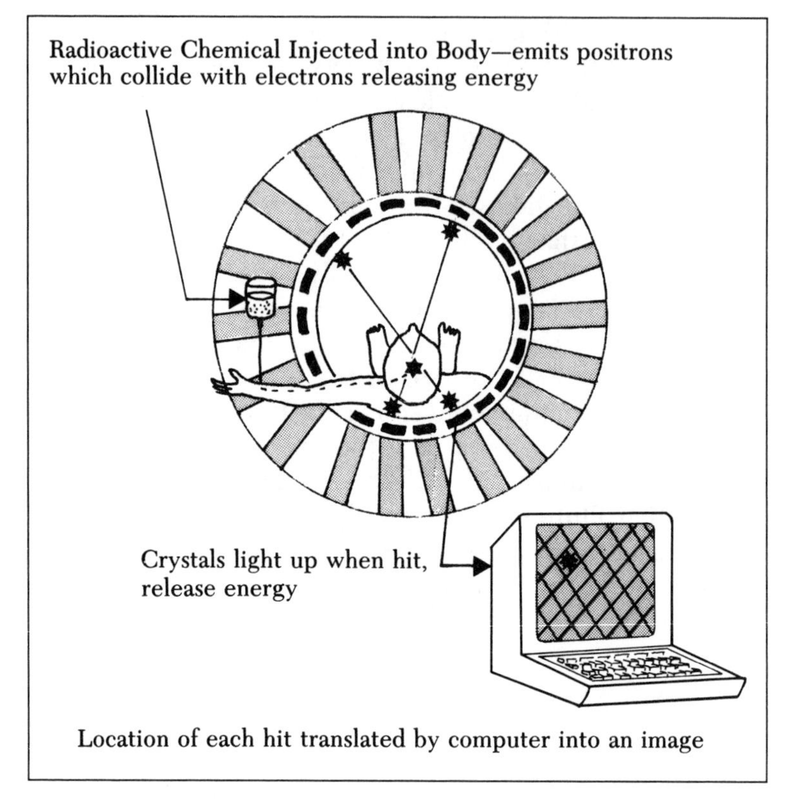

terial, the patient is positioned in a scanning device shaped like a large donut. The radioactive material emits POSITRONS, which collide with ELECTRONS and in so doing release ENERGY in the form of GAMMA RAYS. The gamma-ray flashes are plotted and a COMPUTER is used to translate this data into an image of the chemical reactions taking place in the human body. See also **COMPUTERIZED AXIAL TOMOGRAPHY (CAT)**, **DIGITAL SUBTRACTION ANGIOGRAPHY (DSA)**, and **MAGNETIC RESONANCE IMAGING (MRI)**.

Precambrian The earliest era of EARTH's history, ending about 570 million years ago. It was during this period that the Earth's crust formed and life first appeared in the seas. Earth's history consists of two chapters: the first 4,000 million years (the Precambrian), when the Earth and life first formed, and the remaining 570 million years (the Phanerozoic), when life blossomed. It is only in the last 13 percent of Earth's history that life became complex. See **GEOLOGIC TIME SCALE**.

Precession In astronomy, the term refers to the slow, conical motion of the Earth's axis of rotation, caused by the gravitational attraction of the SUN and MOON on the equatorial bulge of the EARTH. Imagine a spinning toy top that wobbles on its axis as it slows in speed. See ASTRONOMICAL CYCLES.

Primates The order or classification of similar organisms that includes humans, lemurs, monkeys and apes, and several other kind of mammals that began to evolve from other mammals less than 100 million years ago. Several humanlike primate SPECIES began appearing and branching about five million years ago, but all except one became extinct. The line that survived led to the modern human species.

Most species of primates still live in trees as did the ancestors of all present ground-dwelling monkeys, apes, and humans. The various designs of primate bodies correspond to the different ways of moving around: clinging and swinging, leaping, four-footed walking, or two-footed walking. The ability to stand and move about on two feet leaves the hands free for other purposes such as throwing objects.

Anthropologists believe that this ability gave humanlike primates an edge (or a leg up, if you will forgive the pun) over rival species of primates. See AUSTRALOPITHECUS, EVOLUTION, and HOMINID.

Probability, Laws of The mathematical rules governing the relative possibility of an event occurring, as expressed by the ratio of the number of actual occurrences to the total number of possible occurrences. Rolling dice is often used as an example of calculating probability. Each die is a cube with six faces, each of which is imprinted with between one and six dots corresponding to the numbers 1 to 6. On each roll or toss of a pair of dice there are 6×6, or 36, possible outcomes. To calculate the probability of rolling a 7 (getting a sum of 7), it is necessary to compare the number of outputs where the sum 7 is possible (i.e., 6) to the total possible outcomes (36). The probability of rolling a 7 is thus 6/36 or 1 in 6. Because there are more ways to get a sum of 7 than there are ways to get any other sum, 7 has the highest probability. Probabilities become more difficult to compute as the number of favorable and possible outcomes become more difficult to count.

Protein Protein MOLECULES may be thought of as the workhorses of life. Some proteins act as ENZYMES whose job is to digest the proteins we eat. Other proteins, such as HEMOGLOBIN in blood, help to carry OXYGEN from the lungs to the rest of the body. Still others form connective and supportive tissue. Proteins constitute a large portion of the MASS of every life-form and are necessary in the diet of all animals. The chief function of all living CELLS is assembling protein molecules according to instructions coded in DNA molecules. Protein molecules are long, usually folded chains made from 20 different kinds of amino acid molecules. The function of each protein depends on its specific sequence of AMINO ACIDS as well as its exact shape. See **CHROMOSOMES**, **DNA**, and **GENES**.

Protista One-celled organisms that are classified as neither plant nor animal. Biologists classify all living organisms on Earth into five kingdoms. The plant and animal kingdoms are confined to multicellular organisms and are called PLANTAE and ANIMALIA. MONERA are BACTERIA and related forms; FUNGI are multicellular and resemble plants but lack chlorophyll and do not use PHOTOSYNTHESIS. The remaining classification, including the one-celled organisms such as amoebas, are called Protista. See **TAXONOMY**.

Proton At one level of understanding, protons are one of the three PARTICLES that make up an atom. ATOMS consist of a NUCLEUS, made of protons and NEUTRONS, around which orbit a cloud of ELECTRONS. The factors that determine an atom's chemical properties are the number of protons in its nucleus and how this number relates to the number of electrons in orbit around the nucleus. Nearly all the MASS of an atom is in the nucleus. Protons and neutrons weigh about 1,800 times as much as electrons.

When physicists probed the innards of the atom more deeply, however, they found that the neutrons and protons are, in turn, made up of QUARKS, three in each type of particle. A major question in physics today is whether quarks are made up of still smaller particles. See **LEPTONS**, and **SUBATOMIC STRUCTURE**.

Psychokinesis The purported ability to move inanimate objects such as chairs, tables, books, lamps, and so on, through mental processes alone. Claims by self-styled psychics or spiritualists to be able to accomplish this have never stood up to objective scientific analysis. See **TELEKINESIS**.

Pulsars Stars that rotate rapidly and generate radio pulses much like a lighthouse emits a powerful beam of **LIGHT** every time it rotates. When detected by radiotelescopes, these stars are called pulsars. Most astronomers believe that pulsars are what is left of a fairly small **STAR** after it has exploded into a **SUPERNOVA**. This is a stage that all stars reach when they exhaust their nuclear fuel and then collapse into a cooler, less energetic form. Because much of the **MATTER** remaining after this collapse has been compressed into neutrons, pulsars are also called **NEUTRON** stars. See **BLACK HOLES**.

Quads When very large amounts of ENERGY are discussed, it is convenient to use a unit called a *quad*, which is defined as a quadrillion (10^{15}) BRITISH THERMAL UNITS, or Btus. A quad is the energy contained in eight billion gallons of gasoline, a year's supply for ten million autos. The United States uses over 20 quads of energy per year to generate its electricity. Total energy consumed in the United States each year now comes to 83 quads.

Quanta Plural of quantum, the fundamental unit of ENERGY. In other words, there is no such thing as "half a quanta." In physics, a quantum is the smallest quantity of radiant energy, equal to PLANCK'S CONSTANT times the frequency of the associated RADIATION. See **QUANTUM PHYSICS**, and **PLANCK, MAX**.

Quantum Physics The theory, originated by Max PLANCK in 1900, that ENERGY exists not in a continuous entity, but rather in separate or discrete units or PARTICLES. Planck called these discrete packets of energy QUANTA, from the Latin for "how much." Physicists had determined that at the molecular level and smaller, MATTER exists in discrete units called ATOMS. It was Planck who showed that the same thing was true of energy. What atomic theory is to matter, quantum theory is to energy. Both concepts are counterintuitive in that they seem to contradict our senses. We perceive the chair we are sitting in, or the ground we are standing on, as solid matter but we know intellectually that both are made up of groups of atoms bound together. In this same way we perceive energy—whether as HEAT, LIGHT, SOUND, electricity, or in any other form—as a continuously variable quantity, which we can adjust by just turning a dial. The truth instead, as Planck demonstrated, is that energy comes only in fundamental, indivisible units called quanta and these units are adjustable only in sequential steps. When the energy of an atom or molecule changes from one value to another, it does so in definite jumps, with no possible value in between.

In 1905 Albert EINSTEIN applied quantum theory to the behavior of light, proving that light does not truly travel in waves (the accepted theory at that time) but in separated particles, which we now call PHOTONS. In 1913 Niels Bohr applied the quantum theory to the atom, creating a model that described the limited number of possible orbits for electrons and thus explained the relative stability of the atom. Understanding quantum effects has made previously incompletely understood phenomena on the atomic scale explainable. According to relativity theory, matter and energy are interchangeable, quantum physics explains the behavior of all fundamental entities, whether they are described as quanta of energy or particles of matter. See PLANCK'S CONSTANT.

Quarks (corks) Considered to be the most basic and fundamental building blocks of all MATTER. Scientists believe that the PARTICLES produced by the BIG BANG were not ATOMS, nor were they even atomic particles such as PROTONS, NEUTRONS, or ELECTRONS. They were, instead, the prime constituents that came together right after the Big Bang to form the particles that we know today. The leading figure in the discovery of the existence of quarks was American physicist Murray Gell-Mann, who first suggested that the protons, neutrons, BOSONS, and some other particles were made up of quarks. No quarks have ever been seen in the laboratory because, according to theory, they cannot exist as free particles. There is no proof that quarks exist but because they are the best way to explain how atoms behave, most physicists believe that these particles within particles do exist. See SUBATOMIC STRUCTURE.

Quasars (short for quasi-stellar radio source) The most distant GALAXIES from EARTH, believed to be the first stage in the evolution of galaxies. Most astronomers believe that quasars are the nuclei of young galaxies, but there are many unexplained mysteries about these far-off points of LIGHT. Because of the way in which the universe is expanding, the farther away an object is from

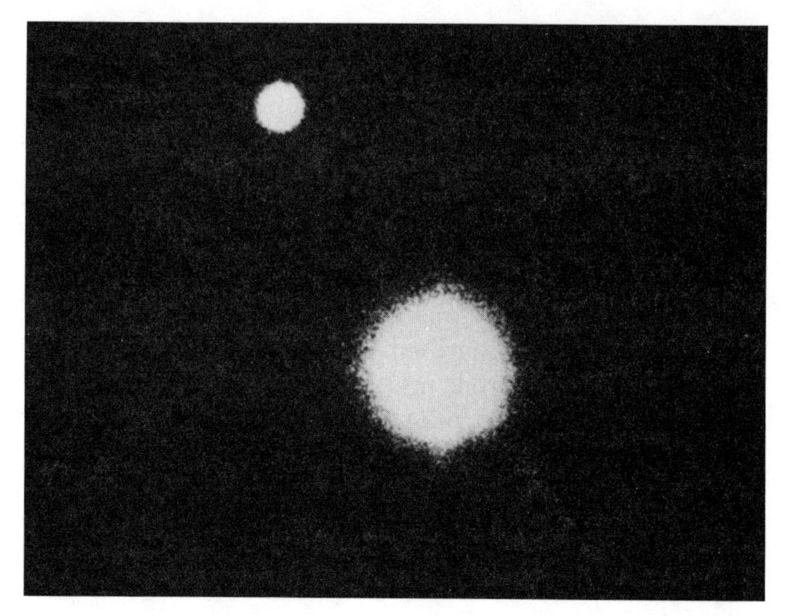

QUASARS are the most distant galaxies from Earth. A single quasar seems to produce energy 100 trillion times more powerful than an ordinary star. Pictured is Quasar 3C-273, which is billions of light-years from our solar system.
Source: NASA photo.

Earth, the faster it is receding from Earth. The nature of the light emitted from quasars (**REDSHIFT**) implies that they are the farthest known objects—billions of light years—from Earth. A single quasar seems to put out as much light as a thousand ordinary galaxies, each consisting of billions of stars. Quasars seem to produce **ENERGY** 100 trillion times more powerful than an ordinary **STAR** and astronomers have no explanation as to how this much energy could be produced.

RAD A unit of absorbed RADIATION. The term *rad* is an acronym for "radiation absorbed dose" and is defined as the energy absorption of 100 ERGs per gram of tissue. Rads measure the radiation that has been absorbed by *something* as compared to REMs, which measure the radiation absorbed by humans. Between 500 and 1,500 rads will kill most people. For all practical purposes, rads and rems are the same for all of us who receive our radiation predominantly from natural background and X rays. Radiation from a single X RAY is about 6 or 7 millirems. See CURIE, and ROENTGEN.

RADAR (*Ra*dio *D*etecting *A*nd *R*anging) An electronic device used to determine the presence and location of an object by measuring the time for the echo of a radio wave to return from an object and the direction from which it returns. Currently, objects are seen on a radar screen as *blips*, or points of LIGHT. Advanced radar systems now under development will have improved resolution to the point where objects on the screen will be recognizable as say that of an L1011 or 747 aircraft.

MICROWAVE-imaging radar is capable of identifying objects as more than blips but Department of Defense research funding is limited because advocates of the Stealth aircraft fear the new radar will render the Stealth bomber obsolete. See STEALTH TECHNOLOGY.

Radial Keratotomy See KERATOTOMY, RADIAL (KT).

Radiation All of the ways in which ENERGY can be given off by an atom. This includes X RAYS, GAMMA RAYS, all charged PARTICLES, and all NEUTRONS. Most ATOMS are stable and therefore nonradioactive. Those that are unstable and radioactive give off either particles or gamma radiation. All radioactive substances form stable sub-

Types		
Natural Radiation	Examples of radioactive elements: uranium, radium	Earth is heated by radioactive elements in its core and rocks
"Unnatural" Radiation	Radiation from unstable or radioactive atoms Substances bombarded by radioactive particles yielding:	
	• alpha particles	positive charge, least penetrating
	• beta particles	high-speed electron or proton moves at about the speed of light
	• gamma rays	usually emitted along with alpha and beta particles, greatest energy and penetration
Radioactivity Measurement		
Curie	Major unit to measure radioactivity; number of disintegrations/second in a gram of radium. 1 curie = 37 billion disintegrations/sec. of 1 gram of radium.	
RAD	Unit of absorbed dose by some thing.	
REM	Unit of absorbed dose by a human; effect equals 1 Roentgen of X rays.	
Roentgen	Unit of exposure dose; Unit = about 84 ergs of energy in a gram of air.	

stances in time. Their HALF-LIFE occurs when they have lost half their activity. Radiation is measured in different ways depending on whether it is occurring in air or is being absorbed. See RADIOACTIVITY.

Radiation, Natural Background We are all subjected to natural background radiation every day. On the average, two-thirds of this natural radiation exposure is due to RADON. According to the National Council on Radiation Protection and Measurements (NCRP), the remaining natural background radiation dose comes from cosmic rays, external gamma radiation (primarily from URANIUM, thorium, and potassium in soil and rock), and other inhaled or ingested radiation. NCRP estimates that the average U.S. total from these sources is 260 and 300 millirems per year. See RADIATION, and REM.

Radioactivity The emission of elementary particles by some ATOMS when their unstable nuclei disintegrate. Materials composed of such atoms are called radioactive. There are a number of naturally radioactive substances such as URANIUM and radium. Other substances can be made radioactive by bombarding them with the particles emitted by radioactive atoms. The three most common forms of RADIATION are: ALPHA PARTICLES (or rays), BETA PARTICLES, and GAMMA RAYS. Alpha particles are the weakest, able to penetrate only air, and can be stopped by a sheet of paper; beta particles can penetrate up to a millimeter of lead; and gamma rays, which are bursts of photons, or very short-wave electromagnetic radiation, can penetrate up to seven inches of lead. All forms of radiation can damage living CELLS, although gamma rays are more likely to do so because of their greater ability to penetrate. See CURIE.

Radio Telescope Astronomers use a variety of devices to observe the universe. Optical telescopes can observe only those objects that emit electromagnetic RADIATION in the visible range. Visible LIGHT is only a small part of the much larger spectrum of radiation, and many objects in the universe produce no visible light strong enough to be detected by an optical telescope. The invention of radio astronomy in the 1940s opened a wider window to the universe. By means of sensitive radio antennas, electromagnetic radiation emitted by distant STARS, GALAXIES, and many other objects can be detected and analyzed. Radio telescopes have detected distant signals that started traveling across space soon after the Big Bang. See ASTRONOMY, BIG BANG, and COSMOLOGY.

Radon A colorless, radioactive gaseous ELEMENT produced by the natural decay of radium and present in soil everywhere. Emissions of sufficient strength are considered a hazard to health. According to the U.S. Environmental Protection Agency (EPA), radon-induced lung cancer is one of today's most serious health issues. The EPA's recommended maximum allowable concentration of radon is 4 picoCuries per liter of air. A pico is one-trillionth of something (see NUMBERS: BIG AND SMALL) and a CURIE (named after Pierre Curie, the co-discoverer of radium) is a unit of radiation approximately equal to the amount of radiation produced by 1 gram of radium of radium) is a unit of radiation approximately equal to the amount of radiation produced by 1 gram of radium per second. The current estimated average U.S. indoor-radon concentration is 1 picoCurie per liter of air.

Emitted by rocks in the soil, radon enters buildings largely through cracks in the foundation. Two 1990 EPA surveys showed that about one out of five homes in the United States exceed a reading of 4 picoCuries. This concentration of radon poses about the same lung-cancer threat to householders as smoking half a pack of cigarettes daily or receiving 200 to 300 chest **X RAYS** annually. By further polluting indoor air, smokers elevate radon risks to themselves as well as to nonsmokers. See **RADIATION, NATURAL BACKGROUND.**

Radwaste Jargon for radioactive waste. More than 24,000 tons of intensely radioactive waste have accumulated around the nation's 111 operating nuclear power plants today. This total is estimated to rise to 40,000 tons within a decade. At least an equal amount has accumulated at the military weapons production facilities. What to do with this mountain of dangerous material is a major national problem. Over the years, scientists have argued about different proposed ways to dispose of radwaste. They have suggested dumping it on the ocean floor, rocketing it into space, lowering it into extremely deep holes and dropping it into glaciers. Ultimately, the U.S. Department of Energy has decided to bury it. The problem is where, since there are technical or political objections to every site that has been nominated. Finding an acceptable solution to the radwaste problem is the major stumbling block to the expansion of nuclear power. See **NUCLEAR ENERGY**, and **YUCCA MOUNTAIN.**

RAM (Random Access Memory) **SEMICONDUCTOR** data storage that can be changed or edited at will but is erased when the computer is turned off. In computerese, RAM memory can be read or written to. That is, information can be taken out of or put into this memory. Early computers had less than 64,000 characters of RAM (called 64K RAM). Personal computers today often have two or four megaBYTEs (2MB or 4MB equal to 2 or 4 million characters) of random access memory.

RAM capability is often used to compare the capacity of one COMPUTER with another. See **BITS**, and **ROM (READ ONLY MEMORY)**.

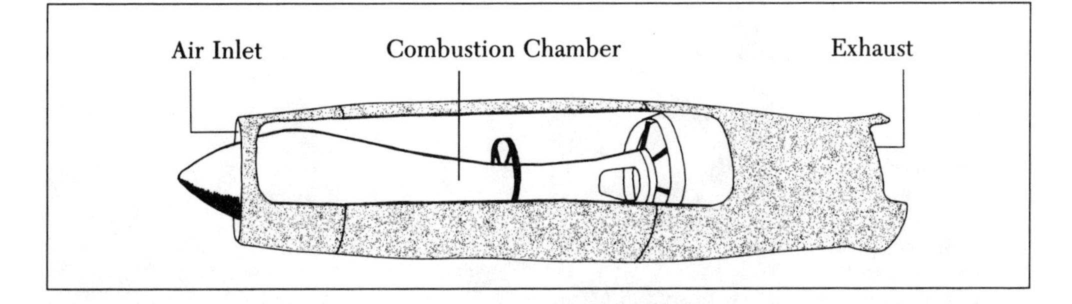

RAMJET is the simplest of all the jet engines in that it does not use turbines or compressors but relies on forward motion to ram air into the combustor. The air, heated and compressed by the ramming at supersonic speeds, then mixes with the fuel and burns providing forward thrust.

Ramjet Ramjets are the simplest of all the jet engines in that they do not use turbines or compressors but rather rely on the forward motion to ram air into their combustors. The air, heated and compressed by the ramming process, then is mixed with fuel and burned. Ramjets have to be brought up to speeds by some other form of propulsion, because their air ramming effect does not take place at speeds below MACH 1.5. The upper limit for ramjets is about Mach 6, where a decrease in fuel efficiency takes place. Combination of TURBOJET/ramjet engines have been considered for future Supersonic aircraft, with the turbojet handling takeoffs and landings and the ramjet used during hypersonic cruising. See **SST**.

Reactor, Breeder A nuclear breeder reactor has two purposes: it generates electrical ENERGY and at the same time it produces fuel for other FISSION reactors. Theoretically, a breeder reactor will extract 60 to 80 percent of a fuel's total energy, compared with about 1 percent efficiency of a conventional fission reactor. Both breeders and conventional reactors rely on splitting ATOMS (fission) to generate HEAT, which ultimately creates steam to drive electric generators. Breeders, however, also use the fission process to convert URANIUM 238 into plutonium. From each fission of uranium in the core, more than one plutonium atom is manufactured. In other words, more fuel is created than is consumed.

Despite the obvious advantages of breeder reactors, not a single commercial breeder is in operation in this country today. Safety is the main cause of concern in the

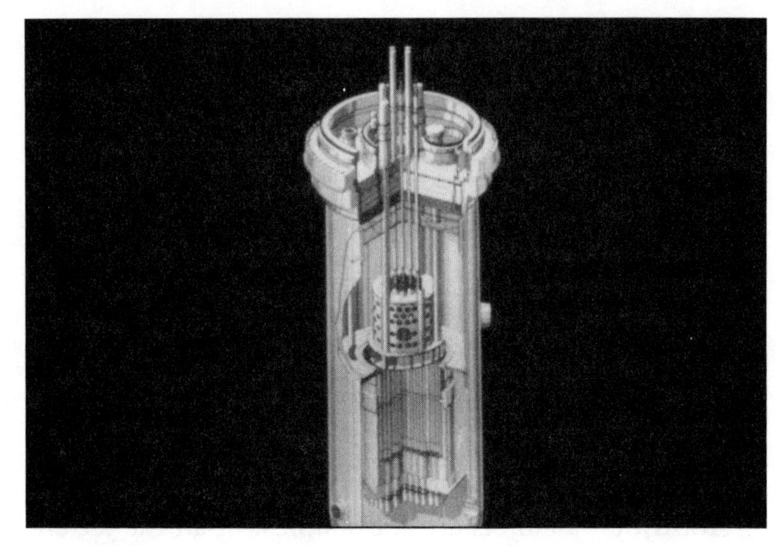

REACTOR, BREEDER. Artist's cutaway drawing of the Clinch River Breeder Reactor Project core. This project, now deactivated, was designed to demonstrate that a breeder reactor could operate reliably and safely in a utility environment. High costs and technical problems led to the deactivation of this project.
Source: U.S. Department of Energy photo.

development of breeder reactors. The same concerns that give us pause in the case of fission reactors are quadrupled where breeder reactors are concerned. Breeders create plutonium, and this material is considerably more dangerous and longer-lived than uranium. Some experts consider plutonium to be the most dangerous material in the world. If even a small amount of this material were to escape into the ATMOSPHERE, it would be a major catastrophe. Another concern is nuclear weapons proliferation. Spent fuel from a breeder reactor must be reprocessed to recover the plutonium created in its operation. The reprocessing is the one stage in the fuel cycle where material directly usable in weapons manufacture would be available. Storage and transportation of this "weapons-grade material" is a cause of concern. See **NUCLEAR ENERGY**, and **FUSION, NUCLEAR**.

Reactor Designs A new generation of nuclear power reactors that would be cheaper to build and easier to operate are now under consideration. The new designs feature *passive* safety systems and are labeled by the would-be manufacturers as "inherently safe." Currently, the nation's 111 nuclear power plants rely on what is called *defense in depth* systems in which components that might fail are backed up by other design features. Existing reactors, for instance, are surrounded by very strong steel and concrete containments, as a last barrier in case multiple failures cause a radiation leak. The new designs rely instead on built-in features for safety. A Westinghouse Electric Corporation design would deliver emergency cooling water to the reactor without the use of pumps. When sensors determine that the cooling water needed, it would automatically be forced in by the pressure of gas stored in tanks. Another concept (from General Atomics) calls for fuel to be encapsulated in graphite instead of metal and cooled by HELIUM instead of water. According to General Atomics, the metal used in existing reactors can melt, but the encapsulated fuel cannot get hot enough to damage the graphite. Also, water can boil away, but helium cannot.

According to the Union of Concerned Scientists, there is nothing inherently safe about a nuclear reactor. They submit that although it is possible to design reactors that are safer, and it is also possible to design a reactor that is economically competitive, they are not convinced one can do both. See **NUCLEAR ENERGY**.

Reactor Fundamentals A nuclear power plant is a complex assembly of steam generators, turbines, cooling systems, monitoring and inspection equipment, and power transmission lines. The heart of the complex is the reactor vessel, which contains a central core. The core contains a group of control rods that are moved in and out of the core to regulate the rate of FISSION. Dissipating the high residual HEAT in the core is the biggest potential problem with nuclear power. In *pressurized water reactors,* steam generators convert the hot water from the reactor vessel to steam, which then drives a turbine and produces electricity. In *boiling water reactors,* the steam is generated directly inside the vessel. See **REACTOR DESIGNS**.

Recombinant DNA The scientific, biological technique used to manipulate or rearrange genetic material to alter hereditary traits. Also known as *gene splicing* or GENETIC ENGINEERING. Recombinant, or recombined, DNA involves taking a GENE from one organism and splicing it into another. If the transfer is successful, the recipient organism will carry out the instructions of the new gene. See **DNA**.

Red Giant Stars Huge, relatively cool stars that appear redder in color than do main sequence or younger stars. The life span of stars is a function of their MASS. Smaller stars last for trillions of years, whereas the giant stars have the least time to live: typically only 10 million to 100 million years. As a STAR goes through its final phases and runs low on fuel, its outer portion expands and cools. The star's color changes from a yellow-white to a deepening red and it becomes a *red giant.*

Redshift The change in the wavelength of SOUND or LIGHT that occurs when the source of the sound or light is moving toward or away from an observer is called the DOPPLER EFFECT. If a source of light waves is moving toward a receiver, the frequency of the wavelengths increases and the wavelength is shorter, producing a bluish light (called blueshift). If the source of light is moving away from a receiver, the frequency of wavelengths become longer and the frequency decreases, producing a reddish light (called redshift). In 1929 the astronomer Edwin P. HUBBLE detected that the light from distant stars was becoming redder. He determined that this redshift meant that stars were receding from EARTH. Redshift indicates that all distant GALAXIES are moving away from Earth and the more distant the galaxy, the faster it is moving away from us. The concept of an EXPANDING UNIVERSE supports the BIG BANG theory for the origin of the universe.

Relativity See GENERAL THEORY OF RELATIVITY, SPECIAL THEORY OF RELATIVITY, and EINSTEIN.

REM (Roentgen Equivalent Man) A rem is a unit of absorbed RADIATION equal to the biological effect produced by one ROENTGEN of X RAYS. The not-to-be-exceeded standard of exposure for workers in nuclear plants in both the United States and Soviet Union is now 5 rems per year. New standards recommended by many concerned scientists would limit annual exposure to 2 rems per year. These new standards would result in action to protect aircraft crew members, some of whom routinely incur exposures of half the new limits from radiation from the SUN and STARS, which is stronger at the higher altitudes.

The current maximum exposure to the general population from an atomic installation must be no more than

.10 rem per year. Radiation from an X ray is about .006 or .007 rem. Government studies have shown the average American can be expected to be exposed to about .36 rem of radiation from various sources annually. See CURIE, and RAD.

REM (Rapid Eye Movement) A stage of sleeping during which rapid eye movement under closed eyelids occurs. Sleep researchers use electroencephalographic techniques to measure brain waves and other physiological signs to indicate dreaming stages of sleep in volunteers. Intense dreaming during REM sleep apparently includes mental processes inherent in cognition and problem solving.

Ribosome The specialized part of a CELL having the specific function of reading the GENETIC CODE (brought by the RNA from the DNA in the nucleus) and assembling the PROTEIN MOLECULE that is called for by the code. Ribosome functions as the site of protein manufacture.

Richter Scale Commonly used scale that measures the magnitudes of EARTHQUAKES. (See diagram on page 262.) Named for U.S. seismologist Charles F. Richter (1900–1985), it is a LOGARITHMIC SCALE, which means that for every increase of one on the scale, there is a tenfold increase in ground motions, or height of the seismic waves. An earthquake measured at 6 on the Richter scale would be ten times the magnitude of an earthquake measured at 5 on the same scale. Magnitude and ENERGY released by an earthquake are not the same thing. The energy released by an earthquake's motion is about 30 times greater for each increase of one on the Richter scale. See ERG, PLATE TECTONICS, and SEISMOLOGY.

RISC (Reduced Instruction Set Computer) A type of COMPUTER CHIP that executes fewer commands than conventional computer chips and thereby streamlines and accelerates the operation of the chip. RISC chips improve computer speed without making the machines more complex. They do this by simplifying the HARDWARE as compared to conventional or so-called CISC (Complex Instruction Set Computers) chips, which can do more but slow down the system and create bottlenecks.

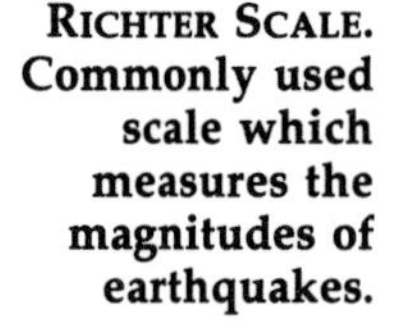
RICHTER SCALE. Commonly used scale which measures the magnitudes of earthquakes.

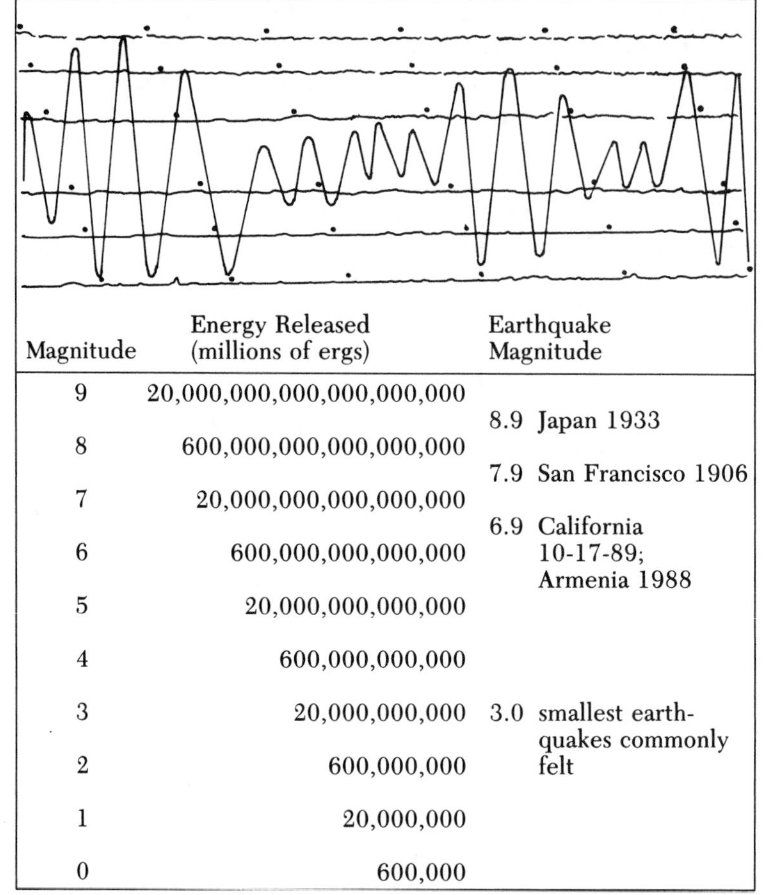

Magnitude	Energy Released (millions of ergs)	Earthquake Magnitude
9	20,000,000,000,000,000,000	
		8.9 Japan 1933
8	600,000,000,000,000,000	
		7.9 San Francisco 1906
7	20,000,000,000,000,000	
		6.9 California
6	600,000,000,000,000	10-17-89; Armenia 1988
5	20,000,000,000,000	
4	600,000,000,000	
3	20,000,000,000	3.0 smallest earth-quakes commonly
2	600,000,000	felt
1	20,000,000	
0	600,000	

RISC chips may be thought of as a football playbook in which 115 plays have been cut down to 25. The team has fewer options, but it can execute each of them better.

RNA (Ribonucleic Acid) A long-chain MOLECULE that plays an important part in the regulation of a CELL by DNA, especially the formation of PROTEINS. There are two kinds of RNA, messenger RNA (mRNA) and transfer RNA (tRNA). *Messenger* RNA transcribes the genetic message from DNA and carries it out of the nucleus to the cellular machinery that follows its instructions in the production of proteins. *Transfer* RNA helps select a specific AMINO ACID and transfers it to the RIBOSOME, the machinery inside a cell that assembles the protein molecules called for by the GENETIC CODE. See **DNA**.

Robots and Robotics The term *robot* is used to designate any artificial device or system that performs functions ordinarily thought to be appropriate for human beings. The term *robotics* refers to the science and technology of the design, construction, maintenance, and use of robots. The science and science-fiction writer Isaac Asimov originated this latter term in a story published in 1942 (later collected into a book entitled *I, Robot*). The replacement of human workers by robots, as is common today in manufacturing, is called *automation*. Industrial robots have found their greatest use today on automobile assembly lines, particularly in Japan. Future computerized robots are in development that have the ability to "see" and "hear." NASA is considering the use of such supersmart robotic systems for assembling space-station components in Earth-orbit. See also **ARTIFICIAL INTELLIGENCE.**

Roentgen (also Rontgen) A unit of radiological dose, defined as the dose that corresponds to the release of 83.3 **ERGS** of energy in one gram of air. Named after the German physicist and discoverer of **X RAYS**, Wilhelm K. Roentgen (1845–1923). A roentgen is equal to the amount of **RADIATION** producing one electrostatic unit of positive or negative charge per centimeter of air. The difference between **RADS**, **REMS**, and roentgens is important in physics. Roentgens are always measured in air. Rads measure the radiation that has been absorbed by something. Rems measure the radiation absorbed by humans. See also **CURIE.**

ROM (Read Only Memory) **SEMICONDUCTOR** data storage that is not erased when the **COMPUTER** is turned off. Data stored in ROM memory cannot be changed. ROM memory is used to permanently store machine operating programs that are frequently used. ROM memory is what is necessary to run a specific computer and therefore is not often used to compare one computer with another, as RAM memory capacity is. See **RAM.**

Satellite A natural body that orbits around a PLANET such as a moon. Also, a human-made device designed to be launched into orbit around the EARTH, some other planet, or the SUN. See also **MOONS**.

SATELLITE, in this case one of Neptune's moons (Satellite 1989N1) discovered and photographed by *Voyager* 2.
Source: NASA photo.

Saturn The sixth PLANET out from the SUN, Saturn is considered one of the most beautiful telescopic sights in the sky, principally because of its system of rings that surround the body of the planet. Saturn's rotation period on its axis is rapid, about 10¼ hours. Saturn moves around the Sun at an average distance of 886 million miles. The planet's orbital period around the Sun is 29 years 167 days. The volume of this large planet is sufficient to contain 740 Earths, but its mass is only 95 times that of Earth. Average density is therefore low, less than that of water. The TEMPERATURE of Saturn is judged to be about −180°C and its ATMOSPHERE consists mainly of

SATURN photograph assembled from *Voyager* 2 images obtained in 1981 from a distance of 13 million miles (21 million kilometers) on the spacecraft's approach trajectory. Three of Saturn's icy moons are evident at left. They are, in order of distance from the planet: Tethys, Dione, and Thea.
Source: NASA photo.

HYDROGEN but also contains METHANE. The most striking feature of this planet has always been its rings, now known to be chunks of ice and rock orbiting the planet. VOYAGER SPACECRAFT discovered that Saturn's rings were really made up of many thousands of thin ringlets. We now know that JUPITER, URANUS, and NEPTUNE also have rings, but Saturn's are the most spectacular.

Another feature of Saturn fascinating to astronomers is its, at least, 18 MOONS. These icy bodies orbiting Saturn display a wide range of geologic evolution. Many of them show craters more than four billion years old, but at least one of them is so new that no craters are visible. See SOLAR SYSTEM.

Scanning Tunneling Microscope (STM) A versatile new tool researchers use for probing the complexities of surface structures, atom by atom. Originally the technique was used to investigate the atomic and electronic structure of SEMICONDUCTOR surfaces. Now scanning tunneling microscopes are beginning to play an important role in studies of the molecular and chemical properties of a wide range of solid surfaces. This instrument, invented in 1981, works by positioning a tiny metal tip within a few atomic diameters of a surface, close enough to allow electrons to leap, or *tunnel,* across the gap and generate an electrical current. As the gap between the tip and the sample surface increases, the current decreases. A scanning mechanism pulls the needle over the sample's surface, constantly adjusting the tip's height to keep the current constant. The tip's bobbing journey produces a sketch of the surface's microscopic contours—a kind of miniature topographic map in which the hills and valleys represent arrays of ATOMS. Scientists are thus able to chart surface bumps and grooves as small as individual atoms.

Scientists knew that the microscope could also be used to rearrange atoms. They discovered that they could reposition the atoms, one by one, by moving the microscope tip close and dragging them along. The first application of this technique was pure showmanship: Scientists at IBM arranged 35 atoms to spell out the company logo in letters

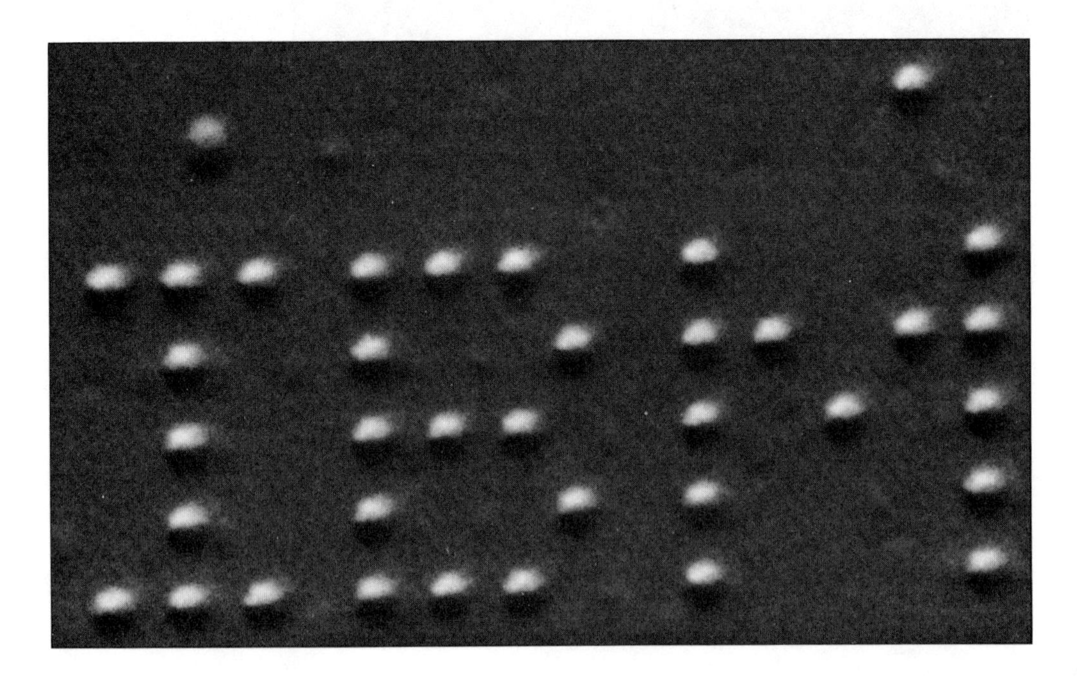

SCANNING TUNNELING MICROSCOPE (STM) used by researchers to probe the complexities of surface structures, atom by atom. Here this microscope is used to spell out the letters "IBM" in atoms. These IBM letters are about 500,000 times smaller than those on this page. The distance between atoms in the pattern is about 50 billionths of an inch (13 angstroms) while the entire IBM is about 660 billionths of an inch (168 angstroms) in length. The image shown here is magnified almost 6 million times.
Source: IBM Corporation photo.

one-atom high. Written on this scale, more than 10,000 copies of your daily newspaper would easily fit inside a typewritten period on this page. The potential scientific applications for this new instrument keep growing. Physicists, engineers, chemists, and biologists are starting to use the scanning tunneling microscope for probing the structure of biologically important MOLECULES such as DNA, monitoring the formation of thin films on metals, and many other applications.

Scientific Literacy, Definition of The ability to understand the scientific vocabulary well enough to follow, or, if necessary, take part in public debates about issues involving science and technology. The American Association for the Advancement Of Science (AAAS) goes a bit further in their definition. The AAAS defines a scientifically literate person as "one who is aware that science, mathematics, and technology are interdependent human enterprises with strengths and limitations; understands key concepts and principles of science; is familiar with the natural world and recognizes both its diversity and unity; and uses scientific knowledge and scientific ways of thinking for individual and social purposes."

SDI (Strategic Defense Initiative) Popularly known as Star Wars, SDI is the proposed advanced missile defense system. The original SDI concept planned an invulnerable space "shield" that would make offensive ballistic missiles obsolete. LASERS were proposed as the basic weapon system including hard X ray lasers which were to have been powered by a controlled nuclear explosion. That early concept did not survive expert scrutiny. After technological reality set in, however, the SDI objectives were changed to more modest ones including blunting a major Soviet attack, preventing small missile attacks by other countries, or providing a measure of defense against accidental or renegade attacks. Costs of SDI research are expected to rise to $12.5 billion by 1997 and Congressional debate about the usefulness of this anti-ballistic missile system will be rancorous. Although many critics support some research as a hedge against a Soviet missile technology breakthrough, deployment they say is impractical and destabilizing. Critics also point out that small countries with nuclear weapons could find means other than missiles for delivering the weapons.

Currently under development as part of the SDI program are three approaches to ballistic missile defense: (1) the space-based "Brilliant Pebbles" concept, (2) the ground-based Exoatmospheric Reentry Vehicle Interceptor System (ERIS), and (3) the ground-based High Endoatmospheric Defense Interceptor (HEDI). *Exoatmospheric* means beyond the Earth's atomosphere and *Endoatomospheric* means within the Earth's atmosphere. Brilliant Pebbles, known technically as kinetic kill vehicles, envisions a fleet of tiny maneuverable SATELLITEs, each only three feet long and equipped with miniature rocket motors, sensors, and guidance systems to home in on enemy missiles. They would carry no explosives but would destroy their targets by crashing into them. As a defense against a massive incoming attack, Brilliant Pebbles—at least in theory—is more efficient than ground-based interceptors, because it destroys missiles in their boost phase, before multiple warheads and decoys break away from launch vehicles. Both of the ground-based systems, ERIS and HEDI, use high-speed, nonnuclear rockets to destroy ballistic missiles on impact. ERIS attacks enemy warheads while they are still in orbit; HEDI attacks incoming ballistic missiles after they have reentered the Earth's ATMOSPHERE. The major technical challenge facing engineers on both of the ground-based systems is how to discriminate between warheads and decoys. A major political/budgetary challenge to all three systems in an era of Washington-Moscow cordiality can be expected.

Seismology The science or study of the action of seismic, or shock, waves generated by EARTHQUAKES or underground explosions. Most of humankind's knowledge of the Earth's interior comes from the study of the action of seismic waves. Seismic waves can be detected far from their source and can thus be used in detecting underground nuclear tests. See EARTH, COMPOSITION OF.

Semiconductor Substances such as silicon or germanium that have electrical conductivity capabilities intermediate between that of a resistor and a conductor; that is, they can either resist or easily pass the flow of electric current. An area of a piece of silicon (called a CHIP) can be chemically treated to readily conduct electricity, whereas a different area of the chip can be made to prevent the flow of current. TRANSISTORS in a COMPUTER are made of semiconducting material and function as switches. Electric current goes from one end of the device to the other, through a gate that sometimes lets the current pass and sometimes does not, providing the on/off signals necessary for a BINARY system.

SETI (Search for Extraterrestrial Intelligence)

Some astronomers have long held that there are probably so many PLANETS in the universe that even if only a tiny fraction were suitable for life, there should be thousands or even millions of planets with life. Recent advances in astronomy and physics have strengthened the theory that there could be many planetary systems hospitable to life. If there are advanced civilizations out there somewhere, how do we communicate with them? One approach has been the NASA-sponsored project that listens for artificially generated electromagnetic signals coming from space on the assumption that any technologically advanced civilization would be producing radio/TV/RADAR signals just as EARTH does. The earliest of Earth-generated electromagnetic signals are by now nearly 100 light-years away in all directions. Attempts to detect alien signals beamed toward Earth have so far been unsuccessful, but the number of stars that have been examined is less than .1 percent of the number that would have to be examined if there were to be a reasonable statistical chance of discovering extraterrestrial civilizations.

Project SETI (rhymes with "jetty") has been limping along on a shoestring budget for three decades. Now NASA has developed new computer and signals-processing technologies capable of a much larger search. The new equipment, scheduled to go into full-time operation in 1992, will provide a search ten billion times more comprehensive than all previous attempts combined. Scientists are divided on the usefulness of SETI and some consider the entire project as being flaky or pseudoscience. Advocates of SETI, on the other hand, point out the huge return on a modest investment that would result with radio contact with other worlds. SETI advocates are careful to keep LIGHT-YEARS of distance between themselves and believers in UFOs, saying that reports of such visits are based more on wishful thinking than on science. See DRAKE'S EQUATION, and EXTRATERRESTRIAL INTELLIGENCE.

Shooting Stars Not really stars at all but rather METEORS or small masses heated to incandescence as they pass through Earth's atmosphere. If a large meteor impacts EARTH, a rare occurrence, what is left of it after the encounter is called a METEORITE. Most meteors are very small particles no larger than grains of sand. They enter our ATMOSPHERE at altitudes between 20 and 50 miles above the ground and, as they fall, they reach speeds of between 7 and 43 miles per second. The HEAT generated by their collisions with air molecules causes the particles to melt and vaporize.

Whenever the Earth in its journey around the SUN passes through the dust left over from defunct COMETS, the dust burns up in the Earth's atmosphere and we see this phenomenon as showers of shooting STARs. Because the Earth travels along the same orbit through the SOLAR SYSTEM each year, showers of up to 50 shooting stars appear at regular intervals. On an ordinary night, however, when there is no shower, an observer can expect to see between three and six meteors per hour. See ASTEROIDS.

Sidereal (sye-dee'ree-al) Time as measured with respect to the STARs as opposed to time measured with respect to the SUN (solar time). Time on our watches or clocks is based on solar time. When the EARTH rotates once with respect to the Sun, one ordinary (solar) day has passed. When the Earth has rotated once with respect to the stars, it has also moved ahead 1/365 of the distance in its path around the Sun. The Earth must then rotate a little farther before an observer on Earth sees the Sun come back to the same place in the sky. Specifically, the Earth must rotate an additional 1/365 of a day, or 3 minutes 56 seconds. Therefore a solar day is longer than a sidereal day. Sidereal time matches solar time at the fall EQUINOX, around September 21. After that, sidereal time grows earlier than solar time by 3 minutes 56 seconds each day. At the spring equinox, sidereal time is 12 hours earlier than solar time. Astronomers use sidereal time to keep track of when certain stars will appear in the night sky, since a given star or constellation rises at the same sidereal time each night.

Singularity The term used by physicists and mathematicians to designate the point in the universe where the equations of the GENERAL THEORY OF RELATIVITY break down. The universe at the first moment of time (BIG BANG) is called a singularity, as is a BLACK HOLE. Physicists believe that at a time in the past (currently estimated at between ten and twenty billion years ago) the distance between neighboring GALAXIES must have been zero. That is, all MATTER in the universe was contained in one single point. At that time, which we now call the Big Bang, the density of the universe and the curvature of space would have been infinite. See EINSTEIN.

Sixth Sense See EXTRASENSORY PERCEPTION.

Skin Cancer Medical and biological scientists are of the opinion that essentially the same spectrum of RADIATION responsible for sunburn is also responsible for skin cancer. Two forms of skin cancer are recognized. One variety (nonmelanoma) is found principally in older people, indicating cumulative effect of exposure over many years. This variety of skin cancer is easily diagnosed and can be successfully treated. The other, more dangerous, variety of skin cancer (MELANOMA) affects a younger segment of the population (30 to 50 years of age) and is a serious and often fatal medical problem. The current depletion of the Earth's protective OZONE layer apparently is contributing to a dramatic increase in melanoma.

Health officials have warned Americans to minimize sunbathing, pointing out that the death rate from melanoma has jumped alarmingly in the past 60 years. By the year 2000, an estimated one in 90 Americans will develop melanoma during their lifetimes, as compared to only one in 1,500 in 1930. There now appears to be a clear cause and effect between ozone loss and increased melanoma mortality. Experts calculate that a 7-percent ozone loss may translate into nearly a 10-percent increase in skin cancer mortality. See OZONE DEPLETION.

Software, Computer The set of programs or instructions that control a computer's functions. As a musical score is to a symphony, software is to a computer. Two types of software are necessary to get the most out of our modern computers: (1) an operating system to control the computer's basic functions, usually built into the machine at the factory; and (2) applications programs of the type we can purchase at a software store to perform a specific task for us (such as word processing or accounting). See **HARDWARE, COMPUTER**.

Solar Constant The average rate at which radiant solar **ENERGY** is received from the **SUN** by the **EARTH**. Sunspots and solar-flare activity cause fluctuations in the rate at which the Sun radiates energy, but for most meteorological purposes solar radiation may be assumed to be unchanging. At the outer limits of the atmosphere, this energy arrives at the continuous rate of 1.35 kilowatts per square meter when the Sun is directly overhead—a figure known as the *solar constant.* More technically, the solar constant is equal to 1.94 **CALORIES** (as a unit of **HEAT**) per minute per square centimeter of area perpendicular to the Sun's rays as measured outside the Earth's **ATMOSPHERE** when the Earth is at its **MEAN** distance from the Sun. See **SOLAR ENERGY**.

SOLAR ECLIPSE occurs when sunlight is blocked by the Moon from reaching a small region on Earth. Though a solar eclipse occurs somewhere on Earth every year and a half, it takes more than 300 years on the average for an eclipse to pass your particular location. The July 11, 1991, total solar eclipse was visible from the Big Island of Hawaii and from the west coast of Mexico, near the tip of Baja, California.

Solar Eclipse When sunlight is blocked by the **MOON** from reaching a small region on **EARTH**. Because of the tilt in the Moon's orbit with respect to the Earth's, the Moon does not pass exactly between the Earth and the **SUN** every month. But about every year and a half, this event does occur. As the Sun, Moon, and Earth move in space, the shadow cast by the intervening Moon sweeps a long, thin path across the Earth, thousands of miles long and as much as hundreds of miles wide—observers within this shadow witness a total solar eclipse. Although an eclipse occurs somewhere on Earth every year and a half, it takes about 300 years for an eclipse to return to a particular location. See **LUNAR ECLIPSE**.

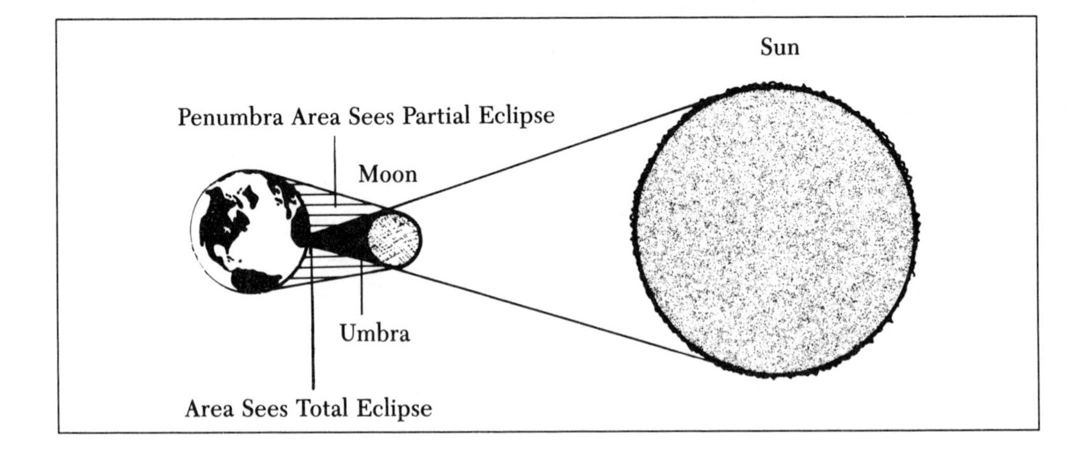

Solar Energy Almost all of the ENERGY we use starts out as solar energy. In the core of the SUN, nuclear energy is transformed into electromagnetic energy, including sunlight. All of the FOSSIL FUELS we use—coal, oil, wood—are really solar energy stored in carbon compounds. Nuclear fuels used in our atomic power plants are solar and stellar energy stored in heavy atomic nuclei. Wind and water power are both examples of transformed stellar energy. Even the internal HEAT of Earth is generated by the radioactive decay of elements that started out in the STARS and was incorporated into primitive Earth. Tidal power is the only form of energy we use that does not originate in stellar energy. Tidal energy results from gravitational pull and the rotational momentum of Earth. This is a minor exception because humankind does not convert much tidal power to usable energy.

The number one problem with solar power, when used to make electricity, is the cost. At current costs, solar or photoelectric cells are too expensive and inefficient to be used for the commercial generation of electricity. Energy for SATELLITES in orbit around the Earth or for space probes is produced by PHOTOVOLTAIC conversion of energy. In these cases, the need for electricity is small and cost is not a factor. The biggest problem with solar heating systems is storing the heat for later use. In some regions of the world, solar water heating is competitive and cost-effective. Israel, Japan, California, and the southwestern areas of the United States are locations where the use of

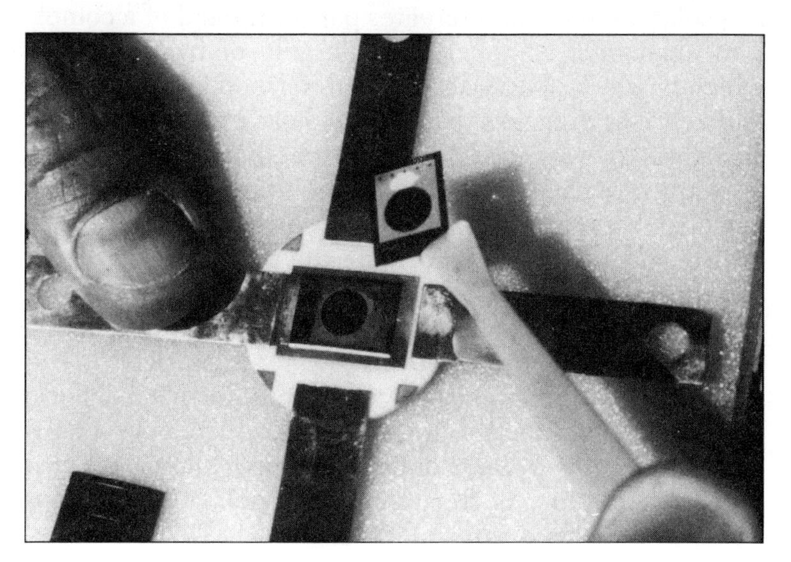

SOLAR ENERGY, when used to generate electricity, requires a photovoltaic cell. Here a solar cell is being stacked in a mechanical arrangement. The top cell converts light from the blue, or high-energy part of the spectrum, while the bottom cell converts the red, or low-energy, part of the spectrum.
Source: U.S. Department of Energy photo.

solar power is growing. Almost 60 percent of Israel's households now have solar-heated water. Some 11 percent of Japanese houses use solar systems, usually for heating water. See SOLAR CONSTANT.

Solar System That region of the universe that includes our SUN, its nine known major PLANETS and their MOONS or SATELLITES, and objects such as ASTEROIDS and COMETS. In other words, the Sun and all the objects that are in orbit around it. The major planets, in order of their average distance from Sun, are MERCURY, VENUS, EARTH, MARS, JUPITER, SATURN, URANUS, NEPTUNE, and PLUTO. The size of this relatively "small" solar system is about 7.3 billion miles across. The approximate distance from the Sun for each planet is:

Mercury	36 million miles
Venus	67 million miles
Earth	93 million miles
Mars	141 million miles
Jupiter	480 million miles
Saturn	900 million miles
Uranus	1.8 billion miles
Neptune	2.8 billion miles
Pluto	3.6 billion miles

Solar Wind The stream of charged PARTICLES emitted by the SUN as it constantly sheds a tiny fraction of its MASS into interstellar space. One manifestation of this solar wind is that it creates part of the tail of a comet. The head of a comet consists largely of frozen water, which is why COMETS are often described as dirty snowballs. As it approaches the Sun, the head becomes warmer and starts to evaporate. The evaporated material is driven outward from the Sun in part by the solar wind. The electrically charged particles of the solar wind excite gas ATOMS and MOLECULES, forming the brighter part of the tail of comets. Another indication of solar wind is the colorful display of the *aurora borealis,* or *northern lights.* When the path of the electrically charged particles in the solar wind are bent and trapped by the Earth's magnetic field, they energize the atoms in the Earth's upper ATMOSPHERE and this process emits the reds, yellows, greens, purples, and pinks of the aurora. The point in far distant space where the solar wind of charged particles is overcome by the interstellar magnetic field marks the point where our solar system ends and interstellar space begins.

Solstices The two days of the year when the SUN shines the longest in the Northern Hemisphere (summer solstice) or the shortest (winter solstice). The dates for these two events, which mark the beginning of summer and winter in the Northern Hemisphere, occur about June 21 and December 22. These are the two times of the year when the Sun has no apparent northward or southward motion, at the most northern or most southern point of the Sun's apparent path among the stars. See also **EQUINOX**.

SONAR (Sound Navigation And Ranging) The use of SOUND waves to locate underwater objects or the depth of water under a ship (RADAR uses radio waves in a similar way). Seawater is opaque to virtually all types of electromagnetic radiation and, therefore, radar won't work under the ocean. Sound waves, on the other hand, can travel over long distances through the ocean and, therefore, are the primary means of detecting, locating, and classifying submerged targets. There are two types of sonar: passive and active. *Passive* sonar functions solely as a listening device. *Active* sonar emits pulses of sound (called *pings*) into the ocean and then listens for the echoes reflected from undersea objects. See **SONARGRAPHY**; and **SOUND, SPEED OF**.

Sonargraphy The use of high-frequency SOUND waves to "see" within the human body. A painless and relatively inexpensive medical imaging technology, sonargraphy is used to examine pregnant women and in the examination of breast, heart, liver, and gallbladder. The system uses piezoelectric (or pressure-sensitive) crystals that convert electric pulses into vibrations that penetrate the body, strike the organs within, and reflect back to the surface, where the crystal functions as a receiver. The returning signals are used to sketch the target's location, size, shape, and even texture for display on a screen. See also **ULTRASOUND**.

Sound A form of ENERGY caused when an object vibrates, creating movement in MOLECULES in elastic mediums such as air. We hear sound when the disturbed molecules moving through air reach our eardrum, causing it to vibrate and send small pulses of electrochemical energy to the brain for interpretation. The human ear can hear sounds with frequencies between 20 and 20,000 cycles (vibrations) per second. Sounds with frequencies below 20 cycles per second or **HERTZ** (Hz) are called *infrasonic*. Those above 20,000 are called *ultrasonic*.

Sound intensity is measured in **DECIBELS** (dB), named after Alexander Graham Bell (1847–1922). The sound of rustling leaves, for example, would fall into the 0 to 20 dB range, while city traffic noises would usually be in the range of 70 to 100 dB. See **DOPPLER EFFECT**; **SONAR**; **SONARGRAPHY**; **SOUND, SPEED OF**; and **ULTRASOUND**.

Sound, Speed of **SOUND** travels through air at about 1,125 feet (343 meters) per second at sea level. Measured in miles per hour (mph), the speed of sound at sea level is about 740 mph. A **MACH** number measures the ratio of an object's speed to the speed of sound in miles per hour. Mach 2 is therefore equal to 1,480 miles per hour (or twice the speed of sound). For a long time, the speed of sound was considered a barrier or limit to speed of aircraft. Because aircraft undergo an abruptly increasing drag force induced by compression of the surrounding air at speeds close to the speed of sound, **MACH** 1 was considered a barrier. This hypothetical sonic limit was broken in the 1950s. Sound travels faster through solids and liquids than through air. The denser the medium, the better the sound conduction. See **DOPPLER EFFECT**, **SONAR**, **SONARGRAPHY**, and **ULTRASOUND**.

Space Exploration Space exploration can be said to have started with the American *Mariner* and the Soviet *Venera* missions to **VENUS** in the 1960s. As of this writing, unmanned U.S. spacecraft have explored all the **PLANETS** in this **SOLAR SYSTEM** except **PLUTO**, the planet farthest away from the **SUN**. The Soviets have reached only Venus and **MARS**. Plans for future unmanned U.S. exploration include missions to Venus, **JUPITER**, the Sun, Mars, and the **KOPFF COMET**. Planned, manned U.S. missions include a new American **SPACE STATION** and, eventually, a manned mission to Mars. Controversy and budgetary squabbles continue to exist in the scientific community over the issue

SPACE EXPLORATION in the future may include a manned Mars mission. This artist's concept depicts hardware that might be involved in the event manned visits ever occur. Hardware seen here includes the Mars explorer, a traverse vehicle, a habitation module, a power module, a greenhouse, central base, and a Mars airplane.
Source: NASA photo.

of manned versus unmanned space exploration. At this writing, the bulk of NASA resources are to be invested in manned space facilities, but critics of this approach have called for the allocation of at least 20 percent of NASA's budget to unmanned scientific missions. See COMETS, GALILEO SPACECRAFT, and MAGELLAN SPACECRAFT.

Space Station The NASA-proposed structure in space designed to house astronauts doing science experiments and preparing for voyages to the MOON and MARS. The space station would be a permanent habitat or base of operations in space. The Soviet Union already has a

SPACE STATION. Proposed structure in space designed to house astronauts over an extended period of time. This artist concept shows a shuttle vehicle off-loading supplies to the Space Station Freedom.
Source: NASA photo.

space station called *Mir*, which they have used as a more or less permanent floating laboratory with rotating crews since its launch in February 1986. Both *Mir* and the American project require assembly of components in space to make a large station. The proposed U.S. space station is far more ambitious than *Mir* or the previous U.S. *Skylab* and would require numerous shuttle flights for the lofting of modules, support structures, solar panels, station equipment, and supplies. The laboratory and habitation modules for the U.S. part of the station, in which Japan, Europe, and Canada also are participating, will be 27 feet in length. The first segment is scheduled to be launched in 1995, with assembly scheduled for completion by 1999. Proponents of the space station consider it to be the next logical step in expanding human capabilities to live and work in space, and, thus, to explore the vast space frontier.

Many space scientists are opposed to building the space station. They feel that the money that would be spent on expensive manned space programs would be better allocated to unmanned space probes and more scientific projects. There is also considerable Congressional

skepticism about the need for or usefulness of the proposed space station. See SPACE EXPLORATION.

SPACE COLONY as imagined by NASA planners for sometime in the 21st century. The two twin cylinders are each 19 miles long and 4 miles in diameter and they are seen in this artist's concept drawing as they would appear from an approaching spaceship. Each cylinder would rotate around its axis once every 14 seconds creating an Earth-like gravity. Solar energy would be the source of power and lunar or asteroid materials would be used for **(continued on page 280)**

Space-Time

EINSTEIN'S SPECIAL THEORY OF RELATIVITY states that space and time are interrelated. For example, the rate of the flow of time depends on the state of motion of the observer: A clock on a moving laboratory appears to be ticking more slowly than a set of identical clocks distributed throughout a stationary laboratory. Another example: Two events or occurrences at two different locations can be seen to be occurring simultaneously by one observer but will be seen as not occurring simultaneously by a moving observer. Einstein considers time as the fourth dimension, more or less equal to the three spatial dimensions—length, width, and height. Einstein sees none of the four dimensions as absolute. Depending on how fast an object moves, the dimensions of space can stretch or shrink, and the passage of time can speed up or slow down. These effects are only

construction. The cylindrical portion is the living area for the several hundred thousand inhabitants. The cuplike containers ringing the cylinders are agricultural stations while the large movable mirrors at the sides of the cylinders would be used to direct sunlight into the interior, regulate seasons, and control the day-night cycle.
Source: NASA photo.

measurable when objects travel at close to the speed of light. See TIME DILATION EFFECT.

Special Theory of Relativity The first of EINSTEIN's two major theories, special relativity considers space and time to be closely linked dimensions rather than two different dimensions as NEWTON had thought. Published in 1905, the special theory contained a number of surprising implications. One of these is that the speed of LIGHT is the same for all observers, regardless of their relative motion. Moreover, the speed of light in empty space is the absolute limit; nothing can ever be accelerated up to that speed nor will anything be observed moving faster.

The special theory also postulates the EQUIVALENCE OF MASS AND ENERGY. Einstein showed that any form of ENERGY has MASS, and matter itself is a form of energy. This is expressed in the most famous equation in the world: $E = mc^2$. In this equation, E represents energy in ERGS, m represents mass in grams, and c represents the speed of light in centimeters per second. Because light travels at 30 billion centimeters per second, it can be seen that the conversion of only a small amount of mass will produce a lot of energy. For instance, 1 kilogram (2.2 pounds) of mass can be converted into 25 billion kilowatt-hours of energy or enough to supply an industrialized nation for several weeks. Also, less than 5 pounds of mass, when the energy is released explosively, can destroy a city and kill a million people.

The special theory also predicted that mass and time change with increase in velocity—as a PARTICLE of matter moves faster it becomes more massive. All of Einstein's concepts and predictions as set forth in the special theory of relativity have been verified many times over by observation and experiment. See GENERAL THEORY OF RELATIVITY.

Species In biology, the term refers to a group of closely related and interbreeding living things. A species is the smallest standard unit of biological classification. Species can be further classified into various subspecies, races, or breeds. For instance, cats, dogs, and snakes are species; whereas Persian cats, collie dogs, or pythons are breeds or subspecies. Species are a type of life that breeds true—that is, a lark gives birth to a lark (and not a sparrow), a monkey gives birth to a monkey (and not a chimpanzee). There is, of course, room for differences within species—blond or dark hair, short or tall height, and so on. See TAXONOMY.

Spectroscopy The science that deals with the use of an optical device called a spectroscope to analyze LIGHT and other electromagnetic RADIATION. The spectroscope is used for producing and observing a SPECTRUM of light or other radiation, consisting essentially of a slit through which the radiation passes, a lens to focus the incoming radiation into parallel lines, and a prism to spread out the white light into a spectrum, or band, of colors.

Every element gives off a specific electromagnetic radiation that can be identified. Thus spectroscopy enables astronomers to determine which elements are present in a distant star or galaxy and the temperature and chemical composition of planets within our solar system. In addition, spectroscopy reveals, by the stretching of wavelengths, whether or not a distant object is moving away from Earth and how fast. See DOPPLER EFFECT, and REDSHIFT.

Spectrum Refers to the entire range of wavelengths produced when a beam of electromagnetic RADIATION is broken up into its array of entities. We are most familiar with the band of colors produced when sunlight is passed through a prism, comprising red, orange, yellow, green, blue, and violet: Red has the longest wavelengths and lowest frequency, and violet has the shortest wavelength and highest frequency.

It was NEWTON who first observed that when sunlight passes through a prism, the colors that make up the LIGHT are dispersed into what he named a spectrum. Today the entire broad band of electromagnetic ENERGY, not just the visible but all radiation, is subject to analysis and study by means of heat effects, photography, and so on. See ELECTROMAGNETIC SPECTRUM.

Speech Recognition COMPUTER-based speech recognition is a well-established technology. Systems that recognize single words or simple phrases from a limited vocabulary are being used in factories for controlling machinery, entering data, inspecting parts, or taking inventory. In some hospitals, doctors and nurses can keep their hands free for working with critically ill patients by wearing microphones so that they can describe their actions to a computer that logs the information and keeps the necessary records.

In 1990, a voice typewriter was developed that could recognize 30,000 words and adapt to individual voices. This system recognizes *discrete utterances*—words with at least a 1/4-second pause in between—and displays a list of possibilities when a word is spoken. It will print the top word unless told to keep looking and waits for the next word to be spoken.

SST (Supersonic Transport) Commercial aircraft that fly faster than the speed of SOUND. The Concorde, developed by the French and British, is an SST. Only 16 Concordes were built and only about six are still flying. Although a spectacular aircraft in appearance, the Concorde has serious shortcomings: It carries only about 100 passengers, is inefficient in the use of fuel, and has a range of only 3,700 miles. The British and French aircraft makers are now considering a new SST that would accommodate between 200 and 300 passengers, use fuel far more efficiently, and have a range of up to 7,500 miles. A flight from Europe to Japan on the proposed super-Concorde would take 5 hours, as compared with 12 hours now. The United States is also considering development of a new-generation SST. Both the proposed British-French SST and the American aircraft are designed to fly in the STRATOSPHERE and therein lies the problem.

The Concorde flies at an altitude of just over 50,000 feet. The proposed planes would have to fly in the thinner atmosphere between 55,000 and 60,000 feet in order to withstand the stresses of supersonic speed. This region of the ATMOSPHERE in which temperature fails to decrease with height is known as a region of *temperature inversion.* The effect of temperature inversion is to inhibit the vertical mixing of air, which means that the engine emissions from SSTs will remain in the stratosphere for years. A fleet of aircraft flying in the stratosphere in the future would leave behind their several years' of accumulation of the effluents they had produced. A fleet of 500 supersonic aircraft using existing engine technology could seriously deplete the OZONE layer, which is already in danger from synthetic CHLOROFLUORCARBONS. The proposed SST fleet could reduce ozone by 15 to 20 percent, almost three times the damage predicted from chlorofluorocarbons (CFCs).

Stalactites Rock structures usually found in caves. Stalactites are formed on the ceilings of caves as water drops down, depositing minerals as it does so. To distinguish stalactites from STALAGMITES, remember that stalactites "hang tight" to the ceilings while stalagmites "might" reach the ceiling some day.

Stalagmites Rock structures found on the floor of caves and formed by water dripping down from above, depositing minerals as it does so. See STALACTITES.

Standard Model In physics, the theories—set of equations—that state that there are exactly four types of FORCES in the universe that control the ways in which objects interact. Of these four, the strong nuclear force and the weak nuclear force act only within atomic nuclei. The third is the electromagnetic force and the fourth is the gravitational force. Taken together, these theories can predict the outcome of every known fundamental interaction. The weak nuclear force controls the process of radioactive decay. The strong nuclear force binds PROTONS and NEUTRONS (known as NUCLEONS) together in the nuclei of ATOMS, and binds the elementary PARTICLES called QUARKS together to form each nucleon. *Electromagnetism* is the force that produces LIGHT and all other forms of electromagnetic RADIATION. Electromagnetism also bundles atoms together as MOLECULES forming all MATTER as we know it. Gravitation is the force that holds each STAR and planet together, and retains the PLANETS and MOONS in their orbits around stars, and stars in their orbits in GALAXIES. See NUCLEUS, ATOMIC; and SUBATOMIC STRUCTURE.

Star A celestial object that generates ENERGY by means of nuclear FUSION at its core and, therefore, it is self-luminous. Our SUN is a star. Other bodies in the universe—PLANETS, ASTEROIDS, MOONS—do not generate

LIGHT of their own but shine only because they reflect the star-light that falls on them. Stars are hot balls of gas held together by their own GRAVITY. Our Sun is an average size star, but large enough that one million Earths could fit inside it. Some stars are 15 times less massive than the Sun, while others are 60 times more massive. Our Sun is a moderately hot star, about 15,000 ° C (27,000 ° F) at its center and about 6,000 ° C (11,000 ° F) at its surface. The coolest stars are the reddest ones, moderately hot ones are yellowish in color, while the hottest stars are blue-white. Stars are often found grouped into clusters, which may number from a few dozens to tens of thousands. The star clusters are part of a giant collection of stars called a galaxy. The current estimate of stars of all known GALAXIES in the visible universe is at least one sextillion (1 followed by 21 zeros). See **RED GIANT STARS**, **SUPERNOVA**, and **WHITE DWARF STARS**.

Star Wars See **SDI** (**STRATEGIC DEFENSE INITIATIVE**).

Statistically Significant Applied to the findings of a scientific experiment, the term indicates that the results are other than what one would expect if chance alone influenced the outcome. In other words, the term indicates confidence in the results of an experiment. This confidence is often expressed quantitatively as in "significant at the 0.05 level of confidence," which would mean that the odds are less than 5 percent that chance alone could have produced a given result of an experiment. Mathematical tests of statistical significance are especially important in medical research, where it is common that both an experimental group and a control group will show improvement even though only the experimental group has received the drug or treatment under test. Expectations and other psychological factors influence reported results and mathematical analysis is necessary to determine statistical significance. See **PROBABILITY**.

Stealth Technology Designed to be able to evade or limit detection by defensive RADAR systems, stealth technology is intended to be the answer to sophisticated ground-to-air missile defense systems that currently deter attacks on heavily defended targets by manned bombers.

attacks on heavily defended targets by manned bombers. Stealth technology is, for the most part, classified, but a combination of sophisticated new materials and a sharp-angled design that presents a minimum radar image work together to make stealth aircraft difficult to detect. In the Persian Gulf War, the F-117 Stealth fighter/bombers were the only Allied aircraft flown inside Baghdad. The F-117's ability to avoid detection by Iraqi radars worked. None of the bat-winged planes were hit in the six-week war. The overall record of the F-117 in the Gulf War was outstanding. The F-117, which carried precision-guided bombs, accounted for only 3 percent of Allied aircraft used in the Gulf War but struck 43 percent of the Iraqi targets that were hit.

Despite the success of the F-117 fighter/bomber, the proposed B-2 Stealth bomber remains a highly controversial issue. The B-2 is the most expensive aircraft ever built. A total of $30 billion has been spent on research and another $31.1 billion has been projected as the cost to build 75 Stealth bombers. To put that $61 billion figure in some perspective, it is more than the Department of Agriculture's annual budget and five times the annual budget for the National Aeronautics and Space Administration. Proponents claim that it will be money well spent—that the Stealth bomber would be able to evade defensive radars and penetrate to any target undetected. Critics call the Stealth's mission marginal and say radar-evading cruise missiles that the Air Force already has pose

STEALTH BOMBER (B-2), shown here taking off on an experimental flight, is the most expensive aircraft ever built.
Source: Northrop Corporation photo.

a serious threat to any radar-defended target at a fraction of the cost of the manned bomber. Also, advanced **MICROWAVE** radar technologies may make the manned Stealth bomber obsolete.

Stratification The process by which materials, usually rock, form themselves into layers. In geology, the term refers to the natural process over long periods of time that formed beds of material, often in parallel layers one upon another. The geological study of stratified rocks, their nature, occurrence, and relationship to each other, as well as their age and classification is called *stratigraphy*.

It was Charles **LYELL**, the father of modern geology, who first analyzed layers of rock and reasoned that younger rock strata usually lay above older strata and that therefore these layers were indicative of age. Lyell also recognized that different layers of rock contained different complements of fossils and that these too represented distinct eras of geological history. See **GEOLOGIC TIME SCALE**.

Stratosphere The region of the **ATMOSPHERE** of **EARTH** above the **TROPOSPHERE**, beginning at an altitude of 5 to 10 miles and extending to about 30 miles. In this region, **TEMPERATURE** rises with altitude and warmer air sits above cooler air. Also, the stratosphere is a stable, virtually cloudless region with very slow vertical mixing of air. The climatic differences between the troposphere and the stratosphere are important in terms of atmospheric pollution. In the troposphere, weather action such as wind and rain as well as the natural vertical mixing of the air permits this region to cleanse itself of most pollutants in about a week. In the stratosphere, on the other hand, cooler air is at the lower altitudes creating a situation known as an *inversion layer*. Pollutants that would be carried out of the troposphere in a week remain in the stratosphere for many years. A 1975 National Academy of Sciences study said, "The stratosphere can be likened to a city whose garbage is collected only every few years instead of daily." See **OZONE DEPLETION**.

Strong Force One of the four known fundamental forces of nature: electromagnetic force, **GRAVITY**, the weak force, and the strong force. The strong force holds together PROTONS and NEUTRONS inside an ATOM. See STANDARD MODEL, and SUBATOMIC STRUCTURE.

Subatomic Structure According to what theoretical physicists call the STANDARD MODEL, all MATTER consists of PARTICLES called FERMIONS. These particles exert attractive or repulsive forces on each other by exchanging force-carrying particles called BOSONS. FERMIONS come in two varieties: QUARKS and LEPTONS. Quarks are the basic constituents of particles such as PROTONS, NEUTRONS and mesons. The lepton category includes charged particles such as ELECTRONS and MUONS, and uncharged, virtually massless particles called NEUTRINOS.

Bosons carry the four known fundamental forces of nature. GRAVITONS carry the force of gravity; PHOTONS carry electromagnetism; GLUONS carry the strong nuclear force that holds together protons and neutrons; whereas the weak force, which is responsible for radioactive decay, is carried by weak bosons.

It should be understood that other forces, and other particles, may yet be discovered. It is possible, for instance, that quarks are made up of particles still more fundamental. At this writing, however, the Standard Model represents the current understanding of subatomic physics. See ATOMS.

FERMIONS Particles of Matter	BOSONS Particles of Energy or Force
Quarks–within protons and neutrons	**Gravitons**–gravitation
names include: up down charm strange top bottom	**Photons**–electromagnetism
	Weak Bosons–weak nuclear force
	Gluons–strong nuclear force
Leptons–include electrons, muons, and neutrinos	

Subliminal Learning Instructional or educational material purportedly delivered below the threshold of consciousness. Vendors of audiotapes claim that the power of subliminal learning can take off weight, improve self-esteem, increase I.Q., improve memory, sharpen up golfing ability, or what-have-you. Psychologists studying the mind's ability to register information directed to the unconscious say that there is no credible scientific information that subliminal messages can persuade or cure. In fact, there is no evidence of any perception of these messages at all, let alone evidence that they are effective. There is still some debate among psychologists on this issue, but most investigators agree that popular understanding of subliminal persuasion is naive, and that lore about subliminal perception is largely nonsense.

Sun The nearest STAR to EARTH, our Sun is a ball of glowing gases, principally HYDROGEN, some 864,000 miles in diameter. The Sun is over a million times larger

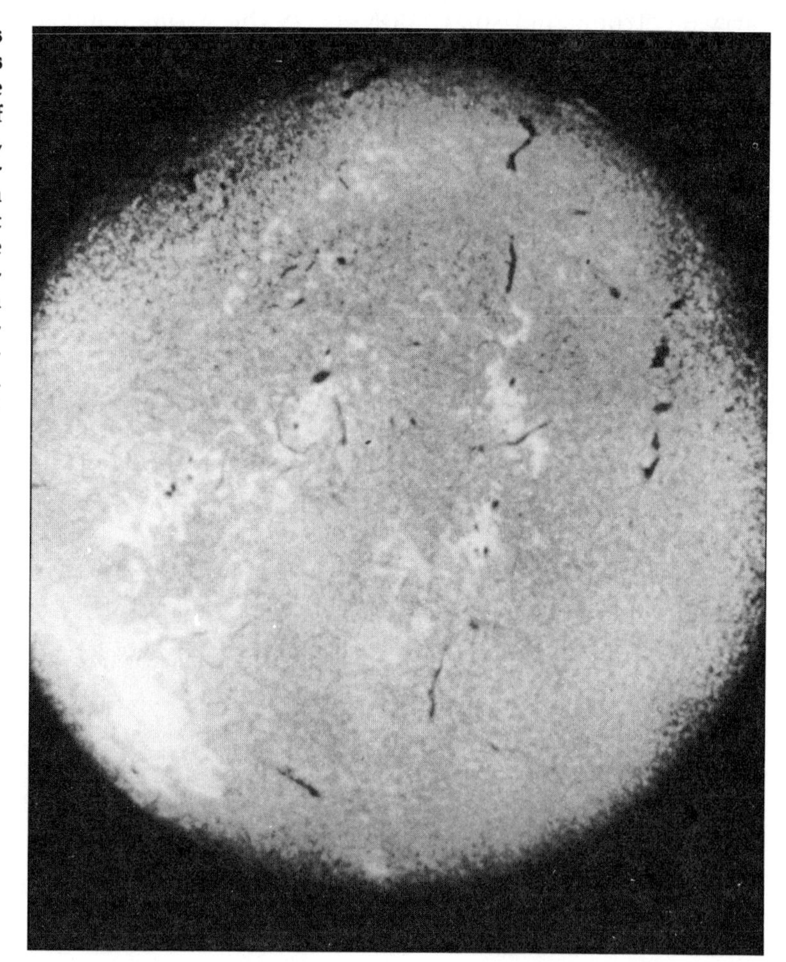

SUN spots, as shown in this NASA photo of the Sun during one of its active phases, are highly magnetized dark areas in the Sun's turbulent surface layer. The total energy emitted by the Sun during active periods appears to increase slightly.
Source: NASA photo.

in volume than the Earth and it is powered by a nuclear process not unlike a hydrogen bomb. Formed about five billion years ago out of a primordial cloud of gas and dust, the Sun is in its middle age. After another five billion years, the FUSION reaction inside the Sun will have deposited so much HELIUM ash in its core that its nuclear furnace will be forced into hotter reactions. When this occurs, the Sun will expand enormously into what is called a RED GIANT STAR. The expansion will cool the Sun's surface, but because of its increased size, the total heat that it will radiate will be far greater than at present. MERCURY, VENUS, and Earth will be consumed in the flames. After another two billion years, the Sun will start to shrink and, in its final phase, will become what is known as a WHITE DWARF STAR. See SOLAR SYSTEM.

Superconductivity A material is considered a superconductor when it loses all resistance to the flow of electricity. This phenomenon occurs when some materials are chilled down to an extremely cold TEMPERATURE. Different materials require different degrees of refrigeration to achieve this state, and research today is focused on finding new materials that can superconduct at higher temperatures. A major breakthrough occurred when a material was discovered that would offer no resistance to the flow of electrons at temperatures above 77 KELVIN (−320.8 degrees FAHRENHEIT). At temperatures below 77 Kelvin, relatively expensive helium must be used as a refrigerant, whereas at temperatures above 77 Kelvin, much cheaper liquid NITROGEN can be used.

Although many technical problems remain, superconductivity has great potential for improving the efficiency of power generation and distribution; advanced electronic systems, including superfast computers; and transportation systems, including levitating trains. See LEVITATING TRAINS, and SUPERCONDUCTOR SUPERCOLLIDER (SSC).

Superconductor Supercollider (SSC) Proposed instrument designed to study elementary PARTICLES and FORCEs. When completed, the superconductor supercollider will be humankind's largest ACCELERATOR/COLLIDER. In fact the SSC will be the largest, and at an estimated cost of $8 billion, the most expensive scientific device ever built. At a site in Texas, 25 miles south of Dallas, the SSC will consist of a racetrack-shaped tunnel

54 miles in circumference, 10 feet in diameter, buried 150 feet underground. Inside the tunnel, two rings of superconducting magnets—9,000 of them—will be spaced along the particle beam pipes to focus, propel, and guide two beams of **PROTONs** traveling in opposite directions. The protons will race around the track, gaining momentum with each circuit until they are traveling at near the speed of **LIGHT**. At special chambers called *interaction halls*, the protons will cross over and collide with an **ENERGY** 20 times greater than has ever been achieved before. Researchers hope to use the great force of this collision to re-create the primordial energy levels of the birth of the universe and in so doing deepen human understanding of the nature of matter.

Each increase in **ACCELERATOR** energy from the earliest **CYCLOTRON** in 1940 to today's **TEVETRON** and **LEP** have increased humankind's understanding of why objects have **MASS** and how the basic forces that govern our universe behave. Accelerators of the 1950s and 1960s revealed the existence of hundreds of different types of subatomic particles previously unknown to science. Using data from even more energetic accelerators, theorists have arrived at the concept that all matter is ultimately made up of **QUARKS**, **LEPTONS**, and interacting forces. This concept is called the **STANDARD MODEL** and it is regarded as the most successful attempt to describe the fundamental nature of matter. There are, however, gaps and unanswered questions in the Standard Model and physicists hope that the proposed SSC will provide insights into the remaining major uncertainties of high-energy physics. Critics of the supercollider point out that studies in high-energy physics are unrelated to any conceivable practical use and that the country cannot afford all of the proposed high-budget scientific projects now under consideration. Proponents of the SSC claim that it is essential to understanding the world around us, and that this endeavor is one of humankind's premier intellectual challenges. The debate can be expected to continue. See **SUBATOMIC STRUCTURE**.

Supernova A supernova occurs when a **STAR** more massive than our own **SUN** transforms all of the **HYDROGEN** in its core to iron and completely explodes. Supernovae are very rare, occurring about once every 100 years in any one galaxy. A few are detected by astronomers each year, but almost always in distant GALAXIES. No supernova has been detected in our MILKY WAY galaxy since the year 1604. The first time that a supernova was seen with the naked eye since that time was in 1987. The 1987 supernova occurred in the large MAGELLANIC CLOUD, a companion galaxy to our own.

All stars die after they have exhausted their nuclear fuel. Stars with little MASS die gradually, but those with relatively large mass die in a spectacular explosion. For a few seconds a supernova may shine as brightly as an entire galaxy. Then it may give off as much LIGHT as 200 million suns for several weeks. It is this extremely bright flash of light that we call a supernova. If the amount of original MATTER in the star was on the low end of the range, what is left over after the explosion will form a NEUTRON star. If the amount of matter is large enough, the collapse that occurs after the explosion will form a BLACK HOLE.

On July 4, A.D. 1054, a supernova exploded that lit up the sky most impressively. It stayed that way, with light enough to read at night, for over three months. Historians tell us that all the world's religions picked up converts. See **PULSARS**.

Symbiosis In biology, the term applies to the living together of two dissimilar organisms in a mutually beneficial relationship. An example of symbiosis is lichen that grows in leaflike or crustlike forms on trees and rocks. Lichen is composed of two different organisms, an algae and a fungus, living together and supporting each other. The fungus provides moisture for the algae and the algae provides food for the fungus.

Synapse The junction, actually a microscopic gap, between two neighboring NEURONS, or nerve CELLS. It is across this gap that signals are transmitted from one nerve cell to the next nerve cell. These chemical signals are called NEUROTRANSMITTERS. An average neuron in the human brain has between 1,000 and 10,000 synapses with nearby neurons. It has been estimated that the human brain contains some ten trillion (10^{13}) synapses, giving the brain an enormous potential capability for information storage and transfer. In his book *Dragons Of Eden,* Carl Sagan compares synapses in the human brain with the microprocessors in a computer system and concludes that the human brain is 10,000 times more densely packed with information than is a modern computer.

Synfuels Synthetic fuels made by combining natural substances with chemical additives to manufacture a fuel. An example is the extraction of crude oil from oil shale. The United States has vast amounts of oil shale reserves in the western states. The technology to extract and convert the oil shale to usable gasoline is too costly at present to make synfuel a commercially attractive alternative to petroleum-based gasoline. The oil shale reserves in the United States are so large that they could supply our demand for fuel for several hundred years. See **FUELS, ALTERNATE.**

Synodic Period The term *synodic* refers to two successive conjunctions of the same bodies. In astronomy a synodic month differs from a **SIDEREAL** month because of two different reference points used to measure the **MOON**'s path around the **EARTH**. The Moon circles the Earth (relative to the stars) in 27.32 days. It turns on its own axis in precisely that same period. However, as the Moon revolves around the Earth, Earth is revolving around the Sun. By the time the Moon has made one revolution about the Earth, the Earth/Moon combination has moved in relation to the Sun. The Moon's revolution about Earth with respect to the Sun is called the synodic month, which is 29.53 days long.

Systolic One of the two components of blood pressure measurement—the first and the highest number in the usual reading. Blood pressure measured when the heart is contracting to pump blood is called systolic as opposed to **DIASTOLIC** pressure, when the heart is resting between beats. Readings are in millimeters of mercury as it rises in a testing device; the higher the pressure, the higher the number. A systolic blood pressure of 140 is considered normal; 140 to 159 is considered borderline hypertension; 160 or higher systolic pressure is called hypertension.

Blood pressure and heart rate, as indicated by pulse, are not the same thing. Blood pressure rises throughout childhood and then continues to rise in later life. Heart rate, on the other hand, slows down as one grows older. The human heart beats roughly 70 times a minute, or four times for every breath. When increased demands are made upon the heart, such as during periods of exercise, the heart beat quickens (and the heart pumps more blood with each beat).

Taxonomy The science or technique of classification. In biology, the taxonomic classification of living organisms according to their resemblances and differences. The original system of taxonomy had only two kingdoms: plants and animals. Some organisms, however, do not fit into either of these kingdoms (e.g., BACTERIA) and, therefore, the number of kingdoms had to be enlarged. Today there are five kingdoms, as follows:

MONERA	Bacteria and related forms
PROTISTA	One-celled organisms with nuclei, such as amoebas
FUNGI	Mushrooms, molds, and other fungi
PLANTAE	Multicelled plant life
ANIMALIA	Multicelled animal life

Tectonics, Plate See PLATE TECTONICS.

Telekinesis The alleged ability to move or deform inanimate objects, such as metal cutlery, through mind power alone. Professional magicians such as Uri Geller claim the ability to bend spoons through mental processes, but skeptics have often demonstrated that Geller's tricks can be duplicated using conjuring techniques. See PARANORMAL, and PSYCHOKINESIS.

Telemetry The automatic measurement and transmission of data or information by means of radio relays from the source (e.g., a spacecraft) to a distant receiver. SATELLITES transmit their data to EARTH by means of telemetry.

Telepathy Alleged communication between minds by some means other than sensory perception. Despite considerable research efforts, mind reading of any sort remains unproven. See EXTRASENSORY PERCEPTION (ESP), and PARANORMAL.

Temperature The measurement of heat or cold of some object or substance with reference to some standard value. Three scales are used to measure temperature: FAHRENHEIT, CELSIUS, and KELVIN. In the United States we are most used to the Fahrenheit scale and we know that a temperature around 70°F is comfortable and that when it gets to be 100°F outside it is hot. Most other countries, however, use degrees Celsius (or Centigrade) and when we travel we have to convert. The Kelvin scale is used almost exclusively for scientific work. The Kelvin scale starts at ABSOLUTE ZERO, the point at which molecular motion stops. See HEAT, and THERMODYNAMICS.

	Absolute Zero	Water Freezes	Room Temperature	Water Boils
Fahrenheit	−459°	32°	68°	212°
Celsius (Centigrade)	−273°	0°	20°	100°
Kelvin	0°	273°	293°	373°

To convert:

Fahrenheit to Celsius	5/9 (F° − 32)
Celsius to Fahrenheit	(9/5 × C°) + 32
Celsius to Kelvin	C° + 273

Terabyte In COMPUTER language a BYTE is eight BITS, or the unit of information that represents one letter or a digit in the computer's memory. *Tera* is the metric designation for trillion; therefore, a terabyte is roughly equivalent to one trillion letters or digits. (Because computers use the base-two or BINARY system, terabyte is not precisely one trillion bytes but rather 2^{40}.) A terabyte contains the information roughly equivalent to 2,000 multivolume encylopedias. In the supercomputers of the future, memory capacity will be measured in terabytes of data. See NUMBERS: BIG AND SMALL, and TERAFLOP.

Teraflop The capability of future large supercomputers will be measured in "teraflops" or trillions of mathematical operations per second. Computer designers predict that by the end of this century, COMPUTERS will be thousands of times faster than today's supercomputers and that this will be achieved by teaming hundreds or thousands of independent processors—silicon CHIPS—to solve problems far too complex for even the most powerful computers today. See also NEURAL NETWORKS, and TERABYTES.

Tevatron Collider Located at the Fermi National Accelerator Laboratory in Illinois, the Tevatron is, at this time, the world's most powerful particle accelerator. The Tevatron shoots PROTONS against antiprotons at collision energies adding up to 1.8 trillion electron-volts. Researchers sift through the debris left over after these head-on collisions in search of exotic subatomic PARTICLES. Specifically, the researchers are looking for evidence of the "top" QUARK, a rapidly decaying subatomic particle and the only member of the quark family not yet detected. See COLLIDER, and SUBATOMIC STRUCTURE.

Thermodynamics The branch of the physical sciences that deals with the study of HEAT. In particular, thermodynamics (from the Greek words meaning "movement of heat") is concerned with the relationship between heat and mechanical ENERGY or work, and the conversion of one into the other. Early studies of the behavior of heat revealed that a certain quantity of work is equivalent to a precise amount of heat, no matter how that work is turned into heat. For example, it was also determined that the phenomena of heat behavior is governed by two main principles (called the first and second laws of thermodynamics), which are the foundation blocks of modern physics. See TEMPERATURE, and THERMODYNAMICS, FIRST AND SECOND LAWS OF.

Thermodynamics, First and Second Laws of The first law simply states that "ENERGY is conserved"; that is, it is indestructible—there is always the same total amount of energy in the universe. Energy can neither be created nor destroyed; it just changes form, such as from chemical energy in fuel to heat or to mechanical energy. The second law is a bit more complex. The second law states that "the ENTROPY of the universe tends to a maximum." Entropy is the measure of the total disorder, randomness, or CHAOS in a system. The effect of increased entropy is that things progress from a state of relative order to one of disorder. With this progressive disorder there is increasing complexity. Expressed in another way, the second law states that HEAT, if not affected by some additional form of energy, will always flow from a warmer place to a colder place.

The first law precludes the possibility of ever constructing a perpetual motion machine, because it tells us that we can only get energy out of a machine by supplying it with *at least* as much input energy in some form or other, either mechanical, electrical, chemical, or heat. Inventors of perpetual motion machines (and there are still a lot of them out there) are victims of self-delusion. Claims that a small amount of energy can be transformed, somehow or other, into large ones are demonstrably false. The second law implies that we cannot convert heat into energy unless we have a difference of TEMPERATURE, and that the bigger this difference the bigger the fraction of input energy that we can utilize. The fraction of input energy that is not utilized will, of course, be dissipated as heat, since there can be no loss of energy as a whole. The laws of thermodynamics have obvious application in engineering and they are also widely applied in physics and chemistry. See THERMODYNAMICS.

Thermonuclear Pertains to devices (such as the HYDROGEN bomb) that use the nuclear FUSION reaction that takes place between atomic nuclei when heated to very high temperatures. (The prefix *thermo* comes from the Greek for hot or heat). See NUCLEAR ENERGY.

Thin-Film A technology used in making INTEGRATED CIRCUITS in which a film of material only a few microns thick is deposited on a substrate. A *micron* is a millionth part of a meter, and a *substrate* is the supporting material on which a circuit is formed or fabricated.

Three Mile Island (TMI) Site of a nuclear power plant in Pennsylvania where in March 1979 an accident occurred that resulted in a partial melting of a significant portion of the reactor core and some RADIATION leakage into the ATMOSPHERE. TMI had a system of four barriers to prevent radiation from reaching the world outside and three of these failed for technical reasons or through human error. In a combination of equipment failure and control room errors, the fuel was allowed to overheat. When a malfunction in the cooling system occurred, the TMI operators misread their instruments and instead of flooding the fuel core, they turned off the backup cooling system. They did exactly the opposite of what was necessary. For 40 minutes the radioactive URANIUM core was almost entirely uncovered by coolant. The fourth and last of the defense barriers, a thick, steel-reinforced containment building, did prevent nearly all of the radiation from escaping. Core damage was major, but the release of RADIATION was minimal—one millionth of what was released at CHERNOBYL.

TMI was close to a major catastrophe. Had much more time passed without proper action being taken to replace the coolant, far more than 52 percent of the core would have melted. In that event, the molten fuel would have reached and penetrated the reactor vessel and come into contact with water lost from the core. A steam explosion would have resulted, breaching the containment dome, and leading to radioactive contamination on the scale of the Chernobyl accident. See CHINA SYNDROME.

Tides Tides in the Earth's oceans are mostly caused by the MOON (the SUN plays a minor role). The Moon's GRAVITY pulls water on the near side of the EARTH toward the Moon, and pulls the solid body of Earth away from water on the Earth's far side. In most places on Earth, therefore, we have two high tides every day. Because the Moon's effect can be calculated, the times and heights of tides for specific dates and locations can be predicted. This information is listed in tide tables. See also TIDES, AS ENERGY SOURCE.

Tides, as Energy Source The periodic rise and fall of the waters of the ocean can be harnessed as a renewable source of ENERGY. In many locations, tides rise to extraordinary heights. In parts of Nova Scotia, Alaska, and Northern France, the tides rise 40 feet or more twice a day. Because the water surges back and forth through narrow channels in these areas, the potential for generating electrical power exists. Hydroelectric power can be produced by damming the basins and using the tidal flow to run turbines. Two-way turbines have been developed that can be activated as the tide flows in either direction. Storing and transporting the generated electricity remain as technical problems.

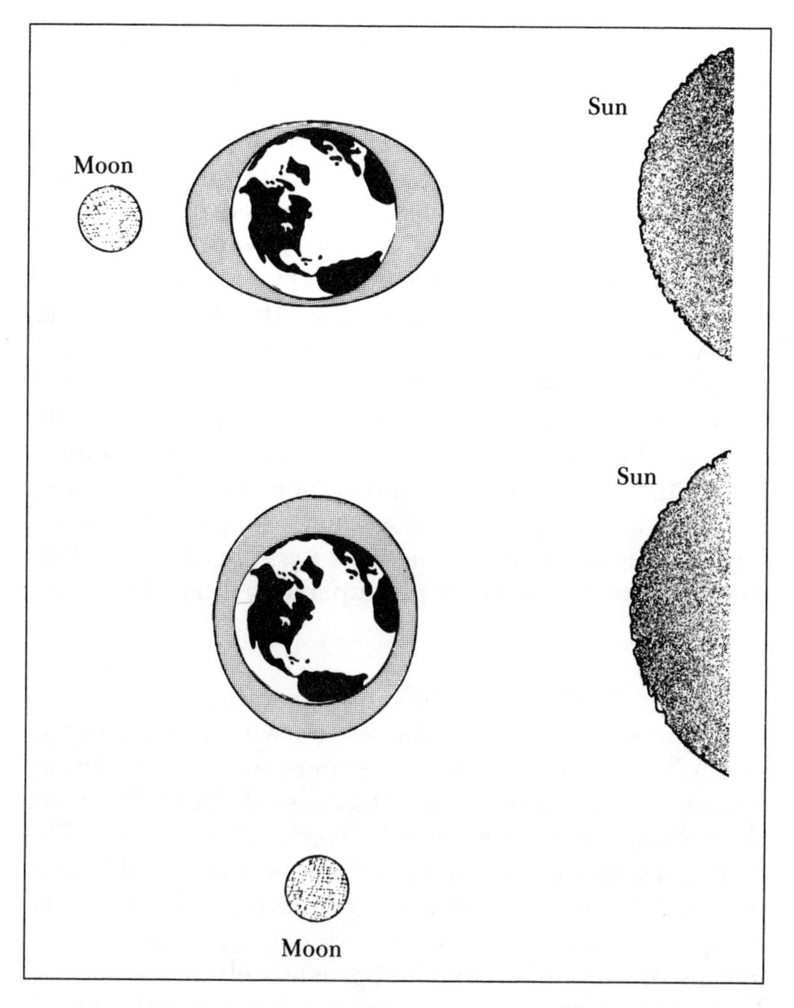

TIDES result from the gravitational pull of the moon and the Sun (mostly the moon). Spring tide, with the greatest deformation, occurs when the moon, Earth, and Sun are aligned. Neap tide, with the least deformation, occurs when the moon, Earth, and Sun are at right angles.

Although tidal power is not a significant source of electrical energy on a worldwide basis today, it is a potential supply that will likely come into use when the costs of alternative fuels—either in dollars or in environmental damage—become too high. The advantages of tidal power include the lack of environmental damage and the constant source of energy. See also SOLAR ENERGY.

Time Dilation Effect In physics, and particularly in Einstein's SPECIAL THEORY OF RELATIVITY, this term refers to the loss of time of a moving clock as observed by a stationary observer. EINSTEIN postulates that not only does time appear slower on a moving system, but all time processes are slowed down. This means that the digestive processes, biological processes, and atomic activity all slow down. At the relatively slow speeds at which we travel today, this effect is negligible, but at speeds approaching the speed of LIGHT, time slows down appreciably. At the speed of light, time would stand still.

We travel today, even with rocket propulsion in space, only at very small fractions of the speed of light. If in some far-off future, space travelers were to move at speeds approaching the speed of light, their rate of time passage would be much slower than that for those of us who waited on EARTH. The space travelers could reach some distant destination and return to Earth in what seemed to them to be only a few years, though on Earth many centuries would have passed. They would, in effect, return to a world of the future. Their friends, relatives, children, and grandchildren would be long dead. This effect is sometimes called the *clock paradox* although Einstein did not see the effect as a paradox at all. See **TWIN PARADOX**.

Time, Standard Time on our watches or clocks is based on the rotation of the EARTH with respect to the SUN. When the Earth has rotated once with respect to the Sun, one ordinary (solar) day has passed. Basically, when the Sun appears to be directly overhead, it is noon. The problem with that system is that any person on a different line of longitude would have a different solar time. To compensate for this effect, a series of 24 regions or divisions of the globe, which approximately coincide with the meridian lines, were set up in 1884, permitting each person in a certain time zone to have the same time. Time jumps, usually by a whole hour, at the edge of a time zone. See also **UNIVERSAL TIME**, and **SIDEREAL**.

Titan (Saturn Moon) One of Saturn's known 18 MOONS and, with a diameter of 3,200 miles (5,150 kilometers), one of the largest moons in our SOLAR SYSTEM. Titan has an ATMOSPHERE that is even thicker than EARTH's, and the surface pressure on Titan is higher than Earth's. Titan is of interest to astronomers because it is encased in a dense atmosphere that could be hiding an ocean or possibly several large seas. It is also believed that the organic-rich nitrogen atmosphere on Titan may nurture chemical processes similar to Earth's before life developed. NASA and the European Space Agency hope to get a much closer look at Titan with the imaging radar of the planned CASSINI mission, to be launched in 1995 to radar-map Saturn and Titan in 2002. See **VOYAGER SPACECRAFT**.

Tomography The word tomography comes from *tomas*, meaning a cutting or section, and *graph*, meaning write: tomography then is the taking of sectional pictures, or radiographs, using medical-imaging devices in which the image of a selected plane of the human body remains clear while the images of all other planes are blurred or obliterated. CAT (Computerized Axial Tomography) scanners are one example of tomography. CAT scanners view a "slice" of the human body from many angles by revolving an X-ray tube around the patient. The resulting pictures are reassembled by a computer to present the data in thin cross-sectional slices. See **CAT SCAN**.

Topology The mathematical field that concerns the fundamental properties of shapes. By studying the steps required to transform one shape into another, topologists establish the relationships between different shapes and learn the differences distinguishing one shape from another. These seemingly abstract efforts are playing an increasingly important role in many disciplines outside of mathematics, from molecular biology to PARTICLE PHYSICS and COSMOLOGY.

Recently a collaboration has sprung up between mathematicians interested in topology, especially KNOT

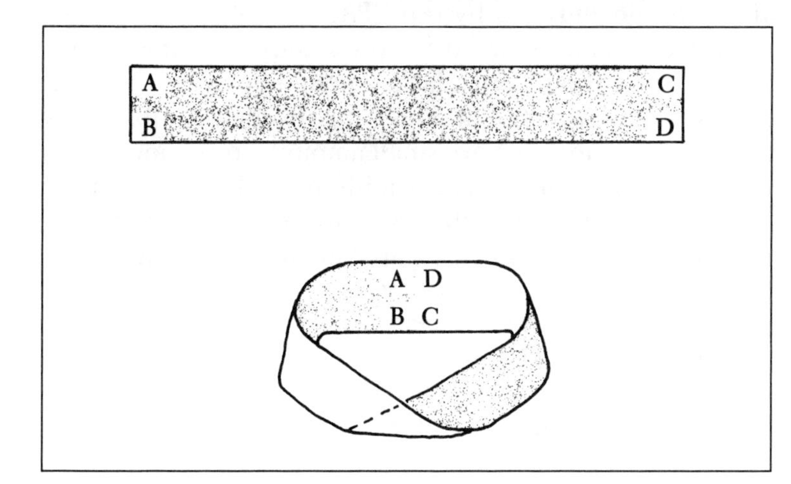

TOPOLOGY is the mathematical field dealing with the study of the fundamental properties of shapes. An example of basic topology is the mobius strip shown here. A mobius strip is a continuous one-sided surface formed by twisting one end of a rectangular strip through 180 degrees about the longitudinal axis of the strip and attaching the ends together.

THEORY, and some molecular biologists attempting to understand the geometry and behavior of DNA MOLECULES. They are investigating topics such as recognizing patterns in PROTEIN sequences, developing new techniques for genetic mapping, tracking energy flows as DNA molecules untangle and shift in position, and analyzing ENZYME actions.

Toxic Waste According to the Environmental Protection Agency (EPA), American industry is pouring more than 22 billion pounds of toxic chemicals into the air, water, and land each year. The chemical manufacturing industry is the biggest source of toxic pollution, with 3,849 plants nationwide emitting 12 billion pounds or about 56 percent of the total release (1987 data). Paper mills and factories discharged 2.8 billion pounds, and primary metal plants, such as smelters and steel mills, released 2.6 billion pounds. Many of the chemicals used in the production and processing of plastics are highly toxic. In an EPA ranking of the 20 chemicals whose production generates the most total hazardous waste, five of the top six are chemicals commonly used by the plastics industry. Disposal of toxic waste is a major national problem. Burying toxic waste is not a completely effective means of disposal because buried waste has a tendency to leach (or perculate through) into the underground water supply. Incinerating toxic waste is more expensive than burning ordinary waste and may result in generating new forms of toxic waste either as ash, or as gases released into the atmosphere. New technologies for disposing toxic waste are being encouraged by the EPA.

Toxic is another word for poisonous and almost anything is toxic if taken in a large enough dose. As the term is used in environmental issues, toxins means substances that are poisonous in very small amounts, often measured in parts per million. Parts per million are difficult to imagine unless we compare this measure to items with which we are more familiar. One part per million is equivalent to:

one inch in sixteen miles
one ounce in 31 tons
one tablespoon in 4,000 gallons.

See also WASTE MANAGEMENT.

Transistors The electronic devices that in the 1950s replaced vacuum tubes in radios, televisions, and COMPUTERS. Transistors, sometimes called solid state devices, are made from SEMICONDUCTOR material. When semiconductor materials, such as silicon or germanium, are combined and altered with impurities such as arsenic or boron, they can be made to pass current in one direction and not in the other (in which case they are called *rectifiers*) or they can be made to increase the current in a circuit (*amplifiers*). The development of transistors was an important step in the miniaturization of electronic circuits.

Faster operating transistors can lead to faster, more powerful computers and development work on new germanium-silicon combinations has led to an experimental transistor that can open and close off a pathway of electrons as many as 75 billions times a second—a rate that nearly doubles the previous record and is seven times faster than silicon transistors in today's mainframe computers. See also INTEGRATED CIRCUITS (ICs).

Transplants See ORGAN TRANSPLANTS.

Triassic Geologic time is divided into three eras, the PALEOZOIC, MESOZOIC, and CENOZOIC (Greek for ancient, middle, and recent life). The eras are divided into 11 periods, most of which are named for the places where rocks from the period were first discovered. The Triassic period, which is in the Mesozoic era, got its name because it was easily divisible into three parts. The Triassic period occurred from 230 to 195 million years ago and is characterized by the advent of dinosaurs and coniferous forests. See GEOLOGIC TIME SCALE.

Triton (Neptune Moon) The largest of Neptune's known eight MOONS, Triton has a fairly circular orbit 204,000 miles above Neptune's cloud tops. Its estimated diameter is 1,678 miles (2,700 kilometers). One of *Voyager* 2's most striking discoveries was a pair of huge geyserlike eruptions towering above Triton's surface. Triton thus joins EARTH and JUPITER's moon, Io, as the only known erupting bodies in orbit around the SUN. Scientists

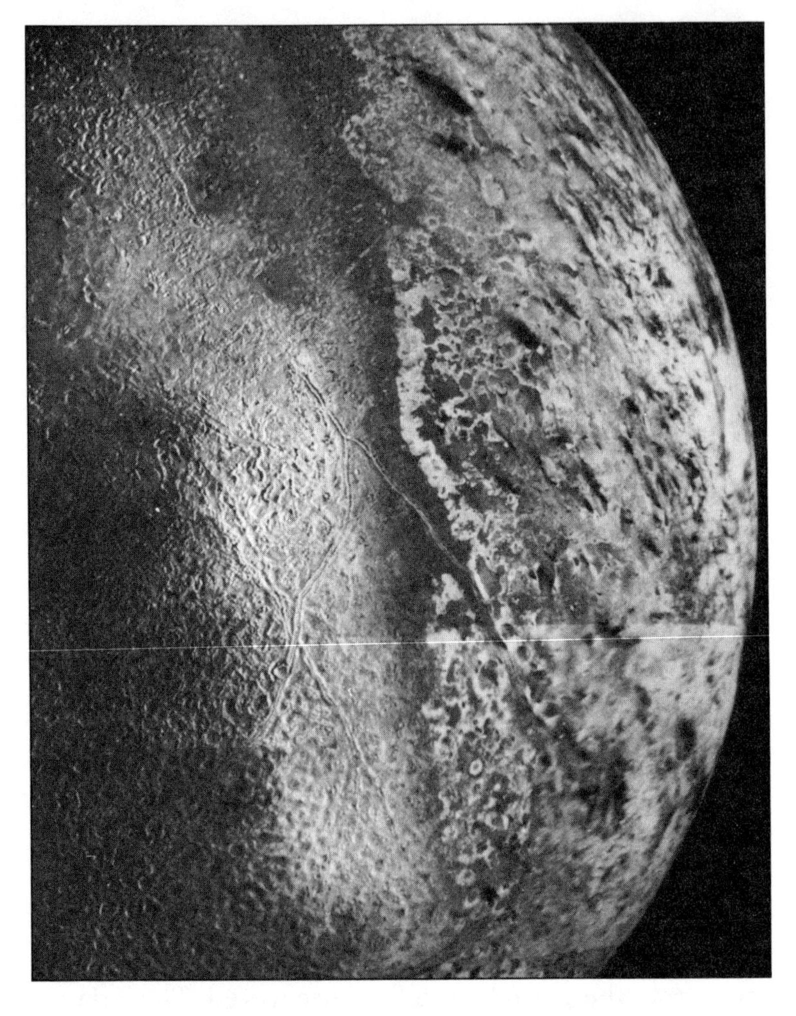

TRITON (Neptune's Moon), shown in this *Voyager* 2 photo is, along with Jupiter's moon Io, the only known erupting body other than Earth in our solar system.
Source: NASA photo.

are still trying to understand the eruption's cause. One theory suggests that a pressurized gas, probably nitrogen, has risen from beneath the surface, carrying up dark particles of some carbon-rich material. From the indications of eruptions, and other evidence, astronomers conclude that this moon is tectonically active; that is, its surface is undergoing constant change. See **VOYAGER SPACECRAFT**.

Tropopause The boundary between the two lowest layers of **EARTH**'s **ATMOSPHERE**, the **TROPOSPHERE** and the **STRATOSPHERE**. The tropopause is the point at which the temperature, which had been falling with increase in altitude, reverses itself. Above the tropopause, the temperature at first becomes roughly independent of height and then increases with height up to a level of about 50 kilometers (160,000 feet). The main reason for the temperature inversion at the tropopause is the OZONE layer.

Troposphere The lowest layer of EARTH'S ATMOSPHERE, extending from the surface to an altitude of from 8 to 16 kilometers (26,000 to 52,000 feet). In the troposphere, temperature falls with increasing altitude above the Earth's surface to a minimum value of about −80 degrees Fahrenheit (average between winter and summer). The troposphere, from the Greek meaning "sphere of change," is the region in which we live, and it is the region that contains winds, rain, storms, and other weather events. It is also a region of relatively rapid circulation and vertical mixing of air. See TROPOPAUSE, and STRATOSPHERE.

Tunneling Microscope See SCANNING TUNNELING MICROSCOPE.

Turbojet The power system used by most of today's jet aircraft. An air compressor in the inlet is used to compress the incoming air, which is then mixed with fuel and ignited in the combustion chamber. The expanding exhaust gases spin a turbine that both produces thrust and powers the compressor. Turbojets are an efficient form of power for aircraft at speeds up to MACH 3 (faster than any commercial aircraft are flying today). At that speed, air ramming into the chamber encounters drag and consequently heats up. The combined effects of aerodynamic heating and combustion raise the internal tem-

TURBOJET, a type of engine that powers most of today's aircraft, uses an air compressor in the inlet to compress the incoming air, which is then mixed with fuel and ignited in the combustion chamber. The expanding gas spins a turbine that produces forward thrust.

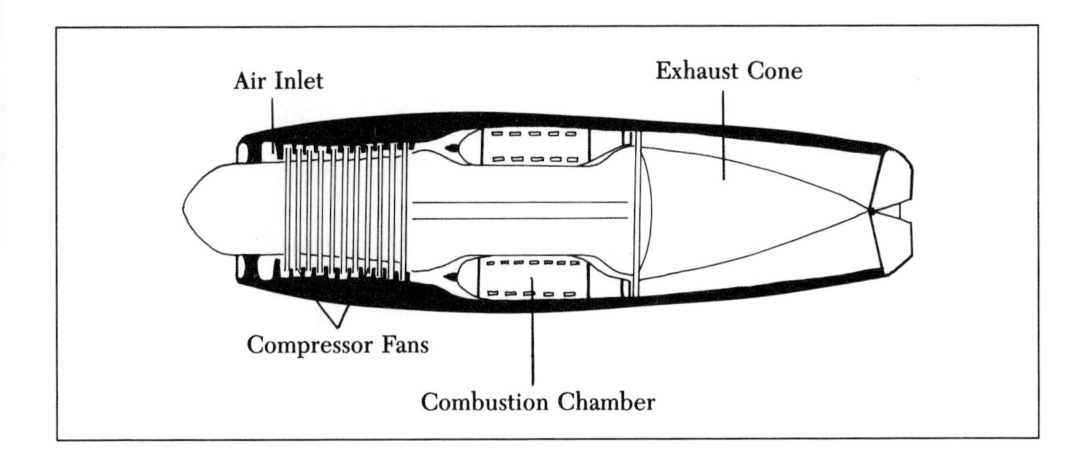

perature of the gases that turn the turbine to a point that is beyond the ability of current state-of-the-art materials and cooling techniques to handle. See RAMJET.

Twin Paradox Refers to the example used to illustrate the TIME DILATION EFFECT in Einstein's SPECIAL THEORY OF RELATIVITY. The example uses two hypothetical twins (or clocks) who are separated, one taking a journey to a distant galaxy and returning at speeds approaching the speed of light. On return to EARTH, the space-traveling twin will find himself much younger than his stay-at-home twin. According to the special theory of relativity, this is no paradox, the results being completely understandable as a consequence of time dilation, or the slowing down of the traveler's clock relative to that of the twin. This effect has been verified experimentally by measurements with atomic clocks. See EINSTEIN.

UFOs (Unidentified Flying Objects) Sometimes called *flying saucers,* these unexplained moving objects are alleged to have been observed in the sky at various times and places and assumed, by those who have claimed to observe them, to be of extraterrestrial origin. Flying saucer accounts are anecdotal and completely free of any verifiable hard evidence. Belief in UFOs is nonetheless popular. There was a wave of sightings in 1952, mostly from rural areas in the southern United States. A new wave of sightings occurred in 1973–74. Three-eyed extraterrestrials in silvery suits were reported in southern Russia in 1990, according to Tass, the once staid Soviet news agency. None have ever been verified by objective scientific observation. None the less, a survey conducted by the Gallup Poll in the summer of 1990 revealed that 47 percent of those polled affirmed their belief in UFOs. One in seven claimed to have seen a UFO.

Ultrasound **SOUND** with frequencies above 20,000 cycles per second or **HERTZ** (Hz). The human ear can hear sounds with frequencies between 20 and 20,000 cycles (vibrations) per second. Sounds with frequencies below 20 cycles per second are called *infrasonic.* Extremely high-intensity ultrasound can be used to kill insects, pasteurize milk, and drill teeth. In medicine, ultrasound is used to destroy diseased tissue in the brain, treat arthritis, and diagnose problems in the gall bladder, kidney, liver, and spleen.

Ultrasound techniques are now being used to shatter kidney stones into tiny fragments that can then be passed from the human body. The principle involved in this treatment is *resonance.* Resonance occurs when an object is put into motion by sound waves from an external source, having the same frequency as the natural frequency of the object. When resonance occurs, the target object is made to vibrate with greater and greater amplitude until, as in

ULTRASOUND is high-frequency sound that produces echoes when directed toward an object, such as the fetus in the mother's womb as shown here. The boundaries between tissues in the fetus produce different kinds of echoes, depending on the relative resistance of the tissues. A computer then converts the received echoes to an image.

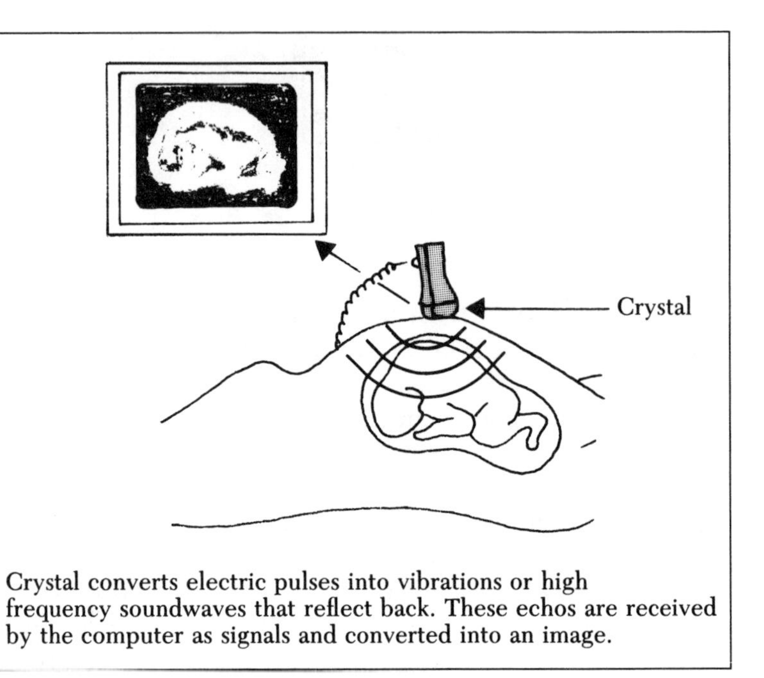

Crystal converts electric pulses into vibrations or high frequency soundwaves that reflect back. These echos are received by the computer as signals and converted into an image.

the case of the kidney stone, it shatters. Ultrasound scans are now accurate enough to take the place of AMNIOCENTESIS for checking fetuses for spinal abnormalities and some other severe birth defects. See SONARGRAPHY.

Ultraviolet (UV) Radiation Short-wavelength RADIATION in that part of the ELECTROMAGNETIC SPECTRUM just beyond the visible violet light but longer than that of X rays. The Earth's OZONE layer blocks most of the biologically harmful ultraviolet radiation and thus protects both plant and animal life. Ultraviolet radiation, particularly in the shorter wavelengths of the spectrum known as UV-B, is the cause of sunburn, SKIN CANCER, and cataracts. The depletion in the ozone layer caused by synthetic chemicals has resulted in increased amounts of UV-B reaching the surface of EARTH. See also OZONE HOLE.

Ulysses Spacecraft Launched in October of 1990, the *Ulysses* spacecraft is on a five-year, 1.86 billion-mile mission to study the SUN's polar regions. *Ulysses* will use the planet JUPITER as a "GRAVITY slingshot" to leave the plane in which all the planets orbit the Sun. *Ulysses'* path

ULYSSES SPACECRAFT will carry nine instruments to conduct experiments at polar regions of the Sun and in interstellar space never before explored. Artist's concept shown here shows the *Ulysses* as it approaches the Sun.
Source: NASA/JPL photo.

around JUPITER will harness that giant planet's gravity to twist the plane of the spacecraft's orbit so that it ends up nearly perpendicular to the solar equator. This will allow the spacecraft to fly almost over the Sun's south pole in 1994 and north pole in 1995.

Uncertainty Principle Simply stated, the principle means that at the level of fundamental particles, one cannot observe a particle's position and its velocity simultaneously. The principle, first enunciated by Werner Heisenberg in 1935, does not mean that the entire field of QUANTUM PHYSICS is uncertain. It means only that it is impossible to measure two properties of a quantum object, such as ENERGY and time or position and movement, simultaneously with exact precision. See **HEISENBERG'S UNCERTAINTY PRINCIPLE.**

Uniformitarianism Not a religion but rather the term given to the geological thesis that the processes that operated to shape the EARTH in the distant geological past were no different than the processes observed today; that is, the Earth has changed geographically over long periods of time and is continuing to change today. EARTHQUAKES, volcanic eruptions, erosions due to wind and water, and, as is known today, the movement of tectonic plates all affect the surface of the Earth and this slow natural process must have been working for a very long time to produce the observed phenomena. The opposite thesis is called CATASTROPHISM, which holds that geological changes resulted from cataclysms occurring during a relatively brief period of history.

The principles of uniformitarianism, first enunciated by Scotch naturalist James Hutton and British geologist Charles LYELL in 1830 to 1840, led to estimates of the age of Earth that far exceeded what had been accepted until that time. As late as the 18th century, the biblical version of the origin of Earth was accepted by the learned world. The age of Earth was thus considered to be only 6,000 or 7,000 years. Understanding the slow geological processes involved has led to the modern scientific estimate of the age of Earth in its present form to be 4.6 billion years. See GEOLOGICAL DATING, and GEOLOGIC TIME SCALE.

Universal Time Astronomers often keep time by the SUN's position in relation to the zero line of longitude, the meridian of Greenwich, England. They use this Universal Time (UT) or Greenwich Time so that data from all over the world can be easily compared. See also SIDEREAL.

Universe, Age of Scientific textbooks and articles have for many years specified the age of the universe to be somewhere between 10 and 20 billion years. This broad range is, of course, less than satisfactory and astronomers have worked to narrow that estimate. Their observations have convinced many, but not all, astronomers that the correct age of the universe is 12 billion years or younger.

The American astronomer Edwin HUBBLE established the fact of an EXPANDING UNIVERSE in 1929. The rate of expansion is a ratio known as HUBBLE'S CONSTANT, which is calculated by dividing the speed at which a galaxy is moving away from Earth by its distance from Earth. To compute the age of the universe, one then inverts the Hubble Constant: Divide the distance by the recessional velocity. This method assumes that the universe has expanded at the same rate since its birth at the BIG BANG (an assumption with which not all astrophysicists agree). Estimates of both velocities and distance from Earth of distant GALAXIES have been subject to a number of uncertainties, and by refining the methods for measuring distance, astronomers have recently taken a step forward in more accurately estimating the age of the universe. See also CLOSED UNIVERSE, and REDSHIFT.

VISIBLE UNIVERSE

Stars

Generate their own light (nuclear fusion at the core); massive. Sun is a star.

Types: neutrons or pulsars
dwarf
giant
nova

Galaxies

Large aggregation of stars; bound by gravity.
Local group is an association of galaxies.

Types: spiral (Milky Way)
elliptical
irregular

Planets

Less massive than stars; shine by reflected light.
Our solar system has nine planets.

Asteroids

Less massive than planets; shine by reflected light; sometimes called "minor planets"

Comets

Lumps of dirt and ice left over from formation of solar system; most believed to be in the Oort cloud.

Our Solar System

Part of the Milky Way galaxy consisting of:

1. Sun
2. Planets—9
3. Moons—satellites around more massive objects
4. Asteroids
5. Comets

Universe, Extent of The dimensions of space, according to the SPECIAL THEORY OF RELATIVITY, are analogous to the dimensions of time; that is, the universe is as big as it is old. Current theory has it that the BIG BANG (the origin of time) occurred 12 billion years ago and the universe started to expand at that time. Astronomers have detected signals in space that started traveling across space soon after the Big Bang. With the passage of billions of years, these signals have changed so that now they are received as radio waves. Astronomers agree that the universe is still expanding, but there is disagreement as to the eventual fate of the universe. If the universe continues to expand forever, as many scientists postulate, then the universe is infinite in size. Will the universe ever contract, thus marking an end to the dimensions of space? Astronomers study this problem but do not know the definitive answer. See CLOSED UNIVERSE, EXPANDING UNIVERSE, and OPEN UNIVERSE.

Universe, Future of According to the BIG BANG theory of creation, the universe began expanding about 10 or 20 billion years ago from a tiny pinpoint to its present enormous size. Scientists agree that it is still expanding but disagree on its final fate. Some astrophysicists suggest that the universe will expand forever. Others think that the forces of gravity will retard the expansion and that at some far-off future date the universe will stop expanding and begin to contract. Perhaps after 50 or 60 billion years, all the matter in the universe will be forced back into a volume comparable to an atomic nucleus. At that time the process could begin all over again with the Big Bang II. See CLOSED UNIVERSE, EXPANDING UNIVERSE, and OPEN UNIVERSE.

Universe, Origin of See ORIGIN OF UNIVERSE.

Uranium A chemical ELEMENT that is naturally extremely radioactive. Naturally radioactive elements disintegrate spontaneously and change into another element. Uranium goes through a series of such disintegrations and ends up as lead. Whenever such disintegrations take place, the radioactive substance emits RADIATION. A particular ISOTOPE of uranium, uranium 235, is so naturally radioactive that it is the main fuel for nuclear REACTORS as well as atomic bombs. A lump of uranium generates enough heat to melt its weight in ice every hour and will continue to do so for a thousand years or more. The unit of radiation that measures the number of nuclear disintegrations occurring per second in a radioactive material is called a CURIE (for Pierre Curie, the codiscoverer of radium). See also **FISSION, NUCLEAR**; and **RADIOACTIVITY**.

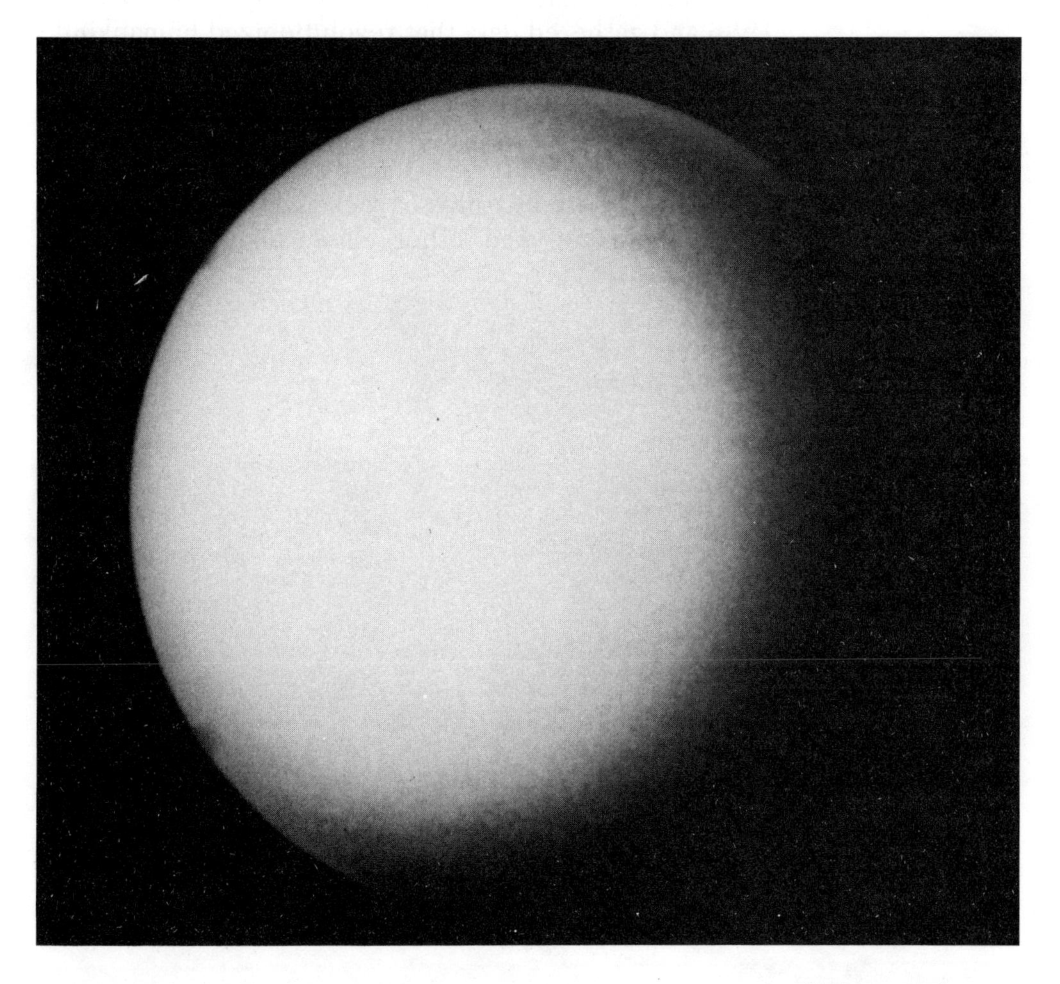

URANUS as shown in photograph obtained by *Voyager* 2 on January 22, 1986, when the spacecraft was 1.7 million miles (2.7 kilometers) from the planet.
Source: NASA photo.

Uranus The seventh PLANET out from the SUN, Uranus is the third largest planet in our SOLAR SYSTEM, JUPITER and SATURN being larger. Uranus, accompanied by its known 15 moons, orbits 1.8 billion miles out from the Sun. Astronomers estimate the temperature of this distant planet to be about −270 degrees FAHRENHEIT (−167 degrees CELSIUS). Recent observations have identified nine rings in nearly circular orbit around the planet. Like

PLUTO, the most distant planet in the solar system, Uranus is tipped on its axis. Unlike the other planets, it does not rotate around an axis reasonably perpendicular to the plane of its orbit. In other words, Uranus rolls around the Sun like a ball on a circular track. This odd orientation means that day and night on Uranus have no meaning in **EARTH** terms.

Voyager 2 spacecraft made a brief flyby of Uranus in 1986 and gathered data that revolutionized humankind's knowledge of this strange planet. Scientists theorize that the newly discovered rings are made of frozen **METHANE**, which reacts under solar **ULTRAVIOLET RADIATION** to produce carbon compounds. Scientists also deduced that Uranus's **MOONS** are mixtures of rock and ices of water, ammonia, methane, and other chemicals. See **VOYAGER SPACECRAFT**.

Vaccines Preparations designed to stimulate the creation of antibodies that recognize and attack a particular infection. Vaccines usually consist of a very mild dose of the disease, dead or weakened BACTERIA or virus, which remain in the system and protect against subsequent exposures. New vaccine technologies developed in the 1990s re-create only parts of a virus to stimulate the immune responses, thus eliminating the risks (usually very small) associated with the use of whole living virus vaccines that can provoke dangerous reactions in some people.

The idea of using one virus to immunize people against another may possibly be the greatest contribution medicine has made to human health to date. Smallpox vaccine was discovered by English physician Edward Jenner in 1798 when he found that inoculating someone with cowpox (a disease that attacks cows) would make them immune to smallpox (an often fatal disease). Jenner called the procedure *vaccination,* from *vaccina,* the Latin name for cowpox. The next advancement was made by Louis PASTEUR who discovered, almost by accident, that a severe form of a disease could be changed into a mild one by weakening the microbe that produced it. The mild form of the disease could then be used as an immunization agent. Pasteur thus expanded upon Jenner's original discovery to develop vaccines for many diseases. See ANTIBODY, and VIRUS, BIOLOGICAL.

Valence or Valency In chemistry, the quality that determines the number of ATOMS or groups with which any single atom or group will unite chemically. In other words, a measure of the combining power of a specific ELEMENT, equal to the number of different chemical bonds one atom of that element can form at any one given time. The term valency is also used in psychology to denote the capacity, or tendency, of one person to interact with others.

Van Allen Belt One of the most spectacular astronomical discoveries of the 1950s was the hitherto unsuspected belts of charged PARTICLES surrounding the EARTH at very high altitudes. These belts are thought to be charged particles trapped in the Earth's magnetic field. The inner of the two regions or belts is located at an altitude of 2,000 miles (3,200 km) and the outer belt is located at an altitude between 9,000 and 12,000 miles (14,500 and 19,000 km). These radiation belts are named after their discoverer American physicist James Van Allen (born 1914) but they are also called the magnetosphere in keeping with the names given to other regions of space in the neighborhood of Earth.

Velikovsky, Immanuel See CATASTROPHISM.

Velocity, Escape The minimum speed needed to escape the gravitational field of a particular body. The FORCE of GRAVITY weakens with height. (To be exact, it weakens as the square of the distance from the Earth's center.) If an object is fired upward at a velocity of 1 mile per hour it will reach an altitude of 80 miles before turning and falling back to Earth (ignoring air resistance). If an object is fired upward at 2 miles per hour it would reach an altitude of 320 miles. It has been calculated that if an object is launched upward at 25,128 miles per hour (6.98 miles per second) it will escape the gravitational force of Earth and never fall back. This speed, 25,128 miles per hour, is Earth's escape velocity.

Venus The closest PLANET to EARTH, and the second planet out from the SUN, Venus is about the same size as Earth. Because it is closer to the Sun than Earth, Venus is much hotter. Its thick ATMOSPHERE traps SOLAR ENERGY inside, making the TEMPERATURE at its surface 900°F (500°C)—hot enough to melt lead. The HEAT is trapped on this planet by a GREENHOUSE EFFECT: Sunlight comes through Venus's clouds and heats the planet's surface, which gives off INFRARED rays, but the water vapor and CARBON DIOXIDE in the atmosphere do not allow the infrared rays to pass back through. The atmospheric pressure on Venus is 100 times that of Earth and the atmosphere is highly corrosive.

VENUS as seen from NASA's *Pioneer* spacecraft.
Source: NASA photo.

Venus has been visited by several spacecraft from the United States and the Soviet Union. It has been found that Venus is mostly covered with a vast rolling plain and has continents. Venus also has volcanoes, which may be active. In fact, the detection by the **PIONEER SPACECRAFT** of major changes in the abundance of sulfur dioxide in the atmosphere above the cloud tops suggests that Venus has been shaken by massive eruptions within the past 15 years. Spacecraft that have landed on Venus have survived for only a few hours because of the high temperatures and pressures. See also **MAGELLAN SPACECRAFT**, and **VIKING SPACECRAFT**.

Very Large Array (VLA) Located near Socorro, New Mexico, and considered one of the most powerful radiotelescopes ever built, the VLA is a field of 27 85-foot diameter dishes linked together. **RADAR** mapping of **MARS** and imaging the rings of **SATURN** have been accomplished by using the VLA as the receiver part of a radar system that utilizes NASA's Goldstone tracking station in California as the transmitter. A great deal about Mars is yet to be learned and maps obtained by this new radar system will help explain surface mysteries such as the indications of "fresh" lava flows on Mars. See **VIKING SPACECRAFT**.

Very Large Telescope When completed, this instrument will consist of four optically linked telescopes, each one containing a LIGHT-gathering mirror 27 feet (8.2 meters) in diameter. When all four telescopes are working together, their combined light-gathering ability will equal that of a single telescope mirror 52.5 feet (16 meters) in diameter, making it the world's largest telescope. Scheduled for operation in the year 2000 and sponsored by eight European nations, the Very Large Telescope will be located on a high mountain (Cerro Paranal) in Chile. See also KECK TELESCOPE.

Video Display Terminals, Radiation from COMPUTER monitors do not emit MICROWAVES. They do emit a small amount of X rays, but these are blocked by the monitors' screens and cases. What is at issue is the RADIATION in the form of weak, pulsed electric fields that create very low frequency and extremely low frequency magnetic fields. Desktop monitors, whether color or monochrome, contain a device called a *flyback transformer* that directs an ELECTRON beam at the inside of the monitor screen. The beam scans back and forth some 15,000 times a second, each time sending off a small electrical pulse. Concern has been expressed over the possibility of adverse health effects from the magnetic fields thus generated. However, government tests conducted in 1991 showed that pregnant women who work at video display terminals are at no greater risk of suffering miscarriages than workers who do not use the terminals.

The best precaution against high levels of exposure to these fields is simply to sit at least an arm's length from the front of the monitor and at least three or four feet away from the sides and backs of nearby monitors where RADIATION is highest. Measurements show that the intensity of the magnetic fields falls off sharply with increased distance from the monitor. At about 24 inches from the front of the monitor, the levels are typically below those cited as cause for concern. Laptop computers use liquid crystal display screens that do not give off the same type of radiation associated with desktop monitors. See ELECTROMAGNETIC FIELD EMISSIONS.

Viking Spacecraft The U.S.-built *Viking 1* and *2* missions to MARS in 1976 and 1977 were the most ambitious space projects to another PLANET ever undertaken. Two identical spacecraft were built, each including an orbiter and a lander. One part of each spacecraft went into orbit around Mars and the other part landed on the surface of the red planet. The two *Viking* orbiters photographed virtually the entire Martian surface. They found giant volcanoes, larger than volcanoes on Earth. They also found vast cratered regions and a canyon as long as the United States measures from coast to coast. Photos taken by the *Viking* orbiters revealed huge sand dune fields that are considered unique among the extraterrestrial surfaces photographed by planetary spacecraft. The *Viking* lander missions were designed to answer questions about the possibility of life on Mars. Each lander contained a miniaturized laboratory designed to look for signs of living organisms. *Viking's* robot arm included a claw able to dig a hole and obtain samples of Martian soil for analysis. No signs of life were found in any of the samples tested by *Viking 1* and 2. See EXTRATERRESTRIAL INTELLIGENCE, SETI.

Virus, Biological An infectious agent, a virus is at the borderline between living matter and nonliving matter. It is not a cell but rather a tiny particle (20 to 300 nanometers in diameter—a nanometer being one billionth or 10^{-9} of a meter) that can reproduce only inside the CELLS of living hosts, mainly BACTERIA, plants, and animals. Viruses invade healthy cells and cause them to synthesize more viruses, usually killing the cell in the process. Generally speaking, antibiotics do not work as an antiviral. There are, however, a number of promising antiviral agents under development. See INTERFERONS, and VACCINES.

Virus, Computer COMPUTER jargon for errors planted in SOFTWARE code intended to cause trouble in a computer system. Not to be confused with BUGS, which are design flaws or unintended errors in software programs. A computer virus is often transmitted by means of borrowed software in the form of floppy disks. A computer virus can cause a computer to stop operations, and/ or crash, dump, or otherwise ruin data. The virus itself is a small chunk of rogue software that is deliberately hidden in a legitimate application program and is designed to spread by reproducing itself. Viruses are rare on home computers but are more common in business or school operations where a number of computers are connected in a network. Programs are now available that screen for viruses. Viruses can be avoided by not using bootlegged software, by not downloading applications from obscure electronic bulletin boards, or by not permitting others to put their disks in your computer.

Voice Synthesis Field of research involving the development of devices that replicate the human voice in words and sentences (as opposed to devices, usually COMPUTERS, that can understand human speech; see SPEECH RECOGNITION). Computers that can translate text into spoken words can facilitate the use of books by the blind.

Also, the deaf can communicate over telephone lines by typing sentences into a computer that are then turned into recognizable speech. Other applications includc computers that can pronounce words from a dictionary or that read out addresses and instructions for people making deliveries and repairs. An English translation device is also under development that will allow a user to type in a word in Spanish, for instance, and hear the English translation.

Voice-synthesis technology involves developing devices that can read words, determine their pronunciation by consulting a set of rules and a list of exceptions, and then synthesize a human voice to pronounce them.

Volt A unit of electromotive force (the ENERGY that drives the current) in an electrical circuit is called a volt, after Alessandro Volta (1745–1827), the Italian inventor of the electric battery. A volt is defined as the difference in electrical potential between two points of a conductor carrying a constant current of one AMPERE. A flow of electricity can be likened to water flowing in a pipe, and voltage can be thought of as the pressure that pushes current through the circuit. See **OHM'S LAW**, and **WATT**.

Voltaic Cells See **PHOTOVOLTAIC**.

Voyager Spacecraft Once every 175 years, the PLANETS line up in such a way that a spacecraft from

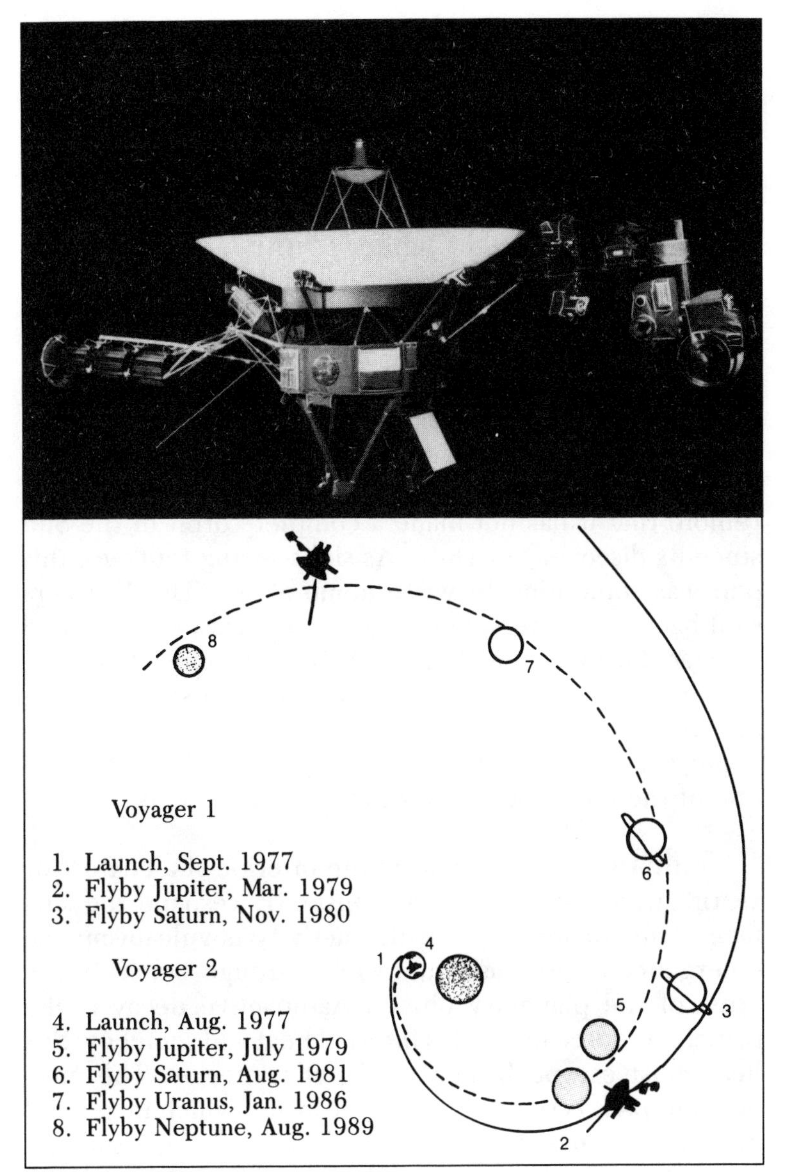

VOYAGER SPACECRAFT and its missions to Jupiter, Saturn, Uranus, and Neptune.
Source: NASA/JPL photo.

EARTH can visit each of the outer planets, in one grand tour. That special opportunity came in 1977, and the United States took advantage of it by launching two unmanned spacecraft, one to make the whole tour, the other to leave the SOLAR SYSTEM after visiting SATURN. By early 1991, *Voyager 1* and *Voyager 2* spacecraft had completed their assigned missions and were beyond the orbits of the known planets. They were, in fact, approaching the HELIOPAUSE, the outer boundary of the SUN's vast magnetic field. Weak signals are still being received from the two spacecraft.

Originally, *Voyager 1* and *2* were designed only to explore JUPITER and Saturn. By 1981, when URANUS was added to the itinerary of *Voyager 2*, both spacecraft had completed all of their mission objectives. *Voyager 1* had encountered Jupiter and Saturn, returning vast amounts of data about the two planets, including evidence of active volcanoes on Io, one of Jupiter's MOONS. *Voyager 2* had arrived at its closest approach to Jupiter and was on its way to follow-up exploration of the Saturnian system. The same gravitational "slingshot" technique that had been exploited to send the *Voyagers* from Jupiter to Saturn enabled *Voyager 2* to be flung in the direction of Uranus and Neptune. The journey from Earth to Neptune, a distance of some three billion miles, took 12 years. Neptune is so remote that it has not made a complete orbit of the Sun since its discovery in 1846. As sight-seeing tours go, this one was something to write home about. The *Voyagers* sent back, not letters, but picture postcards, startling and spectacular images of Jupiter, Saturn, Uranus, Neptune, and an amazing assortment of moons. Almost everything that is known today of Uranus and Neptune comes from *Voyager 2*. The *Voyager* (*1* and *2*) program is considered one of the most successful of all NASA projects.

Vulcanism HEAT from the interior of a PLANET or MOON drives processes that constantly reshape the surface. A major indicator of this activity is vulcanism: the emergence of hot gases and rocks through cracks in the crust of the planetary object. Radioactive decay is the principal source of Earth's internal heat and somehow this heat must escape from the planet's interior. The most efficient path is some form of vulcanism, which in general means any convective flow of hot material to the surface. Other than Earth, vulcanism has been detected only on NEPTUNE's moon, TRITON, and JUPITER's moon, Io, but some astronomers believe that volcanic eruptions may also be occurring on VENUS.

Wallace, Alfred Russel (1823–1913) Codiscoverer (with Charles DARWIN) of the concept of NATURAL SELECTION. Wallace was an English naturalist and explorer who conceived the idea of natural selection independently of Darwin but who agreed to coauthor a paper on the subject with Darwin when it was revealed that they had both been working on the same idea for some time. See also EVOLUTION.

Waste Isolation Pilot Plant (WIPP) A Department of Energy (DOE) facility located underneath the desert near Carlsbad, New Mexico, intended for the disposal of NUCLEAR WASTE by-products from U.S. nuclear weapons manufacturing programs. The DOE facility contains 56 rooms carved out of salt deposits located 2,100 feet below ground. The DOE had planned to start placing waste in this facility by 1990, but regulatory and safety problems have delayed its opening. See RADWASTE, and TOXIC WASTE.

Waste Management The problem of what to do with the nation's garbage output is a technical challenge to engineers. U.S. households will be sending over 170 million tons of waste to municipal landfills annually by the year 2000, according to estimates made by the Environmental Protection Agency. Although this projection shows an increasing demand for landfills, space for landfills is growing harder to find. A 1989 Office of Technology Assessment report estimated that 80 percent of existing landfills will be filled and closed within 20 years. The skyrocketing costs of waste disposal reflects this mismatch of supply and demand for landfill space. Between 1982 and 1988, the average cost to dump municipal waste has more than doubled—from $10.80 to $26.93 per ton.

There are two main approaches to this problem: one, accelerating landfill biodegradation to make the landfills last longer, and two, recycling a higher percentage of our throw-away material. Accelerating biodegradation can be accomplished by adding moisture and other chemicals to make buried wastes decay quicker—from 40 to 50 years to just 5 or 10 years. More than 50 percent of U.S. trash is recyclable and many communities are recognizing the need for this approach. Japan and West Germany have provided examples of what can be done in recycling. Both countries recycle over 60 percent of their waste as compared to 10 percent in the U.S. One example: It has been estimated that if we all recycled our Sunday newspapers, we could save over 500,000 trees every week. See **TOXIC WASTE**.

Watson, James D. The codiscoverer—along with Francis CRICK—of the structure of DNA, and cowinner—again along with Crick, of the Nobel prize in 1962. Prior to the findings of Watson and Crick, nobody knew exactly what a gene was, what it looked like or how it worked. The story of the discovery of the structure of DNA is told in Watson's candid and sometimes abrasive book, *The Double Helix*. Watson is currently (1991) head of the GENOME PROJECT, the 15-year effort to map all the GENES on every human CHROMOSOME. See **DNA**, and **GENETIC ENGINEERING**.

Watt A unit of power equal to the rate of work represented by a current of one AMPERE under a pressure of one VOLT. Named after James Watt, Scottish engineer and inventor (1736–1819). See **OHM'S LAW**.

Wave Theory When energy is propagated by means of coherent vibrations, such as is the case with radio or SOUND, it is often described as a wave. In QUANTUM PHYSICS, wave theory postulates that PARTICLES of MATTER and ENERGY exhibit many of the characteristics of waves and may best be described in this manner. This *wave-particle* duality is a basic feature of nature, and whether LIGHT, for instance, needs to be described as a particle or as a wave depends on the nature of the experiment being performed.

Weak Nuclear Force The fundamental FORCE of nature that governs the process of radioactive decay. According to the STANDARD MODEL in physics, there are four types of forces in the universe. The STRONG FORCE and the weak nuclear force act only within the atomic nuclei. The third force is electromagnetism, and the fourth is the gravitational force. See SUBATOMIC STRUCTURE.

Wegener, Alfred Lothar German meteorologist and geophysicist (1880–1930) who originated the theory of CONTINENTAL DRIFT. See also PLATE TECTONICS.

Weight The phenomenon resulting from gravity acting on mass. Units of weight (pounds, ounces, etc.) are commonly used as a measure of heaviness or mass. See GRAVITY, MASS, and WEIGHTLESSNESS.

Weightlessness The apparent disappearance of GRAVITY that occurs inside any orbiting or freely falling craft. Because we have all seen orbiting astronauts on the nightly television news, we take the idea of weightlessness for granted and think of it mainly in terms of its physiological effects on astronauts. In 1907, however, the concept of weightlessness, or *vanishing of gravity,* led Albert EINSTEIN along a thought path that culminated in his GENERAL THEORY OF RELATIVITY. See also EQUIVALENCE PRINCIPLE.

White Dwarf Stars Some stars in the closing stage of their life become white dwarfs. When a star uses up the nuclear fuel at its center, it collapses. How it ends up depends on how much MASS it had. When an ordinary STAR like our SUN runs low on fuel, it goes through two phases. First its core contracts while its outer portion expands and cools. The star's color changes from a yellow-white to red. At that stage it has become a RED GIANT STAR. Eventually, the outer portion boils away into space leaving the core, a massive, dense sphere only about the size of EARTH—a white dwarf star. So dense are they at this stage that each teaspoonful of white dwarf contains tons of MATTER. See also SUPERNOVA.

WIMPS (Weakly Interacting Massive Particles) Theoretical particles postulated by cosmologists to help explain the discrepancy between theory and observational data with regard to the total mass of galaxies. Studies of orbiting galaxies indicate masses ten times greater than could be explained by simply adding up the mass of all the stars in the galaxy. Astrophysicists speculate that this unseen matter may be BLACK HOLES or BROWN DWARF stars, but it might also consist of subatomic particles that they have come to call WIMPs. No one has ever seen a WIMP—they are invisible to telescopes—but theoretical physicists have good reasons for thinking that they may exist. See DARK MATTER.

Wind Power Wind power generators constitute a form of SOLAR ENERGY in that winds originate because of differing amounts of solar RADIATION received on the surface of EARTH. The Sun warms the spherical Earth more at the equator than at the poles. This uneven warming creates a temperature difference. As the warm tropical air rises, the cold air from the poles moves toward the equator to replace it and this action results in wind.

Wind power is economical for some sites and markets. Integration of this highly variable power source into utility grids is a technical and economic problem. The variations in wind power output require both a backup source of power as well as a method for storing electrical power during those times when demand is low. Sites for wind generation are limited by wind conditions and environmental considerations. The amount of land required per unit of electrical capacity is much larger than for most other forms of solar energy. Interference with television and microwave signals can be a problem unless the wind farms are located in remote areas.

WIND POWER. Scientists at the U.S. Department of Energy are conducting structural testing and performance measurements of commercial wind turbine systems. The goal is to produce wind power systems that generate electricity at less than 6 cents per kilowatt hour. *Source: U.S. Department of Energy photo.*

Despite the many problems involved, wind power is still an attractive alternative way to generate electricity. Between 1981 and 1988, more than 15,000 wind turbines, with a total peak electric capacity of almost 1,500 megawatts, were installed in California—where the vast majority of U.S. wind-powered development has taken place. Wind-powered turbines capable of generating electricity at about 5.3 cents per kilowatt have been developed that, with a rise in oil prices or increased concern about the burning of fossil fuel, will make wind power competitive in many areas. Estimates of the total potential vary, but it appears that at least 20 percent of current U.S. electricity demand could be met by wind turbines installed on the Great Plains, along coasts, and in other windy areas.

Windows, Computer SOFTWARE that creates a graphical operating environment permitting faster and easier ways to accomplish tasks on a COMPUTER. A window is a rectangular area on the computer monitor screen that contains information from a software application, or document file. Windows can be opened or closed, resized, and moved. A user can open several windows on the computer monitor at the same time and can shrink windows to icons or enlarge them to fill in the entire screen. An *application* is a computer program used for a particular kind of work, such as word processing or accounting. An *icon* is a graphical representation of various commands or program items. The use of windows software, along with a computer MOUSE, facilitate the *point and click* method for controlling computer operations, as opposed to having to enter commands through the keyboard.

Workstations COMPUTERS that are linked to one central processing unit and data bank as well as to other workstations in a network. A personal computer (PC, Macintosh, or whatever) is independent whereas a workstation is part of a network. What distinguishes a workstation from a personal computer is not the exterior appearance but what's inside. Generally, workstation features include increased power, the ability to run several programs simultaneously, advanced networking capability, and enhanced graphics. The workstation network gives rapid access to other users (and the material in their systems), to massive data storage, and sometimes, to large mainframe or even supercomputers. Consequently, workstation users have at their fingertips computing power exceeding that of a personal computer by several orders of magnitude. See **LAN (LOCAL AREA NETWORK)**.

Wormhole In **COSMOLOGY**, the highly imaginative and speculative concept of shortcuts through **SPACE-TIME** that would, in effect, act as time machines. According to present theories, when a star several times more massive than the **SUN** collapses to form a **BLACK HOLE**, under certain circumstances, the black hole can turn inside out and poke itself into another part of space-time. The result would be a tunnel through space-time linking two black holes. For most theorists, wormholes do not seem worth thinking about because they squeeze shut before there is any time for a spacecraft or even information to pass through. Some theorists suggest that if a wormhole could be kept open by some means, an advanced civilization could use a wormhole to transmit messages and travel across the universe. Other theorists consider the entire concept as science fiction and point out that wormholes violate the law of conservation of matter and energy. The argument hinges on whether the laws of physics as presently understood permit the creation of traversable wormholes.

WYSIWYG (What You See Is What You Get) **COMPUTER** jargon meaning a **SOFTWARE/HARDWARE** capability that provides the user with the capability to see on the monitor screen all the fonts and graphics exactly as they will print out and the ability to be able to edit on screen. Critics of this system refer to it as "What You See Is All You Get," implying that there are limitations to the WYSIWYG approach.

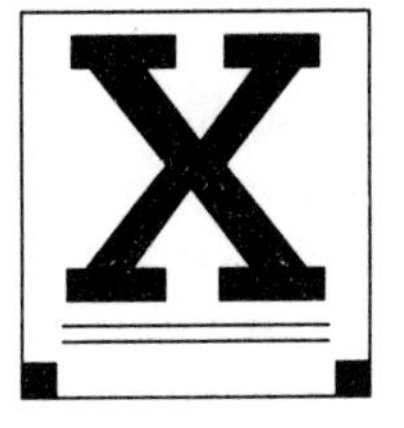

X Chromosome The sex chromosome of humans and most mammals that determines femaleness when paired with another X chromosome. The X chromosome occurs singly in males. All normal women have 44 CHROMOSOMES in each CELL plus two X chromosomes. All normal men have 44 chromosomes in each cell plus one X and one Y chromosome. Sex-change operations cannot change chromosomal makeup, and the difference between a chromosome arrangement containing two Xs and one containing one X and one Y is clearly visible under the microscope. During Olympic competition, for example, an athlete's sex is determined by a simple microscopic examination of a small amount of saliva.

Most color blindness (and hemophilia) is inherited in association with the X chromosome. For this reason more men than women are afflicted; since males possess only one X chromosome, any genetic defect upon it is likely to manifest itself. Females have two X chromosomes, and a defect in one can be masked by the normality of the other. She will not show the disease. She will, however, be a carrier of a chromosomal-linked disease. See **Y CHROMOSOME.**

X Ray A form of electromagnetic RADIATION, shorter in wavelength than LIGHT and capable of penetrating solids and of ionizing gases. X-ray radiation has wavelengths in the range of 0.1 to 10 nanometers (a nanometer is one billionth of a meter). These short wavelengths are able to pass through body tissue making X-ray photography an important medical diagnostic tool. An X-ray photograph resembles a negative of an ordinary photograph, with dense tissue such as bones showing up as white shapes. X rays with very short wavelengths, which can penetrate tissue deeply enough to destroy them, are used in radiation therapy. See **CAT SCAN, MAGNETIC RESONANCE IMAGING (MRI)**, and **POSITRON EMISSION TOMOGRAPHY (PET).**

X-ray Laser The weapon system originally envisioned as the heart of the Star Wars missile defense system. At the weapon's core was a nuclear bomb, which would explode and energize the ELECTRONS in a bundle of LASER rods to emit powerful X-ray beams that would destroy incoming missiles. But the vision proved hard to achieve in reality and X-ray laser research has been significantly scaled back. See **SDI** (**STRATEGIC DEFENSE INITIATIVE**).

X-ray Lithography See MICROLITHOGRAPHY.

Xeriscape (zeer-eh-scape) A modern approach to landscaping which has become popular due to water shortages in many parts of the United States. Xeriscape (from the Greek word *xeros*, meaning *dry*) involves planting drought-resistant plants and low-maintenance grasses that require water only every 2-3 weeks. Other Xeriscape principles—drip irrigation, heavy mulching of planting beds, organic soil improvements that allow for better water absorption and retention—are all applicable to most garden designs.

Xeroradiography A medical diagnostic tool used to detect breast cancer. Xeroradiography combines X-ray imaging with Xerox copying technique. For a time, xeroradiography produced clearer pictures than conventional MAMMOGRAPHY. More recently, with improvements in the mammographic technique, many experts believe that the two procedures may be comparable. With both techniques available, a patient with a problem that cannot be diagnosed adequately with mammography may require an additional view with xeroradiography. See **X RAY**.

Y Chromosome A sex chromosome in humans and most mammals that is present only in males and is paired with an X chromosome. Within the Y chromosome are coded all the necessary instructions for determining maleness. All normal men have 44 CHROMOSOMES in each CELL plus one X and one Y chromosome. All normal women have 44 chromosomes in each cell plus two X chromosomes. The Y is much smaller than the X chromosome and plays a lesser role in inheritance. Males produce two types of sperm: a 22 plus X type and a 22 plus Y type. The Y makes for maleness; its lack makes for femaleness. It is this reason that the male determines an offspring's sex. Ancient monarchs who divorced (or beheaded) their wives because they did not produce male offsprings were deficient in their understanding of genetics.

A type of sex-chromosome abnormality that has drawn attention is the one in which a male ends up with an extra Y in his cells. So-called XYY males are said to be difficult to handle and to have a tendency to rage and violence. XYY combinations may occur in as many as 1 man in every 3,000. At one time, screening of newborns was suggested; however, what society's options were after the screening has never been completely worked out. See GENES, and X CHROMOSOME.

Yucca Mountain Located 100 miles northwest of Las Vegas, Yucca Mountain, really just a 1,500-foot hill, is the site chosen by the federal government for burying the nation's NUCLEAR WASTE—the fuel rods and radioactive debris that has been accumulating at the 111 nuclear power plants around the United States. The repository is intended to hold up to 70,000 metric tons of waste. Once filled to capacity, which will occur around the year 2030, the repository will be sealed off and its entrance shafts refilled. The idea is that the sheer bulk of Yucca Mountain and the artificial barriers of the repository will isolate the waste and protect the surface environment.

YUCCA MOUNTAIN. Artist concept of the proposed high-level radioactive waste repository at Yucca Mountain. The site is located approximately 100 miles northwest of Las Vegas and was chosen because of its good physical and chemical characteristics for nuclear waste containment and for preventing radiation from reaching the biosphere. *Source: U. S. Department of Energy.*

Doubts remain about the safety of this repository. It should be understood that the *spent fuel* is extraordinarily dangerous stuff. **URANIUM** assemblies are removed from commercial reactors after three years' burnup. They are not removed because the radiation is "spent," but rather because they have become *too radioactive* for further efficient use. The term *spent fuel* is a gross misnomer. At issue in the debate over the safety of Yucca Mountain is the possibility of water coming in contact with the canisters of radioactive waste (which are very hot both physically and radioactively). Some scientists have argued that groundwater under the mountain could eventually well up, flood the facility, and prompt a major calamity. The environmental impact, some experts estimate, could exceed that of a nuclear war. The dispute over the safety of Yucca Mountain is yet to be resolved. An independent panel and a group from the National Academy of Sciences (NAS) are separately reviewing the data. The NAS final report is due in 1992. See also **WASTE ISOLATION PILOT PLANT (WIPP)**.

Z Particle or Z° (zee zero or zee naught) Recently discovered subatomic particle that carries the WEAK NUCLEAR FORCE between neighboring nuclear PARTICLES. The weak FORCE, which is responsible for some forms of RADIOACTIVITY, is one of the four basic forces of nature. Z particles are short-lived massive BOSONS (100 times more massive than PROTONS and nearly as massive as an ATOM of silver) thought to have been abundant in the early universe.

In the past decade scientists have postulated a theoretical arrangement known as the STANDARD MODEL, which purportedly shows how the fundamental particles of all MATTER are related. So far, all known particles included in the model can be grouped into three, and only three, families or generations. Exact measurements of the masses and lifetimes of Z° particles will enable scientists to determine whether there are more than three families of basic particles. Understanding the Z° will give theoretical physicists an important key to understanding the master plan by which all matter is organized. Scientists, however, need thousands of Z° particles to carry out their precise measurements and one of the principle functions of the world's large ACCELERATORS today is to produce Z° particles by slamming together negatively charged ELECTRONS and positively charged POSITRONS. See SUBATOMIC STRUCTURE.

Zero Gravity The condition in which the apparent effect of GRAVITY is zero, as in the case of a body in orbit or in free fall. Astronauts find it disturbing to live in zero gravity, with no up and no down. Most astronauts get spacesick—they feel dizzy, sometimes vomit, and exhibit other symptoms similar to airsickness. In space, with no gravity for muscles to work against, the body gets weak, and calcium and minerals begin to leave the bones, making them soft and fragile. Astronauts who have spent months in space resemble hospital patients who have spent a similar amount of time in bed. See **WEIGHTLESSNESS.**

Zodiac An imaginary picture of the night sky, usually in the form of a circular diagram, containing 12 divisions or constellations called signs of the zodiac. Believers in **ASTROLOGY** think that the grouping or relative position of stars influence events, particularly at the time of one's birth. This superstition is solidly held by large numbers of people in the United States despite the lack of any scientific basis.

People have long pretended that the sky is divided into groups of stars, each group having its own name and story or legend. Most of the constellation names used today come from the ancient Greeks because it is the northern sky that was visible from the Greek empire. Dividing the sky according to the zodiac is of little scientific use to astronomers. The position of the **EARTH** has changed since ancient times, so dates for each sign of the zodiac no longer represent the time when the **SUN** is in the related constellation. See **ASTRONOMY**.

Zoo Hypothesis The conjecture that life on **EARTH** has been detected, possibly even visited, by intelligent extraterrestrials who prefer not to disturb us in any way but, rather, prefer to observe our development as if we were animals in a zoo. See also **EXTRATERRESTRIAL INTELLIGENCE**, and **SETI.**

Zygote In biology, a fertilized egg—the cell produced by the union of two sexual reproductive **CELLS** before it divides. In its early stages of development in the womb, a fertilized egg is properly called an **EMBRYO** up to the end of the second month. After that, and until birth occurs, it is called a **FETUS**.